Small-Scale Processing and Storage of Tropical Root Crops

Also in this Series

Westview Tropical Agriculture Series
Donald L. Plucknett, Series Editor

Small-Scale Processing and Storage
of Tropical Root Crops
edited by Donald L. Plucknett

Tropical root crops--basic staples for millions of
people--are highly perishable, and tremendous losses
occur after harvest because of the lack of storage
and processing technology. This book is the first to
fully describe small-scale processing and storage
methods for these root crops, particularly taro,
sweet potato, and yams. The authors emphasize methods
of handling and preserving the crops that require
little in the way of energy or technology, and they
discuss traditional methods of storage and processing
in Africa, Asia, and the Pacific. They also describe
small machines suitable for processing and highlight
examples of higher-level technology. The book is a
milestone in the search for ways to appropriately
modernize traditional agriculture and food systems.

Donald L. Plucknett, professor of agronomy at the
University of Hawaii at Manoa, is currently working
for the Board of International Food and Agricultural
Development. He was chairman of the National Academy
of Sciences Vegetable Farming Systems delegation to
the PRC in 1977, and is president of the International Society of Tropical Root Crops.

Westview Tropical Agriculture Series
Donald L. Plucknett, Series Editor

Small-Scale Processing and Storage of Tropical Root Crops

Edited by Donald L. Plucknett

Routledge
Taylor & Francis Group

NEW YORK AND LONDON

Westview Tropical Agricultural Series, No. 1

First published in paperback 2024

First published 1979 by Westview Press

Published 2019 by Routledge
605 Third Avenue, New York, NY 10158

and by Routledge
4 Park Square, Milton Park, Abingdon, Oxon OX14 4RN

Routledge is an imprint of the Taylor & Francis Group, an informa business

Library of Congress Card Catalog Number: 79-4712

Publisher's Note
The publisher has gone to great lengths to ensure the quality of this reprint but points out that some imperfections in the original copies may be apparent.

ISBN 13: 978-0-367-28743-6 (hbk)
ISBN 13: 978-0-367-30289-4 (pbk)
ISBN 13: 978-0-429-30618-1 (ebk)

DOI: 10.1201/9780429306181

Contents

Section III
Post-Harvest Handling and Storage

Section IV
Processing Techniques and Products

Tables

Figures

Preface

This book is the final product of a workshop on "Small-Scale Processing and Storage of Tropical Root Crops" held in Honolulu, Hawaii, on June 19-23, 1978. It is the first book on processing and storage of taro, sweet potato, and yams.[1] Sponsored by the College of Tropical Agriculture of the University of Hawaii and with financial support from the United States Department of Agriculture from funds provided by Section 406 of the 1966 Food for Peace Act, the workshop provided a technical forum for more than fifty participants from thirteen countries.

The workshop was primarily designed to gather and evaluate the state of knowledge concerning small-scale, low energy-requiring storage and processing of tropical root crops, particularly taro and other aroids, sweet potato, and yams.

The workshop was organized in two phases. The first involved presentation of invited and contributed papers on three major topics: Post-Harvest Handling and Storage; Processing Techniques and Products; and Economic Aspects of Stored and Processed Products. The second phase took the form of working sessions on each of the aforementioned major topics. The working sessions provided an opportunity for world authorities to review the state of knowledge in each of the major topics and

---

[1] A book on processing and storage of cassava was published in 1974 (E.V. Araullo, Barry Nestel, and Marilyn Campbell, <u>Cassava Processing and Storage; Proceedings of an Interdisciplinary Workshop, Pattaya, Thailand, 17-19 April 1974</u>, International Development Research Center, Ottawa, Canada, IDRC-031e, 125 pp.).

to prepare a synthesis report and recommendations for each subject.

Because of the perishability of tropical root crops, successful storage and processing methods for them are especially needed. This book presents information on specific, successful practices that could be used in a number of tropical countries. It is hoped that this pioneering workshop will lead to further efforts to study and evaluate storage and processing techniques that can be adopted by small-scale producers in the tropics.

I would like to acknowledge the unstinting work of Bryan Begley and James Moy, who assisted as members of the Organizing Committee for the Workshop, and Lynne Kobayashi, who prepared and typed the final manuscript and performed myriad tasks in bringing this book to completion.

Donald L. Plucknett
College of Tropical Agriculture
University of Hawaii
Honolulu, Hawaii

An Overview:
Systems Approach to
Root Crop Research

An Overview: Systems Approach to Root Crop Research

J.-K. Wang
K. K. Otagaki

ABSTRACT

Much of the research on taro and sweet potato production at the College of Tropical Agriculture and Human Resources, University of Hawaii, is supported by Grant No. 12-14-5001-121, from the U.S. Department of Agriculture under Public Law 89-808, Section 406 (The Food for Peace Act). One of the few on-going research programs supported under the Food for Peace Act, the comprehensive project was initiated in 1975 and is now completing its third year. The crops studied are taro and sweet potato.

The project is currently supporting systems research activities, either partially or wholly, in the following areas of emphasis:

1. Bibliographic services
2. Production systems analysis and modeling
3. Production of taro
4. Production of stable food forms from taro
5. Crop and stored food protection
6. Economic and marketing analysis
7. Production of sweet potatoes

In addition, the State of Hawaii is funding a separate project on the development of mechanical harvesting equipment for paddy taro, and a substantial research effort in taro diseases is being supported by a CSRS/USDA special grant under PL 89-106.

In this paper, a summary of research activities on taro is given to illustrate the general approach in developing and managing a comprehensive research program on tropical root crops.

INTRODUCTION

Taro (<u>Colocasia esculenta</u> (L.) Schott), a member of the Araceae, is indigenous to South Asia. It is grown throughout the tropics and subtropics for its edible corms. Chung (1958) believed the plant was introduced from India to Burma, China, and Indonesia. Today, it can be found in the Pacific Basin, Africa, the West Indies, and South America. It was introduced to the southern United States in the early 20th century by the United States Department of Agriculture as a supplement to potatoes (Young, 1936). It has become a major source of staple food on many Pacific islands, and it is also an important food source in Africa, Southeast Asia, and the Far East. Ruben Villareal (1975) of the Asian Vegetable Research and Developmental Center has stated that millions of hectares are planted to root crops, and yet the potential of root crops to help alleviate food shortages on a worldwide basis is still not recognized. It is clear that taro has been and will remain an important food crop in the subtropic and tropic regions of the world.

Except in places such as Hawaii, Nigeria, Indonesia, the Philippines, Egypt, and a host of Pacific and Caribbean Islands, where it has developed into a commercial crop, taro is generally produced as a local source of staple food. As such, a steady supply is necessary to satisfy the needs of the population. In underdeveloped countries, the supply source must be close to the population, as high technology transportation and storage are not available or feasible.

Grains such as rice, wheat, and corn can be easily dried and have attributes such as high bulk density, transport capability, stability, and ease of preparation that facilitate delivery to and utilization by consumers. In contrast with the grains, fresh taro cannot be easily dried and does have problems with spoilage, transportation, processing and utilization due to its high water content, low bulk density, and high tissue matter density. In order to reach a broader market, the development of small-scale processing technology for taro becomes important. Simultaneously, increasing attention must be paid to the study of production and

4

processing economics, existing and potential marketing structures, and demands for taro and its processed products.

The importance of ethnic preferences in the
consumption of staple foods is well known. When a
root crop such as taro is used as a supplemental
staple food supply, the question of social acceptability becomes important in an increasingly affluent
society. Thus, a comprehensive systematic approach
must be taken in order to develop and improve production and utilization of taro in the tropical
regions of the world.

The relatively steady demand for taro in areas
where it is used either as a staple or as a supplement, the lack of storage technology, the lack of
processed products that are acceptable to a large
portion of the affluent urban population, and the
need for the development of well organized marketing
structures for taro and its products are all problems common to Hawaii and other taro-producing
areas. An international gathering such as this
workshop is exceedingly useful, as it establishes a
forum to exchange information, to discuss common
problems, and to plan a better coordinated approach
to the future.

There is a strong possibility that taro may
become industrially important in the future. Taro
starch can account for up to 40 percent of such
plastics as polystyrene, polyethylene, nylon, and
polyvinyl. Starches, when incorporated into plastics, can be progressively removed by selective
degradation when the material comes into contact
with bacteria or fungi. Once this degradation process is begun, the removal of starch will increase
the surface area, thereby bringing about an accele-
ration of the degradation process. Taro starch,
because of its extremely fine size, is superior to
other starches such as corn, tapioca, and potato as
a basic ingredient for biodegradable plastics. The
Hawaiian taro food, poi, is one of few foods that
have been shown to be hypoallergenic and can be used
extensively in the production of baby food.

The importance of taro as a staple food, and
the possible development of taro into an export crop
have caused a re-evaluation of taro by a number of
governments. There is a general consensus among
developmental planners and agricultural scientists
that breakthrough developments in the major grains,
like those that occurred during the last two decades, are unlikely to be repeated in the near

5

future. Therefore, for the next decade there is need to examine the potential of root crops. It is our contention that, together with other tropical root crops, taro will be rediscovered through the joint effort of the groups represented at the workshop.

SUMMARY OF ACCOMPLISHMENTS

Many of our current taro research results will be presented elsewhere in the workshop. Therefore, only a very brief listing will be given below to summarize the accomplishments of the project during the last three years.

Bibliography

Approximately 750 taro and 4000 sweet potato citations have been indexed according to subject matter and author categories and have been computerized for quick search and listing. These bibliographies (Rotar, Plucknett, and Bird, 1978; Rotar and Bird, 1978) have proven to be very useful to researchers and developmental project planners. Copies of these cross-indexed citations have been made available to a number of research institutions around the world.

Management

This multidisciplinary project is characterized by quarterly meetings at which various components report on the recent progress and plans. The continuing interchange of ideas and information between components of this project is vital to its continued success. A network approach has been employed in the management of the research project. The general objective, and the interrelationship between sub-objectives, are clearly identified by this technique. The network establishes a schedule for research activities; it also indicates the effect of delay in one segment on the project as a whole.

Agronomy

A collection of germ plasm, consisting of 250 accessions from the Pacific and Asian region, has

been established at the Kauai Branch Station,
College of Tropical Agriculture and Human Resources.
Twenty-three varieties have been field-tested to
date. These materials are being made available to
the cell biology research group at the University of
California, Irvine, so that a methodology for fast
propagation of virus-free planting materials can be
developed. The available varieites are: Apowale,
Haokea, Hapuu, Iliuaua, Kai Uliuli, Lehua Keokeo,
Lehua Maoli, Lehua Palaii, Manini Kea, Manini Owale,
Maui Lehua, Nihopuu, Ohe, Papapueo, Piialii, Piko
Eleele, Piko Kea, Piko Keokeo, Piko Uaua, Red Moi,
Uahiapele, Ulaula Kumu, White Moi.
　　The Iliuaua variety is an upland variety. All
other 22 varieties can be grown in either upland or
paddy conditions.

Production and Evaluation of Stable Food Forms from Taro

　　It was found that using heat to destroy the
acridity of taro also gelatinizes the starch,
resulting in difficulties with drying. A more effective method is to remove the toxicants by settling. It was found that stable forms of taro
products such as noodles and rice can be formed by
adding 10 to 20 percent soy flour. Public Use
Patents for these developments are being processed.
A 0.4 to 0.5 pH unit change at a storage temperature
of 60°C versus less than 0.15 pH unit change at
storage temperatures of 21°C and 38°C over a 12-
month period suggested a possible quality change
affecting protein and lipids, when taro products
were stored above 38°C.
　　Long-term storage of taro products also will
bring pigment degradation. Tests indicated a loss
of anthocyanin, a reddish-purple pigment, and a more
gradual loss of brown pigment over time.
　　When taro flour is processed into pre-cooked
rice or noodle-like products, only minimum preparation is required before consumption. In general,
four minutes in boiling water is sufficient. The
composition of these products can be varied to
include up to 35 percent soy flour and other nutrients. The taste of these products can be altered as
desired. The energy value of taro products measured
3.8 to 4.0 calories per gram, which is quite comparable to the energy value of other cereals.

Crop and Stored Food Protection

In Hawaii, taro is relatively free of serious insect pests. However, potentially serious pests are known to occur in other tropical areas.

A. Insects attacking taro in Hawaii

1. Taro leafhopper, _Tarophagus proserpina_ (Kirkaldy)
2. Cotton aphid, _Aphis gossypii_ Glover
3. Water lily aphid, _Rhopalosiphum numphaeae_ L.
4. Root infesting aphid, _Pemphigus_ sp.
5. Spider, _Tetranychus cinnabarinus_

No chemicals have been cleared for insect control on taro. However, introduced natural enemies have been successful in controlling the taro leafhopper. The other pests occur only occasionally.

B. Diseases

1. Root and corm diseases
   a. Pythium rot, _Pythium_ sp.
   b. Sclerotium rot, _Sclerotium rolfsii_ Sacc
   c. Loliloli taro
   d. Hard rot or "guava seed"

2. Leaf and petiold diseases
   a. Phytophthora leaf blight, _Phytophthora colocasiae_ Rac
   b. Cladosporium leaf spot, _Cladosporium colocasiae_ Sawada
   c. Phyllosticta leaf spot, _Phyllosticta colocasiophila_ Weedon
   d. Viruses

Taro in Hawaii is attacked by several diseases, mostly of fungal origin. Wetland taro is particularly susceptible to the various corm rots, whereas upland taro is practically free of these diseases. On the other hand, wetland taro is nearly free of the leaf and petiole diseases, while upland taro can be severely damaged by leaf diseases. Up to 100 percent of a crop can be lost if diseases are not controlled. Recommended methods of control entail a combination of fungicide applications in concert with cultural practices designed to inhibit the growth of pathogens. Among the most effective cultural practices are: (1) maintaining water flow through the paddy such that water temperature is below 25°C; (2) allowing the paddy to dry out for one to two months between crops; and (3) removal of all plant material from the paddy after harvest. There is also some indication that certain varieties have a higher resistance to corm rots than others.

C. Storage pests

No published information is available on stored product pests which may attack processed taro products. Five species of stored-product beetles were colonized to provide specimens for testing for potential pests of dry processed forms of taro.

Lasioderma serricorne is thought to be incapable of surviving on taro flour. However, this beetle can reproduce on taro chips, indicating it could be a serious pest.

Sitophilus oryzae could be a minor problem for stored taro flour and chips, where the beetles breed in other foodstuffs and migrate to the taro. The beetles can survive on taro chips for some time but do not reproduce.

Alphitobius laevigatus is incapable of surviving or breeding on processed taro under normal conditions (room temperature and humidity).

Triboleum castanum is incapable of breeding in processed taro. Unlikely to be a problem in stored nonenriched taro products.

Cryptolestes sp.--insufficient adults for feeding test.

Economic and Marketing Analysis

Social acceptability and ethnic bias have been identified as two of the most important problems which must be confronted if taro is to become an important energy food source in an increasingly affluent society. Improved marketing and transportation structures have been identified as a key element in the delivery of taro and its products to a large segment of the population in a country, and to generate more income for marginal farmers. The

9

production of root crops such as taro has been found to be highly labor intensive, both in Hawaii and elsewhere. A survey of cultural practices and costs of production for wetland taro has been compiled from the major production areas in Hawaii. An economic analysis of taro processing has also been completed.

Production Systems Analysis and Modeling

The production of taro in the sub-tropic and tropic areas in the world is typified by these major factors:

A. Production activities can be carried out continuously throughout the year and a monoculture can be practiced continuously;

B. A steady demand for the product exists in places where taro is consumed as a staple food, or as a major supplement to staple foods;

C. Unlike cereal grains, the storage technology for taro has not been developed; and

D. Taro production is generally labor intensive, and peak labor demand is directly related to yield.

Considering these factors, the system analysis group has developed a procedure for the design of taro production systems which are capable of delivering steady year-round production while maintaining a high level of production resource utilization efficiency, thereby providing steady employment opportunities to the landless poor.

Engineering

Engineering research has concentrated on the development of mechanical harvesting and field clearing equipment for paddy taro (Krishnan, 1976; Smith and Shen, 1972; Smith, 1975). Continued improvement of the prototype equipment will lead to commercial application in the near future.

DISCUSSION

Taro is a very versatile crop and can be cultivated under widely varying water availabilities, from paddies with continually flowing water to

10

upland cropping with supplemental irrigation. It is a crop with a long but flexible growing season. Root crops, like taro, do not have a definite maturation time. Because of this, the harvest operation can be either delayed or advanced within limits without causing serious yield decline. These characteristics make taro a safe crop under varied environmental conditions.

Compared with high yielding paddy rice varieties, paddy taro does not have much advantage on annual per hectare yield in terms of dry matter; however, its yield is achieved from only one crop, whereas in rice several crops would be required to achieve the same yield. It is much more difficult to put taro into a form to give it long-term storability. Given the established preference for rice, it is unlikely that taro will ever replace rice. However, since taro is less demanding than rice, and is more easily managed because of its long growing season and flexible harvest schedule, taro will remain an important crop.

A steady supply of taro can be achieved by production and marketing scheduling. With a given land area, highly efficient utilization of labor and other inputs can be realized. Most agricultural production requires a seasonal peak supply of labor, with severe underemployment at other times of the year. The ability to utilize a constant supply of labor efficiently by taro production systems has social implications which may far exceed its purely economic importance.

The development of simple processing technology for taro is important, for it may open up an avenue to allow taro products to reach untapped consumers. This development will most likely complicate the already difficult study of taro marketing problems.

The potential of taro starch as a key component of biodegradable plastic material is an exciting alternate use which could be successful provided that a sustained supply of product at competitive cost can be achieved.

Unlike most other major root crops, taro can be produced under a wide range of water availability. This characteristic of taro can be extremely important to agricultural planning, when increasingly the world must look toward the development of agriculture using less dependable water sources.

Taro is a desirable crop. It has been important in the past to many peoples. In many cultures it is still a staple. Its potential as a food

11

source is limited only by the imagination of the scientists studying it. At present, the potential industrial uses of taro starch are just emerging.

REFERENCES

Chung, T.K. 1958. Dispersal of taro in Asia. Ann. Assoc. Amer. Geog. 48:255.

Krishnan, Palaniappa. 1976. Evaluation of auger plow for digging wetland taro. Master's thesis, University of Hawaii, Honolulu, Hawaii.

Rotar, Peter P. and Bird, Barbara K. 1978. Bibliography of sweet potato (Ipomoea batatas). College of Tropical Agriculture and Human Resources, University of Hawaii, draft photocopy.

_____, Plucknett, Donald L., and Bird, Barbara K. 1978. Bibliography of Taro and Edible Aroids. College of Tropical Agriculture and Human Resources, University of Hawaii, Misc. Pub. No. 158.

Smith, M. Ray. 1975. Mechanical harvesting of wetland taro. Am. Soc. Agr. Engr. Paper No. 75-1024.

_____ and Shen, H. 1972. Pickup mechanism for harvesting wetland taro. Trans. Am. Soc. Agr. Engr. 15:1005

Villareal, Ruben L. Sweet potato, its present and potential role in the food production of developing countries. Paper presented at the Regional Technical Meeting on Root Crops, Fiji.

Young, R.A. 1936. The Dasheen: A Southern Root Crop for Home Use and Market. Farmers' Bull. 1924, U.S. Department of Agriculture.

Jaw-Kai Wang and Kenneth K. Otagaki: College of Tropical Agriculture and Human Resources, University of Hawaii, Honolulu, Hawaii

Section I
Synthesis Reports of
International Working Groups

1
Working Group Report: Handling and Storage

D. G. Coursey, Chairman
G. V. H. Jackson, Cochairman
R. S. de la Pena, Rapporteur

A. O. Adenuga   *I. E. Nwana*
S. Chandra   *B. F. Siki*
P. Ching   *M. S. Strauss*
G. H. de Bruijn   *J. Tanaka*
N. Hrishi   *T. K. Tupuola*
S. G. Miller   *J. Watson*

INTRODUCTION

All root crops fall within the category of
perishable staples which have high water contents
and remain metabolically active after harvest. The
problems of their post-harvest biotechnology are
thus fundamentally different from those of the more
familiar durable crops, such as grains. Root crops
are, in some situations, processed into dry food
products, but this document will restrict its
attention to storage and handling of the fresh
commodities.

The principal tropical root crops are cassava,
sweet potato, yams, and aroids. Cassava is the
most important within the developing world, but as
this crop is now receiving serious attention from
several organizations (e.g., Centro Internacional
de Agricultura Tropical, Colombia; International
Institute of Tropical Agriculture, Nigeria; Tropical Products Institute, United Kingdom; Central
Tuber Crops Research Institute, India; and National
Root Crops Research Institute, Nigeria), the concentration here will be on the other crops.

Storage losses of fresh root crops are mainly
due to physical, physiological, or pathological
causes or various combinations of all three.

Mechanical injury occurring at harvest or as a result of bad handling can cause considerable losses both directly and indirectly through enhanced physiological and pathological effects. Damaged produce always has an inherently shorter post-harvest life than undamaged produce.

In addition to mechanical damage, exposure to both high and low extremes of temperature can affect storage life. Some tropical root crops, particularly yams, are subject to chilling damage at temperatures below about 10 to 12°C, resulting in internal discoloration and tissue breakdown followed by rapid decay and loss of quality. Exposure to very high temperatures, such as those caused by direct insolation, can also lead to physiological breakdown, and consequent reduction of storage life.

Root crops, being living organs, are metabolically active, and losses due to wilting and respiration always occur. The rate and magnitude of such losses are greatly influenced by the storage environment, particularly temperature and relative humidity, while, as already mentioned, damaged produce loses water, and frequently respires faster, than sound produce.

Additional physiological loss can also occur as a result of sprouting, which in general renders the produce less marketable and also accelerates respiration and water loss and can predispose the produce to pathological invasion. Sprouting normally occurs at the end of the natural dormancy period, which may vary considerably among species and cultivars, and is also influenced by the storage environment.

In the particular case of cassava, initial loss of quality arises from vascular streaking or primary deterioration, the exact cause of which is not yet known but may be regarded as physiological, as it is not directly associated with specific pathogenic attack. This rapid internal discoloration usually starts at the sites of damage and in the vascular tissues but later spreads throughout the root. It is followed by massive pathogenic/saprophytic invasion.

Biochemical changes in quality may occur during storage in all root crops, e.g., changes in the starch/sugar equilibrium which are influenced by storage temperature, storage at low temperatures leading to increased sugar content.

Phytopathological attack by microorganisms is probably the largest cause of post-harvest loss in root crops, although both physical and physiological damage can predispose produce to pathogenic invasions. Many post-harvest fungal pathogens are wound parasites and only capable of attacking storage organs at sites of injury. Quantitative pathogenic losses result from a massive breakdown of the tissues of the organ, the usual pattern of attack being an initial infection, either pre-harvest or through a post-harvest wound, by one or a few specific pathogens, followed by a secondary invasion by a broad spectrum of organisms which grow on the moribund and dead tissues. Qualitative pathogenic losses are typically the result of blemish diseases which render the produce less attractive and so reduce its market value, without necessarily damaging much of the edible tissue.

Insect pests are not commonly the cause of serious losses in the storage and handling of fresh root crops except in the case of sweet potato. Considerable loss may be caused by rodents, particularly rats, and other animal pests such as monkeys and pigs.

The magnitude of post-harvest losses caused by the factors discussed above varies greatly, but on a global basis has been conservatively estimated at 25 percent of produce actually stored.

CASSAVA

It is well known that fresh cassava normally has a storage life of only a few days. Traditionally the crop is left in the ground and harvested only when needed or when it is to be processed into a dried product of longer storage life. Although the lack of well-defined maturity period and thus the possibility of in-ground storage can be advantageous, this system results in large aread of land being unnecessarily tied up, which is undesirable in those regions where there is considerable pressure on land. Furthermore, susceptibility to pathogenic losses increases, and palatability and extractable starch content decline, when harvesting is delayed far beyond the optimum time.

Small quantities of harvested cassava roots can be preserved for short periods using such simple traditional techniques as reburial in moist soil or coating in mud.

17

It has recently been shown that it is possible to extend the life of cassava roots considerably by storing them in clamps in the field or in containers packed with moist material. While neither of these techniques has yet been widely applied, they obviously merit further attention. Clamps are constructed by placing the roots in cone-shaped heaps and covering them with straw and soil but at the same time providing some ventilation. Roots may be packed in materials such as moist sawdust, moist coir dust, or moist peat in wooden or cardboard boxes and kept successfully for periods of between two and six weeks. These simple techniques provide conditions which promote curing of the roots and reduce moisture loss.

More sophisticated techniques such as refrigerated storage at temperatures below 6°C, and coating washed roots in paraffin wax have also been shown to be successful in maintaining roots in acceptable condition for periods of up to a month, but may not always be economically feasible.

SWEET POTATO

While a large body of work is available regarding post-harvest storage of sweet potato, this is primarily applicable to temperate regions of North America, Japan, and New Zealand. In tropical areas the crop is marketed or consumed shortly after harvest and is not normally held for long periods.

Preharvest

Weevils (Cylas spp.) are a major constraint in sweet potato storage. In regions where they are a problem, the tops are removed no more than three days prior to harvest, while before planting and during growth of the vines, insects can be controlled by application of insecticides. Where weevils are not a problem, the tops can be removed and tubers cured in the ground for a week before harvest. In most areas tubers are harvested and consumed as needed and no curing or ground storage is done. Study of cultural practices, especially crop rotation and development of insect resistant varieties, is needed to allow for increased ground storage.

18

Harvest Time

Harvest time is not primarily dependent upon biological maturity, which is ill-defined. Some cultivars become excessively fibrous and crack if left too long in the ground: others can be ground-stored for long periods without appreciable quality loss. Weevil infestations necessitate early harvest.

Harvest Methods

Sweet potatoes are highly susceptible to damage or bruising during harvest on account of their thin tender skins. Curing in the ground before harvest somewhat reduces susceptibility to damage. Tubers are easily damaged by puncture or cutting regardless of the implements used, and this damage accelerates post-harvest deterioration.

Packaging and Transport

These are primarily convenience oriented. In many areas the crop is grown for local consumption and harvested only as needed. Thus, in these areas packaging is not a consideration.

Storage Methods

Although sweet potato is not generally stored for extended periods in the tropics, a variety of structures--sheds, pits, caves--have been used. In such conditions consideration must be given to temperature and humidity to minimize losses due to evaporation and respiration. A serious problem is rapid weight loss, which is increased by mechanical damage. Storage losses can also be caused by sprouting or by weevil infestation.

Research Emphasis

Research emphasis should be given to weevil and post-harvest decay, but there is also a need for fundamental research on the post-harvest physiology of sweet potato tubers with the aim of extending storage life.

YAMS

Yams, particularly the species D. rotundata, D. cayenensis, and D. alata, have the greatest potential for storage among all the tropical root crops, presumably associated with the fact that the tuber is an organ of dormancy. Some storage methods have been practiced by ancient cultures for a very long time. These include methods varying from storage in heaps and clamps to storage in elaborate barns. Considerable storage losses occur in all these traditional methods.

Preharvest

The storage quality of yam tubers varies considerably between cultivars and is influenced by various preharvest factors. Only sound tubers will store for any extended length of time. Infestation by nematodes, especially Scutellonema bradys, and infection with various rot-causing pathogens, occurs before harvest and results in increased storage losses. Preharvest damage by yam beetle (Heteroligus spp.) and termites may also render tubers more susceptible to storage losses.

Harvest Time

Tubers which are mature at the time of harvest generally store better than immature ones. Time to maturity varies between cultivars and is influenced by soil and by climatic factors during growth. With many species and cultivars, only those yams harvested after the vines have completely senesced are considered suitable for storage.

Harvest Methods

Mechanical damage to the tuber at the time of harvest leads to increased losses in storage, so that methods of harvest which minimize mechanical damage should result in reduced storage losses. Frequency of mechanical damage can be influenced by cultivar characteristics such as tuber size, shape, and distribution by soil conditions, and by harvesting techniques.

Any mechanized system developed to harvest yams destined for storage must minimize mechanical damage, both through the development of appropriate machinery and selection of cultivars with small, regular, shallow placed thick-skinned tubers.

Packaging and Transport

Mechanical damage and temperature extremes
must be minimized during transport. Traditionally
yam tubers are transported and marketed without
benefit of packaging, and little work has been
carried out on packaging either for transport or
shelf life during marketing. Individual wrapping
of tubers in fibrous material which gives both
protection and ventilation has proven successful in
long-distance shipping of yam.

Storage Methods

Fungal and bacterial rots, sprouting, and
weight loss due to respiration have been identified
as important factors contributing to storage losses.
Damage by rodents and insects such as scale and
mealy bug are secondary, but still of some importance. Mealy bugs and scale insects can cause
serious damage to yam in storage, to the extent
that planting setts prepared foom infested tubers
may not germinate.

Botryodiplodia theobromae and several species
of Fusarium and Penicillium are the most important
storage rot pathogens, although a number of other
pathogens have been found in association with
decaying yams.

Time of sprouting is determined primarily by
the length of endogenous dormancy, which varies
with cultivar. Sprouting can be retarded by low
temperature, gamma radiation, and certain chemical
treatments.

A curing treatment of freshly harvested
tubers using temperatures between 32 to 40°C and
relative humidity between 90 and 95 percent for 1
to 4 days to heal wounds and promote suberization
has been shown to improve storage quality of yams.
Traditional methods of storage sometimes tend to
encourage curing during the early period of
storage. However, a specific curing period becomes
imperative if storage takes place in subambient
temperature. Many traditional systems use wood ash
to treat wounded tubers before storage. It is
necessary to cut out bruised areas of the tubers,
as cut surfaces can heal, but bruises rot.

Temperatures of 15 to 20°C have been found to
prolong the storage of yam tubers. However, chilling injury which results in complete breakdown of
the tubers and greatly increased susceptibility to

decay occurs at temperatures below 10 to 12°C.
Adequate ventilation is important in prolonging
storage life.

Rodent damage can be reduced by building metal
guards into barns and other storage structures.

Research Emphasis

1. Selection and breeding of new cultivars for
   improved storage life, enhanced nematode
   resistance, and geometric adaptability to
   mechanized harvesting.
2. Investigations of the biochemical and physiological bases of dormancy.
3. Development of small-scale system of on-farm
   disinfection and curing of tubers.
4. Surveying on an international level of the
   storage life of various species and cultivars
   of yam.
5. Methods which enable movement of yams between
   countries are required, especially from those
   countries where cultivars of superior quality
   are reported.

AROIDS

There is often confusion in the terminology
used to describe the various species of edible
aroids: in this document the nomenclature of
Purseglove (1972) is followed.[1]

Preharvest Techniques

No well-established preharvest techniques
exist which relate to improving methods of storage.
In Hawaiian wetland cultivation fields are drained
three to four weeks before harvest. This reduces
root volume and facilitates removal of the plants,
thus reducing the possibility of damage before
harvest. In countries where Phytophthora colocasiae leaf-blight is important, removal of infected
leaves two weeks before harvest could prevent contamination of corms by spores washed from leaf
lesions.

Harvesting Times

Xanthosoma and Colocasia esculenta var. antiquorum are usually harvested when cormels have gone
into the dormant stage (6 to 12 months) as

indicated by leaf senescence. Colocasia esculenta var. esculenta should be harvested when main plants have declined to the three- to four-leaf stage (5 to 15 months), to prevent regrowth.

In Samoa early harvesting is practiced when plants become infested with high populations of taro leafhopper (Tarophagus proserpina) to prevent a deterioration of corm quality.

With some cultivars of Colocasia esculenta esculenta, time to harvest can be extended for up to 2 years when, as practiced in Samoa, soil is successively heaped around the bases of the growing plants. Yields of 5 to 10 lb (2.27 to 4.54 kg) per plant are achieved by this method.

Harvesting times of Alocasia and Cyrtosperma are very flexible, as taste and texture are not appreciably impaired if plants remain in the soil for several years. In contrast, Amorphophallus is harvested at 9 months, after the leaves have senesced. However, as with Colocasia and Xanthosoma, cultivars of Cyrtosperma differ in their time to maturity. There is no information available regarding optimal time of harvest with respect to storage life and food quality.

Harvesting Methods

The chance of injuring corms at harvest occurs when they are pried or dug from the ground, while further damage can occur when cormels and petiole bases are removed. Wounds made at this stage facilitate the entry of saprophytic fungi and bacteria, for instance, Botryodiplodia theobromae and Erwinia spp.

In many areas, petiole bases are left attached to facilitate handling and to reduce the area of exposed tissue available for infection.

Packaging and Transport

Traditionally no formalized methods of packing for transportation are used. Sundry containers such as wooden cases, sacks, or baskets are used according to local convenience.

Experience gained from shipments of corms to New Zealand from Pacific Islands has underlined the care needed in handling corms to reduce damage and subsequent decay. Corms with 50 to 100 mm of tops attached are now shipped in wooden crates placed in refrigerated containers where temperatures are held

23

at about 5°C. This method has reduced the incidence of corm rot from the previous situation where
boxes were placed in the holds of ships at uncon-
trolled temperatures and where handling and loading
times were far greater.

Storage Methods

In general corms or cormels are not stored,
as they rapidly become unfit for human consumption
within a few days from harvest. Infections by fungi and bacteria and weight losses (more than 10
percent in 7 days) due to respiration and/or trans-
piration are the most important factors inhibiting
long-term storage. Botryodiplodia theobromae and
Fusarium solani are major pathogens infecting
corms of Cyrtosperma, Alocasia, and Colocasia. In
addition, corms of Colocasia are decayed by Pythium
splendens and Sclerotium rolfsii, and in some
countries, Phytophthora colocasiae. Corms of Xanthosoma store for longer periods than those of
Colocasia although within the genus Colocasia, the
storage potential of corms of some antiquorum cvs.
greatly exceeds that of esculenta corms.

Corms of Colocasia may be stored for up to 4
weeks in shaded, leaf-lined soil pits dug in well-
drained soil. Best results are achieved if corms
are unscraped and only the largest cormels are
detached. Under pit conditions, corms remain
physiologically active and root and shoot production may occur, while conditions within the pit may
enhance suberization of wounded surfaces. Storage
life is extended by this method as fungal rots are
suppressed, although bacterium rots caused by
Erwinia chrysanthemi sometimes occur.

Cormels and corms of both C. esculenta var.
antiquorum and Xanthosoma are stored for up to 4
months in trays or heaps at ambient conditions in
both Ghana and the Cameroons. Investigations into
these traditional methods, however, have shown that
losses are considerable.

Corms shipped to New Zealand from the Pacific
islands are packed in wooden boxes and placed within refrigerated containers cooled to 5°C. Satisfactory storage during shipment and in market cold
storage lasts for periods up to 6 weeks. When
stored in this way, they are sprayed with water
twice weekly to prevent desiccation and to maintain
corm color.

When airfreighted to New Zealand, corms are
packed in polyethylene bags, but can be subject to

24

overheating in transit which necessitates cooling upon arrival. If shipments are fumigated, subsequent storage life is impaired.

Soil-pit conditions may be simulated when corms are placed in polyethylene bags. Rots caused by Botryodiplodia theobromae are completely controlled, but in countries where the leaf-blight fungus Phytophthora colocasiae is present, rots may still occur when inoculum is high. Trials have shown that these can be prevented by pre-treating corms in 1 percent sodium hypochlorite for 2 minutes before placing them in the bags.

Research Emphasis

1. Survey of the storage behavior of all the aroid crops, with the aim of selecting cvs. of good storage life.
2. Investigations of biochemical, physiolocial, and pathological parameters for post-harvest loss of the aroids.
3. Development of small-scale systems of on-farm disinfection and curing of corms.

NOTES

1. Editor's note: Purseglove (1972) followed the concept that Colocasia is divided into two major types, C. esculenta var. antiquorum and C. esculenta var. esculenta, and applies the West Indian common names and uses to these botanical varieties. Hence, the "eddoe" of the West Indies is C. esculenta antiquorum (C. esculenta var. globulifera of Hill (1939); also called "dasheen" or "oriental taro"), and the West Indian "dasheen" is C. esculenta esculenta (C. esculenta var. antiquorum of Hill (1939) and commonly known as taro, "Polynesian" taro, or "old cocoyam" elsewhere).

REFERENCES

Purseglove, J.W. 1972 Tropical Crops. Monocotyledons 1. London and New York. p. 58.
Hill, A.F. 1939. The Nomenclature of the Taro and its Varieties. Bot. Mus. Leaflet, Harvard University 7:113.

2
Working Group Report: Processing Techniques and Products

M. R. Villanueva, Cochairman
J. H. Moy, Cochairman
W. K. Nip, Rapporteur
L. Jakeway, Rapporteur

G. J. L. Griffin   C. Pedrana
M. Lambert   E. Trujillo
T. O. M. Nakayama   C. C. Tu
E. U. Odigboh   N. Wenkam
C. Williams

PURPOSES OF ROOT CROP PROCESSING

This section is concerned with the transforma-
tions (via processing) of fresh tubers, from the
time the material is delivered to the processing
area until the product is utilized by the consumer.
It is recognized that processing must be interfaced
with economics, environmental concerns, and human
factors in order to realize its purposes. Among
these purposes are: (1) to make available a
larger quantity of food; (2) to provide a greater
variety of products; and (3) to improve the quality
of life.

Processing can increase the quantity of food
available by decreasing loss. It can provide a
vast array of products which are stable and may not
necessarily duplicate the fresh material. The
product may be better or worse than the original
material; it will certainly be different.

Processing can improve lifestyles by increasing the efficiency of the food delivery system,
thus making available more time for commercial
enterprises or leisure for those concerned with
supplying food.

In particular, the implementation of processing in tropical root crop areas holds promise of

increasing discretionary use of women's time, a
necessary condition for the advancement of any
society.

APPRAISAL OF FOOD VALUE OF ROOT CROPS

A particular opportunity for a gathering of
multidisciplines on a limited topic is to agree on
a single system of appraisal. This is especially
needed for root crops which are frequently deni-
grated because some are not high in protein, nor
are they regarded as "complete" foods.[1]
For tropical root crops which are principally
used as food, it is suggested that the utilitarian
basis of food energy (calories, joules) be used.
It has the advantage of being directly related to
population and is also amenable to material and
energy studies.
A suggested example is the megacalorie
(million Calories, 4,184 million joules). This
unit which represents the approximate caloric
requirements of one person per year (2,740 Cal/day)
is what might be reasonably expected from one
metric ton of roots (assuming a 25 percent yield
and a 4 Cal/g).

Basis for an Energy Unit

A megaCalorie would provide:

$$\frac{1,000,000 \text{ Cal/yr/individual}}{365 \text{ day/yr}} = 2,740 \text{ Cal/day/individual}$$

To obtain a megaCalorie with dry root crops at 4
Cal/g dry weight:

$$\frac{1,000,000}{4} = 250,000 \text{ g d.w.}$$

$$= 250 \text{ kilos}$$

Assuming 25 percent yield of dry weight from fresh
taro, 1,000 kilos of fresh tubers will yield 250
kilos of dry material. A scale of 5 metric tons/
day of fresh roots mean 5,000 kilos/day, or
5,000,000 Cal/day in dry products. Working 250
days/year would yield 1,250,000,000 Cal/year. This
would feed 1,250 people for a year, or more
reasonably at 25 percent of total caloric intake
would feed 5000 people. Thus the unit could be
related to population as well as material (roots)

and energy (joules) relationships. A critical ratio is cost/megaCal, and it is a challenge to economists to delineate and enumerate the numerator.

ASSESSMENT OF STATE-OF-THE-ART

1. Definition of "small scale": "Small scale" processing is defined here as the processing of up to 5 metric tons of raw root crops per day at home or village level, assuming 250 working days per year.
2. Available information on processing technology and data on product quality: Very few data on processed product quality are available. An appreciable amount of information on processing technology is available (Figure 1, Table 1).
3. Missing, but needed, information and knowledge to enable increased utilization and improved processing and storage of root crop products:
   a. Characteristics of the physical, chemical, biochemical, and organoleptic quality of root crops and their products.
   b. Toxic factors in root crops.
   c. Nutritional values of root crops as related to processing.
   d. Effect of post-processing handling and storage on quality of, and possible damage to, the finished products.
   e. Detailed information on processing techniques and equipment not presently available.
   f. Information on sources of equipment: e.g., Tropical Product Institute, South Pacific Commission, College of Tropical Agriculture, University of Hawaii.

PROBLEM IDENTIFICATION

Problems and bottlenecks in small scale processing and storing of root crops are many and varied, depending upon geographical area and conditions. Some of the more general problems are:

1. The availability and acceptance of the raw material. Seasonal changes can affect yield, thus affecting availability. The acceptance of the product can depend largely on cultural practices and beliefs of the consumer.

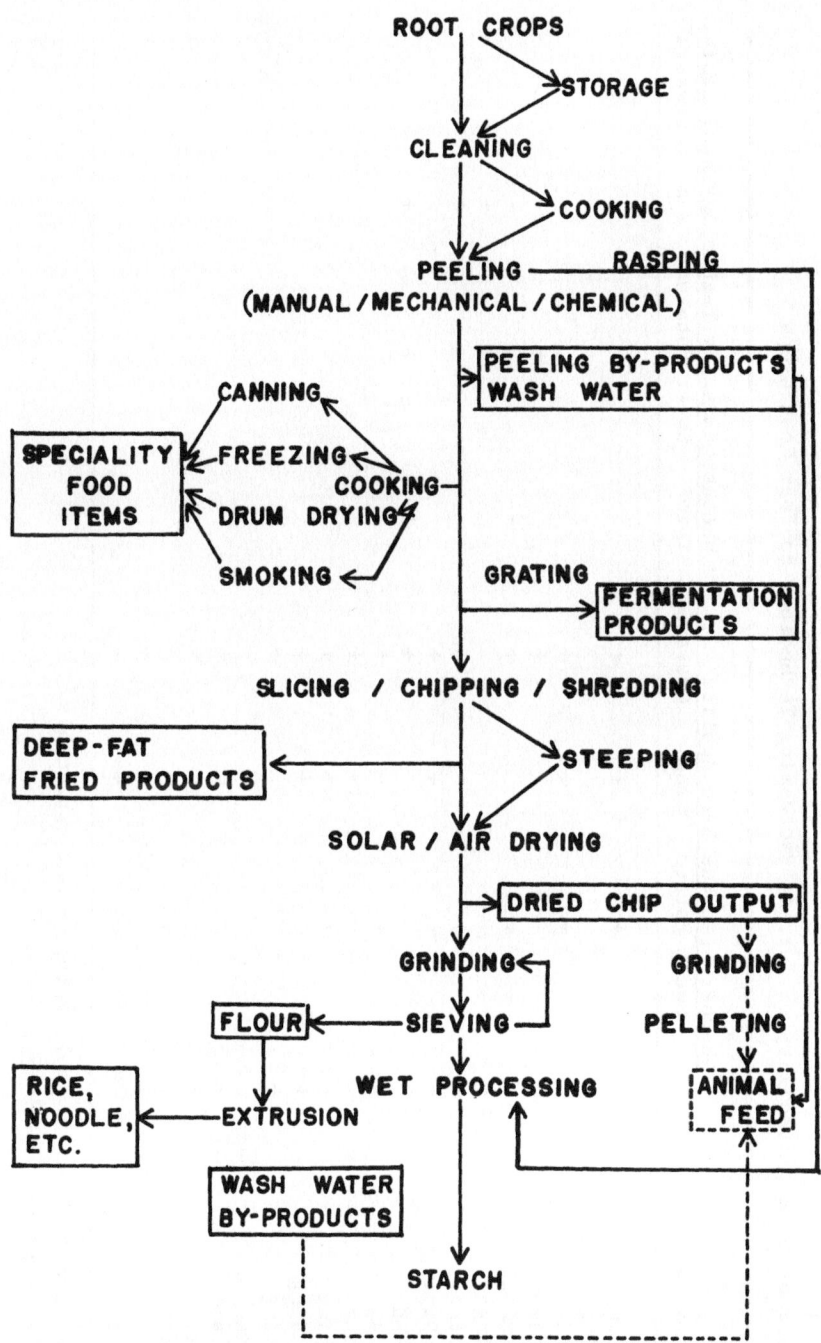

Figure 1. Processing scheme for various root crop products.
Each step can be mechanical or manual.

TABLE 1
Assessment of Applicability of Current Processing Technology to Four Root Crops

Processing Steps	Whether Step Has Been Applied				Whether Existing Technology Applicable			
	Taro	Sweet Potato	Yam	Cassava	Taro	Sweet Potato	Yam	Cassava
Cleaning	+	+	+	+				
Peeling								
Hand	+	+	+	+				
Mechanical	+	+		+	–	+	+	
Chemical	+	+		+			+	
Canning	+	+		+			+	
Freezing	+	+		+			+	
Drum-Drying	+	+	+	+				
Smoking				+	+	+	+	
Grating	+	+	+	+				
Slicing	+	+	+	+				
Chipping	+	+	+	+				
Shredding	+	+	+	+				
Steeping	+			+				
Drying								
Solar	+	+	+	+				
Air	+	+	+	+				
Grinding	+	+	+	+				
Sieving	+	+	+	+				
Wet processing	+*	+	+	+				

+ has been applied or applicable; – not applicable.
* Centrifugation required.

2. The processing procedure is important in affecting the final outcome of the product. Some important factors contributing to this are:
    a. Changes in the drying process (e.g., browning, case-hardening, and change in color of product).
    b. Equipment involved in the processing.
    c. Control of moisture content/water activity during drying.
    d. Simple way of measuring moisture content/water activity.
    e. Sanitary conditions during the process.
3. Packaging and storage are next steps in the general scheme of utilization of root crops. In order to prolong the shelf life of the product, an appropriate technology must be developed that achieves the following desired results:
    a. Prevention of microbial growth and insect, rodent infestation during storage.
    b. Minimizing the effects of light and oxidation by using more efficient packaging techniques.
    c. Finding economical packaging and warehousing to prevent undesirable effects from moisture and temperature.
4. The end products from root crops must meet certain criteria for acceptance; these are usually set by the local populace. Root crops have great significance in traditional and religious ceremonies of many cultures, and perhaps the key element in successful marketing is to retain the root crop in a form that has been recognized throughout the history of the specific region.
5. In many regions cereal starch products are imported when root crop starch could be a good substitute, provided that technology to produce the needed starch existed. One example could be the substitution of grain starch noodles with taro starch noodles, with little change in the final appearance of the end product. The technology for manufacturing these products exists today, but little has been done to apply it at a local level.

31

Additionally, some specific problems in various regions that have been identified are:

1. <u>Transportation and marketing</u>. Transportation problems exist in some regions because of increased production and consumption of root crops. It is necessary to improve market situations in local and metropolitan areas. Also, a processed product should reconstitute into its original state such as the specialty food items shown in Figure 1.

2. <u>Utilization</u>. Animal products are in short supply. To increase meat production root crops should and can be used as animal feeds. This will reduce import of animal feeds. New technology is needed to adapt and improve the utilization of root crops in this capacity.

3. <u>Need to increase production</u>. Increased production of root crops is needed in some regions. Technology of processing root crops should be improved. Current supplies of proteins are limited; this may be supplemented by other sources.

RESEARCH NEEDS

1. Research needed in the immediate future includes the following:

   a. Characterize the physical (mechanical, rheological, and thermal), chemical (cyanide in cassava, acridity in taro, and gums in general), biochemical, nutritive, and organoleptic properties of root crops such as taro, sweet potato, yam, and cassava.

   b. Study the toxic factors in root crops such as cyanide in cassava and the acridity factors in taro. There is an urgent need to determine the varietal differences of these toxicants and to develop simple methods for their quantitative determination.

   c. Assess systematically the effects of processing on nutritive values of root crops.

   d. Improve post-processing handling and storage and, thereby, quality of processed products.

32

e. Evaluate consumer acceptance of various processed products made from root crops.
f. Increase utilization of other parts of the plants.
g. Study the processing economics, especially on energy and labor requirements.
h. Relate processing feasibility of various root crops in terms of some kind of international scales, such as calorie requirements per year.
i. Develop appropriate equipment and test the adaptability of other processing equipment used in other crops to root crop industries in various countries.
j. Assess the economical and social impacts of small scale processing at the village level.

2. Future research needed includes the following:

a. Increase uses of alternate energy sources for processing and storage of manufactured products.
b. Breed the appropriate root crop cultivars for efficient processing and direct consumption.

3. Regional collaboration. The group recognizes the need for collaboration among researchers in various regions who have an interest in the utilization of one or more root crops. The following kinds of collaboration have been suggested:

a. To participate in the South Pacific Commission Regional Technical Meeting in 1980 (tentative) on the "Progress of Root Crop Research in the Pacific Islands," as a continuation of this workshop.
b. To participate in existing research or technical assistance programs in the South Pacific (e.g., USAID- or FAO-supported projects).
c. To establish cooperative studies on the utilization of root crops as alternate energy sources, such as energy from low temperature fuel cells.

4. Possible sources of research funding. The following sources of funding have been suggested for future collaborative work:

a. USAID

b. U.S.--Title XII
c. UNDP
d. Various foundations
e. Local governments

RECOMMENDATIONS

The working group recommends the establishment of an International Research Center on Taro (or Aroids) in Hawaii, or other appropriate countries, to assess and disseminate research results and to coordinate the exchange of information.

It is also recommended that collaborate work on the adaptation of technology be carried out in the appropriate countries.

NOTES

1. Considerations of nutritive completeness should recognize that a total balanced diet is what is desired. It need not be assumed that an individual food be completely balanced. Indeed, it may be much less desirable to have a single balanced food for reasons of monotony and storage problems, rather than two complementary foods which by themselves would be less attractive to insects.

3
Working Group Report: Economic Analysis of Tropical Root Crop Products

B. W. Begley, Cochairman
William Jones, Cochairman
Heinz Spielmann, Rapporteur

H. Miura	T. P. Phillips
M. Muench	G. R. Vieth
P. Phillip	B. Williams
P. Wilson	

INTRODUCTION

The charge given to the members of the economic section was to look critically at the economic aspects of stored and processed products, which need to be examined in assessing the potential of products developed from root crops. This report sets out an approach and a methodology to carry this out. In doing this it is recognized that economic analysis must take into account all of the cost components which determine the competitiveness of a product, and which include and go beyond the costs of processing and storage.

While recognizing that foreign exchange problems or national defense needs can override economic considerations and cause government to deliberately influence supply and consumption, this report has focused on the more normal situation where economic considerations are paramount.

Briefly, this report emphasizes the need to examine closely what products are currently being produced, or that may be developed, which can be competitive in the current marketing system. Critically needed to carry out this task are cost and price data in each marketing situation analyzed.

35

AGRICULTURAL MARKETING AND PRODUCTION

Agricultural marketing, including transporting, storing, processing, and distributing farm products, is usually thought of as separate and distinct from agricultural production. It is not always easy, however, to distinguish when crop growing leaves off and crop marketing begins. Farmers may sell before the crop is harvested, or they may retain title until their crops are sold in a distant city. They frequently hold crops in storage for extended periods before selling them, and in many places may carry out part or all of the processing. In some elemental sense, too, the production of a food crop is not completed until the product is in the hands of its ultimate consumer. Production and marketing activities form a continuum; all are the object of agricultural economic analysis.

This workshop was held specifically "to gather and evaluate the state of knowledge concerning small-scale, low energy-requiring, storage and processing of tropical root crops." However, the economic aspects of storage and processing require that forces affecting the value of various root crops and their products be considered as well, particularly the prices of competing crops and the nature of alternative markets.

The purely economic function of agricultural marketing is the allocation of scarce products over time, space, and form and among consumers. Storage is done by warehousemen, transporting by hauliers, and processing by millers, canners, freezers, and bakers. How well the marketing system performs these allocative functions determines to a large extent the profitability of growing, storing, transporting, and processing of tropical root crops. Marketing efficiency is a matter of achieving optimum allocation at minimum cost. The marketing system is not performing its principal task if shortages and gluts occur side by side, or if seasonal prices rise by more than the cost of storage, or if prices between different locations differ by more than transport costs, no matter how small the spread between farm price and retail price.

COMPETITIVE PRODUCTS WITH ECONOMIC MARKETS

The challenge is to develop products that are competitive with existing products and that can

find a market large enough to justify their production and handling. Analysis of the potential of
root crop products must concern itself with convenience, packaging, and taste factors, the price
of alternative products, and specific market segments.

If a product is to compete with rice, for
example, it should be packaged as attractively and
conveniently as rice; it must be easy to prepare
and be acceptable in terms of taste; and it must be
competitive with rice in terms of price per pound
or per calorie. Most importantly there must be a
large enough group of potential buyers with the
discretionary income to buy the potential root crop
product.

DETERMINANTS OF COMPETITIVE POSITION

Before pushing the first button or kicking the
first starter of a grinder, peeler, or slicer
designed to turn out a new root crop product, one
should carefully examine what products are currently being produced from taro, sweet potato, yam, and
cassava, and of these, which appear to have the most
potential.

Other questions could also be asked relating to
other food products that might yet be developed.
But at this point it would seem to be wiser to look
at what has been or is being developed and work
from there. With a range of products including
bread, sweet potato "rice" (Papua New Guinea), sweet
potato and taro flour and starch, poi, kulolo
(Hawaii), and ice cream (Philippines), there would
appear to be plenty to focus on without venturing
into the world of what might be developed.

While the process of bringing a product from
the producers to consumer is a continuum, we will,
for ease of analysis, break the process into some
identifiable elements. There are three areas which
determine the competitiveness of a product: cost of
marketing activity, demand, government policy.

Cost of Marketing

Supply and prices of raw materials. Among the
critical factors determining the supply of any
(root) crop are the physical production parameters
such as variety, climate, soil characteristic and
water availability.

37

A function of price is to elicit the supply of raw material in quantities which the market requires. Price measures the cost of raw material to the merchant or manufacturer and to other elements of the marketing channel, and it takes into account costs of production and competing demands for the raw material.

Cost of storage. Storage meets the requirements of time utility. For some root crops, e.g. sweet potato, cassava and taro, harvest can be delayed, sometimes for as much as 18 months. The cost of this "live" storage depends upon the alternative uses to which the land can be put (i.e., upon its opportunity cost). This opportunity cost can be very low or zero during the dry season or in cases where the land will revert to fallow. On the other hand, it can be substantial where no alternative to such storage is available.

Costs of post-harvest storage depend upon the quantity stored, duration, marketing channels, and the technology used. Calculation of this cost has two components, product losses and the physical cost of storage. These costs may be borne by the farmer himself, the transporter, the processor, or by a combination of the above.

Cost of transportation. Transportation meets the requirements of place utility. Transportation costs are incurred in the movement of the raw material from the farm to the assembler or to the site of processing or to the final market. It is possible for transportation to substitute for storage by allowing products to be collected from regions with different harvest dates. Transportation costs may fall into several categories: capital costs, operating costs, and labor costs (including opportunity costs of household labor). An additional, and in some cases quite major, component of costs may result from damage and spoilage incurred during transportation.

Processing costs. Processing satisfies form utility and transforms materials into products of higher value. It may at times be a substitute for storage. Frequently special processing of tropical root crops is required to remove toxic principles. Processing costs vary with the degree of transformation, the scale of processing and alternative opportunities for employment.

Marketing costs. The preceding three costs relate to the physical handling of the products. Additional marketing costs pertain primarily to service costs which are required to bring the

various elements of the channels of distribution
together. These include such costs as stall
charges, license fees, payments to market intermediaries, and interest on funds tied up in physical
stocks.

Demand

The amount of root crop products which can be
expected to be bought depends on:
1. The price of the product
2. The disposable income of consumers
3. The prices of competing (substitute)
   products
4. Consumer preferences

To establish the relationship between prices
and amounts sold of a root crop product, we need to
study the price elasticity of demand at various
price levels. Price elasticity of demand relates
change in purchases associated with a change in
price.

We also have to measure the change that occurs
in the consumption of our root crop product with
a change in consumer income which is the income
elasticity of demand.

The cross elasticity of demand measures how
purchases of a given product are influenced by
changes in prices of competing products. It is
also a measure of the substitutability of one crop
for another or one product for another and a measure of the intensity of competition between them.

The price elasticity of demand for all of
these commodities as staple foods is negative, that
is, purchases vary inversely with price, although
for "superfoods" like taro and yams price elasticity tends to zero. Income elasticity of demand for
taro and yams is high but less than one, for sweet
potatoes it is probably low but usually positive,
and for cassava products it tends to be negative
except for the poorest parts of the population.
In general income elasticity of demand for these
products declines as incomes rise.

Cross elasticities among the cereals and the
tropical roots are high and substitution among them
is easy. However, there tends to be a general hier-
archy of preferences. Cereals are usually preferred to root crops and the "white cereals," wheat
(flour) and rice are preferred to the coarser
grains. Among the root crops taro and yams sometimes partake of the characteristic of super foods

but there may also be strong local preference for sweet potato and cassava.

Consumer tastes and preferences affect the demand for a root crop product. Taste and preference are influenced by such factors as:
1. National and local cultural factors and religious beliefs;
2. Social factors relating status to consumption of the product; habit; and miscellaneous factors such as time (season, work day versus holiday, etc.), climate and weather, locality.

Finally, over time the effects of change in tastes and preferences for root crop products through such influences as education, changes in product quality, reliability of supply, merchandizing, and sales promotion need to be investigated.

Government Role in Agricultural Marketing

A recent OECD/FAO report on "Critical Issues on Food Marketing Systems in Developing Countries" refers to the general "over-expectation of the contribution made by cooperatives and state enterprises which could perform effectively only under well-defined conditions, of which policy makers are often not aware," and of the tendency to underrate considerably the potential contribution made to small-scale farmer development by private marketing enterprises.[1]

Farm marketing is relatively complex, especially when commodities must be collected in small losts from many producers, assembled in large lots for transport, and broken once again into small lots for sale to consumers in various urban areas. This requires a multitude of unhesitating on-the-spot decisions that are best made by highly-motivated merchants who can enjoy the rewards but must also bear the costs and risks of their buying and selling decisions. If there are many traders in the markets, and if access to markets and to supporting services, including credit and market information, is on purely economic terms, the market will afford optimum prices to producers and consumers alike, and optimum allocation of goods as well.

Government can contribute most directly to agricultural marketing by keeping interference in the free movement of goods within its boundaries to a minimum and by reducing the costs of marketing functions. Government responsibilities should include the provision of roads and ports since private investors find it difficult to accumulate funds sufficient to this task and even more difficult to recapture the benefits they generate.

Government has a responsibility to provide an economic environment in which there are marketplaces with regular market meetings, communication is rapid and reliable, information about market conditions is available generally, access to markets is not restricted by uneconomic considerations, credit facilities are widespread, and security of contract, including certification of quality and quantity, is assured. Some of these services can be provided by merchant associations or organized marketplaces, but all are proper concerns of government.

Before steps can be taken to improve marketing efficiency, it is essential that the performance of present marketing systems for tropical root crops and competing products be examined and appraised. Preliminary evaluation is possible simply through price analysis in countries and areas where a record of prices over time and space is available. When such data are not available, crude first approximations can be made on the basis of ad hoc price surveys extending over a few weeks. For proper analysis, however, more elaborate price and market cost surveys and interviewing are essential. Unfortunately, such studies are expensive, but they are likely to be less expensive than ill-informed governmental intervention.

CONCLUSIONS

For the economists the major concern when evaluating root crop products is their competitiveness. While this is an expected conclusion, it is often overlooked in the evaluation of storage or processing methods. Such evaluation should begin with an examination of existing tropical root products (locally produced or produced in other countries) and an assessment of domestic market performance. A critical part of such a study is the work carried out on the "determinants of competitiveness" which was discussed in the Competitive Products section.

APPENDIX

Some questions were asked and comments made following the presentation of the economics report. As this discussion was not recorded it is only possible to set out some of the questions that were raised in a formal and somewhat condensed form, and then list some of the comments that were made.

QUESTION 1: Could your report have been more specific? As it stands this report would appear to be as applicable to fish products as to root crop products.

ANSWER: The approach is the same whatever the product one is dealing with. The particular problem is location and time specific. One looks at the competitiveness of a root crop product in a particular region or country at a particular time using the data available in situ.

QUESTION 2: Do we have some data available?

ANSWER: Yes, we have some available. In a paper prepared for this workshop,[2] Truman Phillips quoted a C.I.F. price for Thai cassava chips of $70 per ton. Bill Jones quotes the prices per thousand calories for various starchy staples in different African nations.[3] Clive Pedrana quoted a cost at the factory door for taro in Western Samoa of 6 to 10 cents (Samoan) as against 2 cents (Samoan) for cassava. John Watson reported a price of 10 cents (New Zealand) that the Samoan shipper receives for his taro if the auctioned price for a 60-pound case is $15.

But in general prices for root crop products and for competitive products must be collected in country, and the component costs that determine the final costs in the store or on the street will most likely differ from country to country and from town to town within a country.

Comments

1. It is important to distinguish between products destined for home use and those destined for the marketplace.
2. Economics would seem to have difficulty dealing with purely subsistence production and assigning a value to what may be the most important use made of root crops.
3. Few people grow root crops purely for their own use, but sell or barter at least some of them.

4. It is important to recognize and examine the continuum of root crop utilization from subsistence to small-scale to large-scale processing and to recognize the costs at various stages in the continuum.
5. Attempts should be made to standardize and measure cost per calorie for various root crops. But while this is acceptable to the scientist it was pointed out that it does not really work in the marketplace. Consumers buy particular foods, not calories; however, some data have been compiled on this basis.

NOTES

1. OECD/FAO, "Critical Issues on Food Marketing Systems in Developing Countries," Development Centre of the Organization for Economic Cooperation and Development in cooperation with the Food and Agriculture Organization of the United States, Paris, 1977.
2. Truman P. Phillips, "The Implications of Cassava Processing and Marketing for Other Tropical Root Crops," University of Guelph, Ontario, Canada, June 1978.
3. William O. Jones, Manioc in Africa, Stanford University Press 1959.

Section II
Processing and Storage in Selected Countries

1
Storage and Processing of Root Crops in the South Pacific

Michel Lambert

ABSTRACT

The South Pacific Commission (SPC) is an
advisory and consulting body in Noumea, New Caledonia, founded to promote the welfare of Pacific
peoples. Research conducted under the auspices of
SPC has led to this overview of root crop storage
and processing throughout the South Pacific.

Yam, taro, cassava, sweet potato, arrowroot,
and Irish potato are discussed in terms of specific
problems associated with each. Yam is cultivated
to the greatest extent throughout all the Pacific
islands. It has very high prestige, particularly
in New Caledonia and the Trust Territory of the
Pacific Islands, to the point of undergoing
ritualized cultivation in both of those cultures.
Taro, also cultivated throughout the Pacific, is
difficult to store due to rapid spoilage.

Less widely distributed but no less significant are cassava, which is rising to prominence as
part of a new animal feed industry in both Tonga
and New Caledonia; sweet potato; arrowroot, used
for cooking and for making alcoholic beverages; and
the Irish potato.

YAMS

Yam has a very high prestige in all Pacific
islands, mainly in the Loyalty Islands, New Caledonia, and Ponape, Trust Territory of the Pacific

47

Islands. In New Caledonia the Chief decides where, when, and how yam is to be planted and harvested. In Ponape, cultivation of yam is a top secret activity for men, and yams of large size are especially prized. Yams are classified according to the number of men required for harvest, hence the terms "two-man yam," "four-man yam," "six-man yam" are used.

Semi-mechanized yam cultivation occurs in the Kingdom of Tonga, where the crop is planted mechanically, and New Caledonia, where planting and harvesting machines are used.

Storage

Yam is harvested very carefully in order not to wound the tuber. If wounded, ash is applied to the wound. Yams are not washed with water, but are cleaned of mud or soil. In New Caledonia, tubers are put on shelves made with bamboos and poles in a well-ventilated shed. Sometimes they are stored in a dark, cool place for up to six months. Shelters are protected against rats and insects using mosquito nets or other types of nets.

TARO

Colocasia esculenta

Colocasia esculenta is cultivated in all the Pacific islands, mainly Fiji, Western Samoa, American Samoa, Tonga, Niue, Cook Islands, French Polynesia, New Hebrides, Solomon Islands, and Papua New Guinea.

In Melanesia, mainly in New Caledonia, it was cultivated in the past on irrigated terraces.

It seems that "dasheen" or taro "de Chine" (C. esculenta var. globulifera), which can be stored up to three months in India,[1] is different from C. esculenta var. antiquorum, which does not store as well. There are many taro varieties in the Pacific: 72 in Fiji, 38 in American Samoa, and 17 in Palau.

Storage. Storage of taro corms is difficult. Three papers presented at this workshop (Jackson, 1978; Tupuola, 1978; and Chandra, 1978) describe storage problems in their respective countries. Plant quarantine regulations must be very strict to protect the islands against pests, diseases, and viruses, mainly Phytophthora colocasiae, Phythium,

48

and the taro beetle (<u>Papuana</u> <u>armicollis</u>). All of
these can affect stored taro.

Xanthosoma sagittifolium

This taro is known as New Hebrides taro in New
Caledonia, Samoan taro in the New Hebrides, palagi
taro in Samoa, Ote Sapan in Truk, and Trukese taro
in the Marshall Islands.
<u>Xanthosoma</u> is becoming more and more popular in
the islands, especially in New Caledonia, New Hebrides, and the Trust Territory of the Pacific Islands.
No trials on storage or processing are being
done at the present stage in the Pacific islands.

Alocasia macrorrhiza

This taro is cultivated as an important secondary staple food in Tonga, Wallis and Futuna, where
it is called "kape," French Polynesia ("ape"), and
Western Samoa ("ta'amu"). In some islands--New
Caledonia and New Hebrides, for example--leaves and
corms are very irritating due to the high amount of
calcium oxalate raphides. For this reason it is
often considered as a wild taro or an emergency food.
The Marshallese have learned to cook it with green
coconut husk juice, which reduces the irritation.
<u>Storage</u>. This taro can be stored in a fresh
room up to approximately three weeks (see Tupuola,
this volume).
<u>Processing</u>. No processing is being done at the
present time in the Pacific islands.

Cyrtosperma chamissonis

This is the most common and widespread taro
species in the atolls. It is known as "ieraj" in
the Marshalls, and as "babai" in the Gilbert Islands
and Tuvalu. It is also grown in the Caroline
Islands and other atolls. This is a giant swamp
taro which is an important food for many Pacific
people (Plucknett, 1976).
No trials on storage or processing are being
done at present in the Pacific islands.

CASSAVA

Cassava is cultivated more and more in the
Pacific islands, mainly in Palau (TTPI), Tonga, and

49

New Caledonia. New Caledonia is planning to create
an animal feed industry for which cassava will be
cultivated, harvested, and processed mechanically.

Storage

There are no storage trials being conducted in
the Pacific islands.

Processing

Tonga. A small-scale industry for animal feed
was operating a few years ago.
Fiji. See Chandra (1978) in this volume.
Western Samoa. See Pedrana (1978) in this
volume.

SWEET POTATO

Sweet potato is very popular in Tonga, the Cook
Islands, some islands in Micronesia, and in the
highlands of Papua New Guinea. There are 278 different types of sweet potato in the collection of
the Agricultural Station, Aiyura, Papua New Guinea.

Storage/Processing

For information on Papua New Guinea, see Siki
(1978) in this volume.

ARROWROOT (TACCA LEONTOPETALOIDES)

Arrowroot is grown widely in the Tokelau
Islands, Niue, Tuvalu, Gilbert Islands, and on the
atolls of the Marshalls. The Marshallese cultivate
arrowroot ("makmok"). They plant from May to
October, and harvest from November to April.

Processing

During processing, men, women, and children are
involved. They start by grinding the tubers and
washing them with salt water. They place a canvas
under the filter box in order to catch the starch in
the water; starch settles on the bottom of the canvas in 3 to 4 hours. After that they drain out the
water from the canvas and keep the starch. The

starch is dried in the sun for one week. When dry, it can be stored for many months.

Arrowroot starch can be cooked in many ways, and is used with coconut milk, tuba (toddy), pandanus juice, etc.

IRISH POTATO

Irish potato is cultivated during the cool season in New Caledonia and the island of Tanna, south of New Hebrides (April to October). The average yield in New Caledonia is 15 t/ha.

Storage

Irish potatoes are stored in cool rooms in Noumea (500 t capacity) and Bourail (200 t capacity) at 12°C for 4 to 5 months. A hormone treatment to prohibit germination is carried out just before placing the tubers in the cool room.

Processing

A small-scale industry exists in Noumea at the family level in making chips and french fries.

SOUTH PACIFIC COMMISSION PROJECTS RELATED TO ROOT CROP PRODUCTION AND PROCESSING

1. A Regional Meeting on the Production of Root Crops was organized in Suva, Fiji, from 24 to 29 November 1975. All working papers are collected in the SPC Technical Paper No. 174.

2. A Food Crop and Home Economics Project at the Village Level has been started and will run for three years (1978 to 1980) in three Pacific Islands: Hapaii (Kingdom of Tonga), Tutuila (American Samoa), and Palau (TTPI).

3. A Regional Technical Meeting on Atoll Cultivation will be organized in 1979.

4. The SPC Handbook, Taro Cultivation in the South Pacific, will be published shortly.

NOTES

1. Dr. N. Hrishi, personal communication.

REFERENCES

Plucknett, Donald L.   1976.   Giant swamp taro
  (Cyrtosperma chamissonis), a little-known Asian-
  Pacific food crop.   Proc. Fourth Symposium of the
  International Society for Tropical Root Crops.
  IDRC, Ottawa, Canada.

*Michel Lambert: South Pacific Commission, Noumea, New
Caledonia*

2
Handling, Storage, and Processing of Root Crops in Fiji

S. Chandra

ABSTRACT

In Fiji, small-scale processing of the tropical root crops--ginger, taro, and cassava--are carried out on a very limited scale by only one or two firms. Some of the products are dried ginger, crystallized ginger, ginger pickles and chutneys, taro and cassava chips, and cassava starch. Some associated problems are high wage rates which reduce the viability of the industries, irregular supply of raw materials to the factories, a high degree of variability in the raw materials which affects the quality of the final product, high prices of some crops which results in high cost production and marketing, unavailability or ignorance of appropriate processing technologies, and lack of basic research in food technology. To resolve some of these problems, a food processing laboratory is to be established with the Department of Agriculture. The object of this laboratory will be to conduct research into ways of utilizing and preserving tropical roots and other crops grown in Fiji. Hopefully, the research results will be used by the private sector to establish new processing industries in Fiji.

GINGER

Production

Green ginger production for overseas export is
a relatively new industry in Fiji, although the
cultivated ginger Zingiber officinale (now called
the local cultivar) seems to have been introduced
to Fiji in the early days of European settlement.
By the early 1950s regular exports of green ginger
were firmly established (Haynes, Partridge, and
Sivan, 1973). In 1966 green ginger export totalled
363 metric tons (t), but by 1977 this had risen to
1853 t worth over $1.5 million in foreign exchange.
The major importing countries are the U.S.A., Canada, and the United Kingdom, with smaller amounts
going to Hong Kong and New Zealand.
Ginger is produced by about 50 Chinese, Indian,
and Fijian smallholders. The main production areas
are the hilly, moderately fertile humic latosols of
the Waibau and Sawani areas of southeast Viti Levu
Island (Twyford and Wright, 1965). The total
annual rainfall in these areas is about 3430 mm,
with a fairly uniform monthly distribution, although
the main rainy season is between November and April.
Yields of green ginger are high by world standards,
and average about 35 t/ha over the whole production
area. Although high cash inputs in the form of
fertilizers and hired labor are necessary to attain
such high yields, the gross margins from ginger
enterprise are also high, averaging about $6000/ha.
Ginger is planted between September and December, and the harvesting of mature ginger normally
begins in August. The crop is ready for harvest
when the foliage dries and separates from the rhizomes. The rhizomes are lifted by hand, using
forks, hence the task is extremely labor intensive.
The larger rhizomes are broken into smaller pieces,
sorted, and the diseased and poor quality ginger
are rejected. The main disease of ginger is tuber
scale (Aspidiella hartii). However, the major
economic losses are caused by the burrowing nematode
(Radopholus similis) (Butler and Vilisoni, 1975) and
the root knot nematodes (Meloidogyne arenaria, M.
incognita and M. javanica).[1] These are the most
important internal parasites of the ginger rhizomes
and can be transported in fresh ginger produce.
After sorting, the suitable rhizomes are packed
in crates and taken to nearby creeks for washing.
Up to this point the major problems are: (1) Some

54

farmers fail to understand that rejected ginger, whether poor quality or diseased, should not be used as planting material but should be destroyed. (2) Hand harvesting of ginger is very costly, with the current rural wage rate ranging between $0.60 to $0.80 man-hours, thus restricting the acreage a smallholder can grow. There are no suitable mechanized systems of harvesting ginger on the steep slopes where most of the production is centered. Research trials have shown that ginger grows poorly on flat, alluvial soils on which harvesting can be achieved by animal drawn moldboard plow. (3) Quarantine regulations restrict the export of green ginger to some countries because of the possibility of nematodes in the consignments. For example, fresh ginger cannot be exported to Japan and New Zealand because of the danger of accidentally introducing the burrowing nematode (Radopholus similis) which would create a serious threat to their citrus, banana, and black pepper industries.

Post-Harvest Handling and Storage

At the wash points, dirt is thoroughly washed off the rhizomes and the larger ones which are suitable for export are repacked in crates and transported to the packing shed in Suva. The practice of using wash point rejects for planting material creates a problem because farmers may be planting genetically inferior ginger.

At the packing shed the rhizomes are air dried and the final selection and grading take place. The rhizomes are packed into wooden or cardboard crates for shipping or air-freighting. During this time the ginger is inspected by government produce inspectors. The consignments are fumigated with methyl bromode at 0.05 kg/m³ for four hours, which should kill all nematodes in surface lesions.

Processing

In Fiji the important commercial products of ginger are dried, ginger, crystallized ginger, and pickles and chutneys. One small factory, located in Suva, processes between 1000 to 2000 kg of ginger per day.

Dried Ginger. There are two forms of dried ginger, namely ginger sticks and powdered ginger. Only mature ginger which has been in the ground for about ten months is used for the manufacture of

dried ginger. Rhizomes are broken into 8 cm pieces
to be dried by the sun or in an oven. Sun-drying
takes 6 to 8 days and can be done on a clean surface
such as a concrete floor. Poor quality ginger
results from sun-drying, as the sticks turn black
with mold unless they are chemically treated,
although there is no loss in essential oils (Sills,
1970). Oven-drying of ginger is practiced by the
above factory. To prevent fungal growth and prolong
shelf life, the moisture content needs to be reduced
to below 7 percent. Powdered ginger is obtained by
grinding down ginger sticks using hand or electric-
ally operated machines. Only a small amount of
ginger sticks and powdered ginger is consumed in
Fiji; most is exported.

Crystallized Ginger. Ginger harvested early,
at about 6 to 7 months of age, which has tender,
non-fibrous rhizomes, is used to manufacture crystallized ginger. Figure 1 shows the processing
diagram utilized at the Suva factory. The tasks
that consume the most labor--peeling and cutting the
ginger into small cubes--are done manually. In Fiji
it has been estimated that a good worker can peel
only 5 to 6 kg of fresh rhizomes per day (Sills,
1959a). However, if the rhizomes are steamed first,
as is done in the above factory, peeling time is
reduced considerably, with four workers peeling 1000
kg of rhizomes per day. No suitable machines have
been designed to peel ginger rhizomes. Rough peeling reduces the value of the ginger, because most of
the oil-bearing cells are located within 0.4 mm of
the surface. Most of the crystallized ginger is
exported to Australia, Western Samoa, and Tonga.

Since 1976 the National Marketing Authority of
Fiji has been exporting ginger preserved in brine
to New Zealand, where the product is turned into
crystallized ginger. The rhizomes are washed,
peeled, and preserved in brine solution in large
vats which are treated with SO_2, derived from sodium
metabiosulphite, for about 3 to 4 months. The
growth of spoilage microorganisms is prevented by
the high concentration of salt and the low pH of the
solution, usually about 3.5 or less. The pH can be
further reduced by addition of natural fruit acid or
vinegar. After the treatment period the material is
packed into drums and exported. In 1977 20,000 kg
of such treated ginger was exported to New Zealand.
After storage in brine, the salt can be washed out
by repeated soaking in clean, fresh water. A final
boiling treatment may be necessary to remove the
last traces of salt.

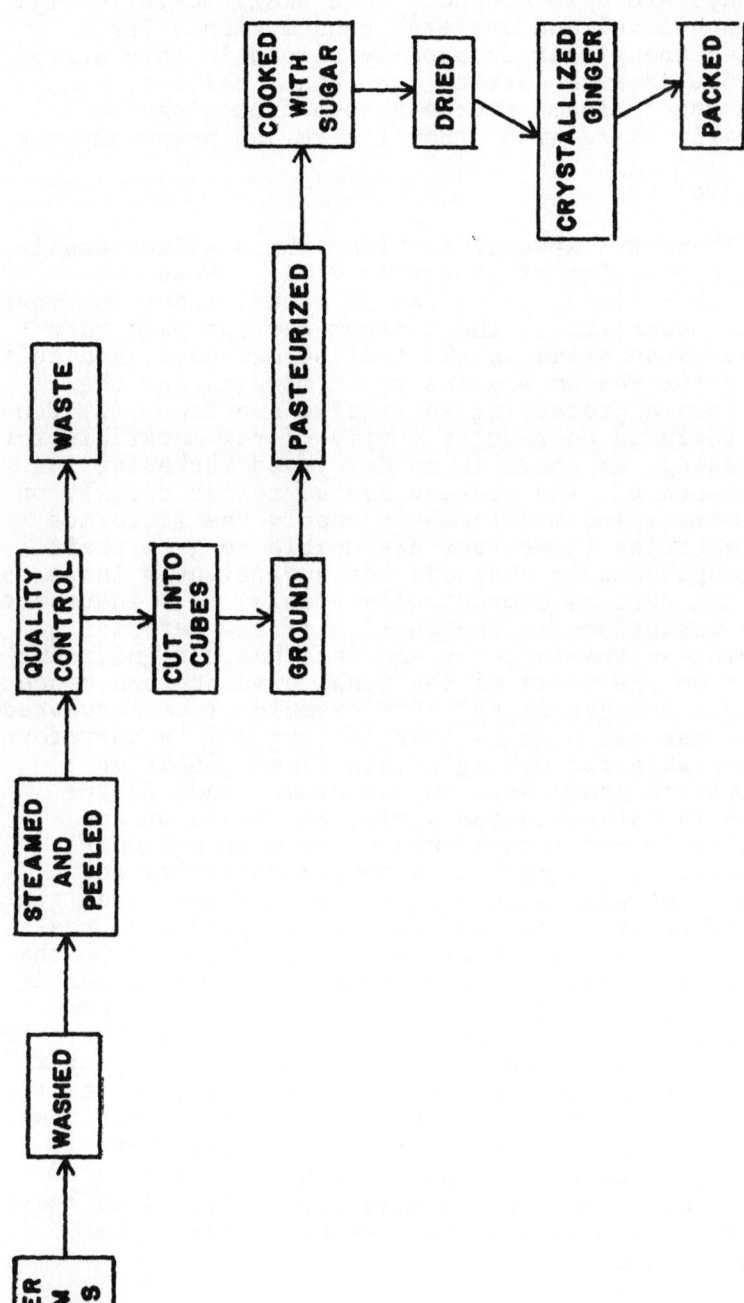

Figure 1. Processing diagram for crystallized ginger.

Pickles and Chutneys. Ginger pickles and chutneys are also produced on a small scale in Fiji for both local and overseas consumption. For pickles the ginger is usually sliced in thin strips and bottled with carrots, gherkins, chilies, and other vegetables. Chutneys made from ginger are generally mixed with other fruits and preservatives.

Problems

There are several problems which affect small-scale processing of ginger in Fiji. These are: (1) High reliance is placed on manual labor for most of the operations. The minimum current wage rate in the urban areas is about $1.00/man-hour, and this is a major reason why the gross margins for the small-scale processing enterprise are low. (2) Usually there is no regular supply of raw materials for processing, as there is no developed marketing infrastructure. The present system relies totally on the farmers and middlemen to supply the factories. The factories themselves are unable to grow their own crops because they are not sufficiently large to make the systems economically viable. (3) There are large variations in the quality of raw material arriving at the factory, and this has a significant effect on the taste of the final product, and hence consumer acceptability. For example, late harvested ginger has too high a fiber content and is therefore unacceptable for making crystallized ginger or pickles and chutneys. On the other hand, if the ginger is harvested too early, the taste of the products is not strong enough for most consumers. (4) There is a large gross margin disparity for farmers between early ginger and late harvested ginger for crystallized and dry ginger products, respectively. Early ginger harvested at 6 to 7 months yields only about 17 t/ha, whereas if the ginger is left in the ground for another 3 to 4 months the yields double. The problem for farmers is that this yield differential between early and mature ginger is not sufficiently reflected in farmgate prices for the two commodities, as the early ginger generates only a 20 percent higher price. Because there is very little extra cost once the ginger as reached 6 to 7 months of age, farmers are inclinded to leave ginger until maturity and thereby derive higher gross margins.

TARO

Production

The processing of taro (<u>Colocasia esculenta</u>) into chips is a new small-scale industry in Fiji, less than two years old. The important varieties of taro grown in Fiji are Samoa, Tausala-ni-Samoa, Vutikoto and Vavai Dina on the hills, and Toakula on the flats. The main production areas are the wetter zones of the main islands of Viti Levu, Vanua Levu, Taveuni, and Kadavu. For most farmers in Fiji all tasks involved in the production of taro are done by hand. Some commercial taro growers carry out the initial cultivation with animal power. A few farmers use animal drawn plows for the harvesting task; the majority use forks. The average yield of taro is about 9 t/ha, and the total national production could be about 17,000 t/yr (Chandra, 1977). Most of the yield is consumed as subsistence products on the farms, while the remainder is marketed in the main urban areas of Suva, Lautoka, Nadi, Nausori, Ba, Sigatoka, and Labasa. About 350 t/yr of taro are processed into chips by one small factory in Suva.

Post-Harvest Handling and Storage

Fresh taro corms can be effectively stored for about two weeks only, at ambient temperatures, after which corm rot results. Storage life can be extended if the corms are stored in cool rooms, but this is not practiced on a commercial scale in Fiji. After harvesting, the taro corms are scraped free of soil and the leaves cut off, but the petioles are left intact to prolong storage life. For urban marketing the petioles are left on the corms, tied, and sold in bundles of three to six.

For processing, the petioles are cut off and the corms delivered to the factory on the day of harvest so as to reduce deterioration. The factory relies on farmers and middlemen for its supply. Usually truckloads of 2-5 t of corms are delivered at a time.

Processing

In Fiji the only processed product of taro is chips. Figure 2 shows the processing diagram for the manufacture of taro and cassava chips at the Suva factory mentioned earlier. Both crops are processed identically. For good eating quality both

59

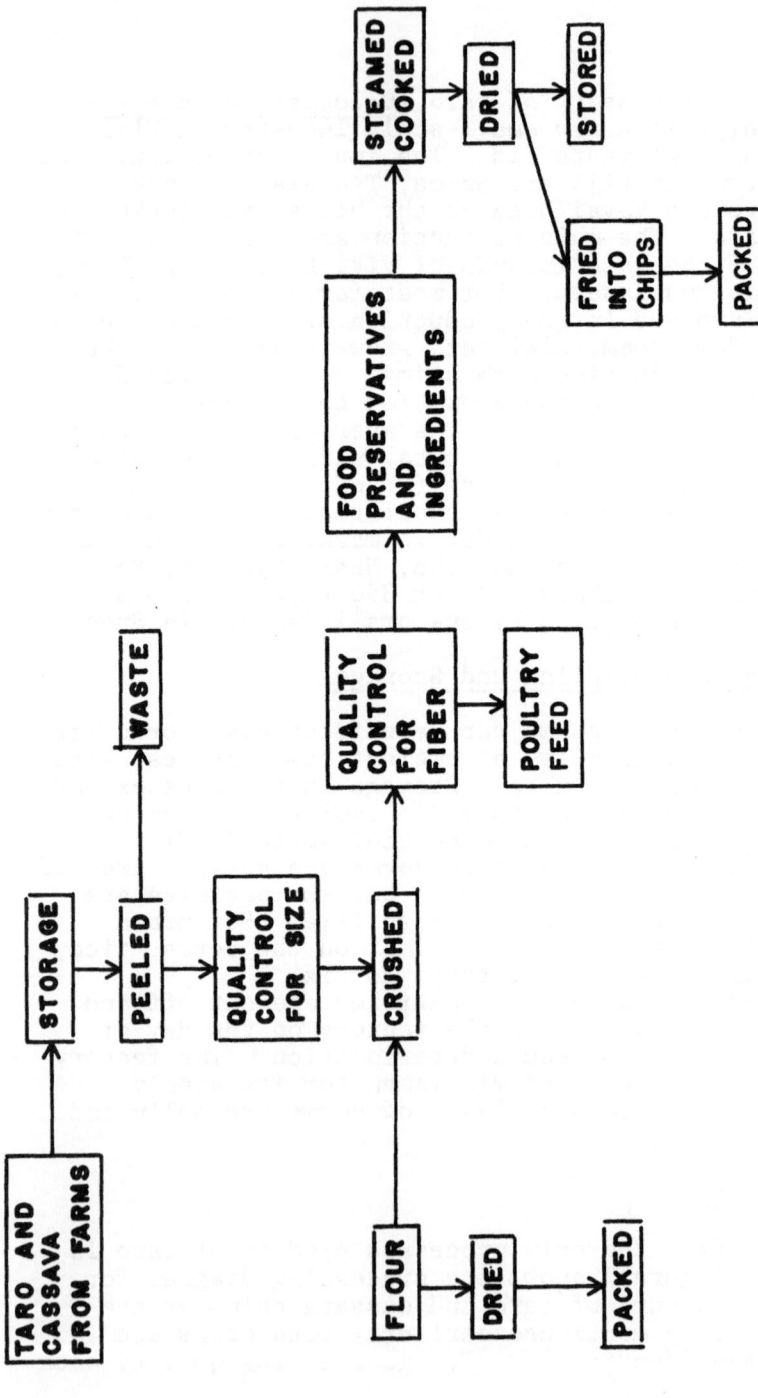

Figure 2. Processing diagram for taro and cassava chips.

crops have to be processed quickly, taro at least the day after harvest, and cassava tubers on the day of harvest. The conversion rate from fresh material to chips, in terms of weight, is 5:2.

Taro chips, which are usually about 2 to 3 cm in diameter, are packed into 57 g cellophane bags. The product is mostly exported to Australia, but there is a growing market locally as consumers become aware of the possibilities of substitution for potato chips.

Problems

Most of the problems in the processing of ginger, such as high wage rates, irregular supply and variability in the quality of raw material, are also applicable to taro processing. A major problem that faces this industry is that taro, being a staple food crop of urban and rural Fijians and thus in constant demand, has a high retail price. Farmers are therefore often unwilling to sell to the factory, which can only offer a lower price. It is possible that mechanized cultivation and harvesting of taro presently being investigated by the Research Division of the Department of Agriculture, may result in taro grown on commercial farms at a lower cost per unit, which would be beneficial for the development of further taro processing industries.

In the factory, quality control is extremely important and, for this reason, all corms have to pass a size and fiber content grading standard before materials are used for processing.

CASSAVA

Production

At present two processed commodities, chips and starch, are derived from cassava (Manihot esculenta) in Fiji. Research is being carried out to evaluate cassava as pig and poultry feed in Fiji, but no definite conclusions have been reached. The important varieties of cassava in Fiji are Vulatolu, Merelesita, Beqa, Yabia Tamu, and Sokobale, and these are grown in all climatic zones. As for taro, all tasks are generally done manually, although a much greater use is made of draft animals for plowing and ridging.

All harvesting is done by hand, although there is one commercial farmer near Suva who uses a

tractor drawn moldboard plow. Currently, considerable research effort is being placed on improving
this method of harvesting. The average yield of
cassava is about 10 t/ha, and the total annual production in Fiji could be about 35,000 t (Casley,
1969). Most cassava is used for subsistence consumption on farms, and together with other traditional root crops, accounts for 51 percent of the
daily kilojoule intake of the rural Fijians (Chandra
and De Boer, 1975).

Post-Harvest Handling and Storage

Fresh cassava tubers at ambient temperatures
have a short storage life, usually about a week, by
which time deterioration in the form of streaking
and later by rotting sets in. Cassava for processing into chips must be sent to the factor in Suva
within a few hours of harvesting.

Processing and Problems

The processing diagram for cassava chips is the
same as for taro and is shown in Figure 2. About
350 t of cassava are used annually to make chips at
the factory mentioned earlier.

Cassava starch was produced on a small scale by
the Fijian Co-operative Market Association, Ltd., at
Nausori, at the beginning of the 1950s (Sills,
1959b). This project was abandoned a few years
later because of (1) irregular supply of tubers to
the factory which relied on crops produced by smallholder smi-subsistence farmers living in the vicinity; (2) too low a purchase price paid to the
farmers; and (3) total reliance on sun-drying, which
meant that production was sporadic because of pre-
vailing weather conditions. At present, some very
small units are producing cassava starch for the
local market, but the bulk of starch consumed in
Fiji is imported. It is likely that this trend will
continue, as the price structure of the raw material
and the total consumption in Fiji are against the
establishment of a large-scale starch factory based
on cassava.

NOTES

1. M.F. Kirby, Research Division, Department of
   Agriculture, Fiji, personal communication.

REFERENCES

Butler, L. and Vilisoni, F. 1975. Potential hosts of burrowing nematode in Fiji. Fiji Agric. J. 37:38.

Casley, D.J.L. 1969. Report on the Census of Agriculture, 1968. Legislative Council Paper No. 28, Govt. Printer, Suva, Fiji.

Chandra, S. 1977. The production, marketing and consumption of root crops in Fiji. Paper presented at the Conference on the Adaptation of Traditional Systems of Agricultural Production to Serve the Needs of a Developing Market Economy, 3-7 October 1977, Honiara, Solomon Islands.

_____ and De Boer, A.J. 1975. Root crops and diets in two Sigatoka Valley villages. Trop. Root and Tuber Crops Newsletter 8:19.

Haynes, P.H., Partridge, I.J., and Sivan, P. 1973. Ginger production in Fiji. Fiji Agric. J. 35:51.

Sills, V.E. 1959a. Ginger products. Fiji Agric. J. 29:13.

_____. 1959b. Cassava starch. Fiji Agric. J. 29:16.

_____. 1970. Note on the preservation of ginger. Fiji Agric. J. 32:33.

Twyford, I.T. and Wright, A.C.S. 1965. The Soil Resources of the Fiji Islands. Government of Fiji, Suva.

S. Chandra: Research Division, Department of Agriculture, Koronivia Research Station, Nausori, Fiji

3
Processing and Storage of Root Crops in Papua New Guinea

Beka F. Siki

ABSTRACT

Traditional methods of storage or storage avoidance techniques for root crops are important in the various subsistence economies in Papua New Guinea, as most people are directly dependent on root crops at the subsistence level. These methods are reviewed for the major root crops (sweet potato, aroids and yams). These include both pre-harvest and post-harvest techniques.

Additional storage methods which are not widely used but which have been evaluated and may be useful at the village level are also discussed. Processing of root crops has not become widespread in Papua New Guinea, although a significant advance has been made with sweet potato. This process is described, together with other possibilities.

INTRODUCTION

In this paper I will briefly mention a few things about Papua New Guinea (PNG) regarding land, climate, people, and social and economic changes, and then go on to discuss storage and processing of root crops under several main headings as follows:

1. Traditional storage methods;
2. Studies on storage methods;

3. Processing of root crops, with particular
    emphasis on processing of sweet potato
    tubers.

PAPUA NEW GUINEA IN BRIEF

Papua New Guinea (PNG) is a land of great
physical variation--from vast swamply plains on the
coast to high alpine mountains and broad upland
valleys with great potential for agriculture. The
highest mountain is about 4500 m above sea level and
the settlement limit about 2500 m. The climate is
tropical and monsoonal. There are only two types of
seasons, "wet" and "dry," which are regulated by
southeast and northwest airstreams, respectively.
The temperature varies very little between seasons
and ranges from hot and humid (average minimum of
20 to 25°C to average maximum of 28 to 34°C on the
coast) to warm days and cold nights in the highlands
(with minimums of 11 to 18°C to maximums of 20 to
25°C depending on the altitude). Temperatures
decrease with increasing altitude. The relative
humidity range is 60 to 80 percent in the highlands
and 70 to 80 percent on the coast (Ford, 1974; Ward
and Lea, 1970).
Papua New Guinea's population is now approaching three million. About 85 to 90 percent of the
people are subsistence farmers living in small rural
villages and hamlets. Most of the population is
heavily dependent upon root crops used as staples in
the diet (Kimber, 1975). The major root crops used
as staples are, in order of importance, sweet potato
(Ipomoea batatas), taro (Colocasia esculenta), yams
(Dioscorea spp.), Chinese taro (Xanthosoma sagittifolium), cassava (Manihot esculenta), and, recently
introduced, potato (Solanum tuberosum). Other
staples are bananas (Musa sp.), sago (Metroxylon
spp.), sugar cane (Saccharum officinarum) and
maize (Zea mays) (Bureau of Statistics, 1963).
There are now considerable changes in the field
of economic and social development in PNG, and this
progress is drawing many people away from subsistence farming into jobs which do not allow them to
grow their own food. This changes some of their
social and cultural habits and, in turn, changes the
food production and consumption patterns in PNG.
Besides these changes, there are natural
phenomena such as periodic droughts and frosts in
the highlands, periodic droughts and flooding in

65

some parts of the lowlands, increasing population growth, and the highly perishable nature of most of the root crops. These factors make it most desirable now to study storage methods for these crops and come up with some solutions to these problems in order to help reduce wastage and make possible the more economic utilization of supplies.

TRADITIONAL STORAGE METHODS

Subsistence farmers through experience know that most root crops deteriorate rapidly once they are harvested. They have learned to counteract this problem by using several cultural techniques to ensure a fresh supply of these crops all the time.

Some of these techniques have been investigated by scientists in some detail. Some are known to be practiced in many places but have never been studied in detail to understand their effectiveness. Some of these techniques are discussed below.

TRADITIONAL METHODS OF AVOIDING THE NEED FOR STORAGE

Year-Round Planting in Different Gardens

Under conditions in some parts of Papua New Guinea where rainfall is fairly well distributed throughout the year, people plant their food crops in different gardens in a regular year-round planting basis. This method helps to provide people with a constant supply of fresh food, thus avoiding the need for elaborate storage techniques.

Progressive Planting in a Garden

Crops such as sweet potato, taro, or Chinese taro are planted progressively so that different maturity periods can be obtained.

Dependence on More than One Staple

In areas where the rainfall distribution throughout the year is fairly even, taro (in the lowlands) and sweet potato (in the highlands) are the basis of the subsistence economy. However, in areas with a more pronounced dry season, it is not possible to plant gardens all year round. People can spread food availability by planting crops with different maturity periods. These are

66

planted using either a mixed cropping pattern or
separate gardens. Thus in lowland areas with a more
pronounced dry season, yams supplement the taro
staple. In other areas where the dry season is still
longer, yams, bananas and taro are all used as
staples.

Mixed Cropping of Different Varieties (Cultivars)

In subsistence agriculture in PNG, a large
number of varieties of the principal staple are
generally planted together in a garden. People
know that there are differences in maturity periods
for different varieties. People employ this method
of gardening so that a continuous supply of fresh
food from one garden is maintained for a number of
months.

PRE-HARVEST STORAGE

Progressive Harvesting of Individual Plants

Progressive harvesting (or "mumuting") of sweet
potato is the most common method of harvesting among
subsistence agriculturists (MacDonald, 1973, quoted
by Rose, in press). This is especially true in the
PNG highlands. "Mumuting" is a Papua New Guinea
pidgin word which is loosely used here (with English
verg endings -ing, -ed) to mean progressive harvesting of crops (Rose, in press).
In progressive harvesting only the mature
tubers are harvested, and the immature ones are left
to be harvested at a later date when these tubers
reach maturity.
Mumuting is considered a useful storage method
which ensures a supply of fresh tubers for human
consumption (Kimber, 1972). Some crops of sweet
potato in the highlands have been reported as being
harvested using this method over three years
(Jamieson, 1968). Rose (in press) reported that
crops could be harvested three or four times over a
period of eight months to one year in the Tari area
of the highlands, depending on variety, soil drainage and rainfall. He compared sweet potato yields
from a crop harvested at six months with those from
a crop "mumuted" at six months and then harvested at
nine months. He found that mumuting increased the
proportion of small tubers (those less than 100 g

in weight) but not total tuber yield production, when calculated on a per day basis.

Using two sweet potato varieties (Gonimi and Serenta), Kimber compared single harvesting and replanting with progressive harvesting and replanting over three plantings (Kimber, 1972a) Apart from the first planting, times of planting were not the same for the two methods because the progressive harvesting always took longer for the crop cycle than did the single harvesting method. Yields also declined markedly over the second and third plantings because the same plots were used throughout. However, results from the third planting were quite reliable. For both varieties, there was no significant difference in yield for marketable tubers (that is, tubers of 100 g or more in weight). Average monthly yield, however, was in favor of the single harvest method (Table 1).

Progressive Harvesting of a Crop

This is a usual technique of spreading the harvest of food crops over a number of weeks or months. It is very common in the lowlands. Small areas of garden are harvested as needed and the remainder of the crop is left in the ground. In this way food is stored as the living plant in the garden.

Jamieson (1968), in his observations on time of maturity of sweet potato in the lowlands, stated that loss of crops through tuber rotting is minor until at least two months after achievement of maximum yield.

TABLE 1
Cumulative and Average Monthly Yields of Marketable Sweet Potato Tubers for Progressive Harvest and Single Harvest (after Kimber, Unpublished Annual Report, 1969-1972).

| Experiment | Progressive Harvest | | Single Harvest | |
	Cumulative (kg/ha)	Monthly (kg/ha)	Cumulative (kg/ha)	Monthly (kg/ha)
1	12,360	950	13,270	1330
2	17,780	740	15,450	740
3	17,280	620	18,040	580

Use of Species with Flexible Harvest Dates

These include swamp taro (<u>Cyrtosperma chamis-sonis</u>), paragum taro (<u>Alocasia macrorrhiza</u>), and cassava. For these three crops, the food is stored in the crop and used as needed.

Swam taro is used as a reserve crop in certain coastal areas. The crop is planted in wet areas and left to grow for several years until it is needed; for example, during droughts or if the staple crop is destroyed. <u>Alocasia</u> is a minor staple in a few lowland areas of Papua New Guinea. As with swamp taro, it is a useful reserve crop. Cassava is also used as a reserve food sometimes, as the harvest date is much more flexible than for the traditional staples (Bourke, 1977).[1]

POST-HARVEST STORAGE

The two most common traditional methods of storing root crops after they are harvested are pit storage and storing in the house. The other common method is storing on platforms in the open. Employment of each method in different parts of Papua New Guinea varies slightly depending on environment, crop, and purpose.

Storage in a House

Storing root crops in the house is the most common method employed, as it is used for day-to-day supply of food for household consumption. Root crops may be stored for up to 2 to 3 weeks, with the exception of yams, depending on the consumption rate and the distance between the family's dwelling and garden.

Sweet Potato

For long-term storage of sweet potato, only good quality tubers are spread on a platform situated in an area of the house that is dark, well ventilated, and free of pests. In most areas in the highlands this platform would usually be in the house where the people live themselves. Fires for cooking and warmth would provide a heat source for curing the tubers. Some tubers are said to be actually smoked lightly over a period of time, and this may have some effect on curing tubers. Sweet

potatoes are placed on the rafter above the fire-
place[2] or in a string bag hanging in or under the
house (Kimber, 1972b). It is well known by the
villagers that storage for a few days improves the
flavor of the sweet potato. This is due to loss of
moisture from the tuber, thus giving a greater
energy value per unit weight and also to the hydrolysis of starch to sugars (Kimber, 1972b).

Taro

This crop is stored in a cooler part of the
house. For long-term storage only good quality
corms are used. Petioles of the corms are not
removed, but are sometimes shortened by cutting to
about 5 to 10 cm above the corm tip.

Yam

This crop can be stored in good condition for
many months after the tubers are harvested. In the
Trobriand Islands and some other areas of Papua New
Guinea, people have developed some elaborate techniques for yam storage. In some areas where yams
are not so important they are stored in ordinary
houses. The main construction materials and a
diagrammatic sketch of the structure of a typical
Trobriand yam house are shown in Figure 1 (after
Gideon, 1975).

Pit Storage or Burying

At Kipu, in the upper Waria area of Morobe
Province, a hole about 1 m deep and 1 m in diameter
is dug on a raised dry area and lined with dried
leaves and grasses.[3] After selecting good quality
tubers or corms, these are placed in the pit, sometimes alternating each layer of corms or tubers
with dried grasses, and finally covered with a
layer of dried grasses and a top layer of soil. The
method varies slightly from place to place. In some
areas there are no linings, and tubers are just
buried in soil or sand, as in certain coastal areas
where taro is said to be buried in sand.[4]
At Kipu, sweet potato can be stored for up to
one month.[5] Yam tubers and taro corms can be stored
better and for much longer than sweet potato. It
was mentioned that they could be stored for more
than six months but this remains subject to scientific testing. Kipu is about 1000 m above sea

70

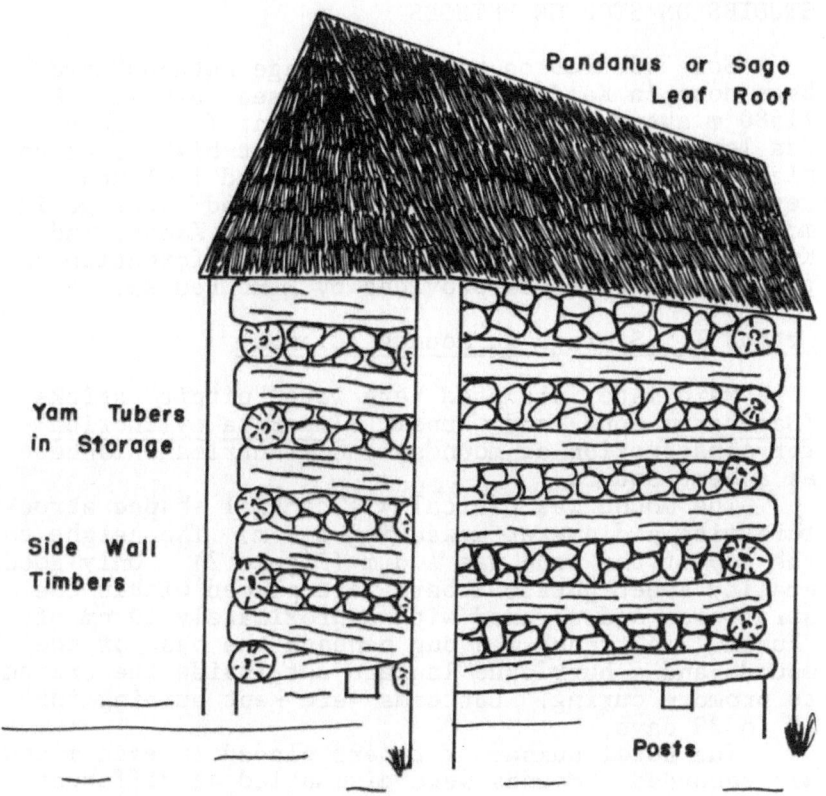

Figure 1. Diagrammatic sketch of typical Trobriand
yam house (after Gideon, 1975).

level. Elsewhere in Papua New Guinea, sweet potato
tubers are said to be stored for only up to 2 to 3
weeks.

Storage on Platforms

Root crops are sometimes stored in the open on
platforms for short periods. An example of this is
when taro is being prepared for a feast. The corms,
with petioles attached, are stored after harvest on
a platform while other food is being gathered and
prepared. This acts as a method of display as much
as storage. Taro may be stored in this manner for
up to a week after harvest.[6]

71

STUDIES ON STORAGE METHODS

Some studies on various storage methods have been done in Kandep (2200 m above sea level), Kuk (1580 m above sea level), and Keravat (20 m above sea level). These stations represent high altitude highlands, mid-altitude highlands, and lowlands, respectively. Three methods were tried: storage in mounds, houses, and pits. The work at Kandep and Kuk was described by Aldous (1976). Information on the Keravat work was provided by R.M. Bourke.

Method 1: Storage in Mounds (Clamps)

Main materials used here were "pitpit" sticks (Saccharum spp.) and "kunai" (Imperata cylindrica) for construction of mounds, and a hurricane lantern as a heat source.

The mound was basically a conical shaped structure with a diameter measuring 1.3 m. The height to the top of the cone was 1.6 m (Figure 2). Only good quality sweet potato tubers were placed within the structures and covered with approximately 10 cm of "kunai." A trench was dug beneath the base of the mound, and a hurricane lantern set inside the trench to promote curing. Lanterns were kept burning for 10 to 20 days.

The total number of tubers placed in each mound was recorded. Mounds were dismantled at different times. As the tubers were removed, they were subjectively placed into three categories as follows:

A--good quality tubers with no visual spoilage.
B--tubers that had one-third or less of the whole tuber spoiled.
C--tubers showing more than one-third spoilage.

The method used at Keravat (Figure 3) was slightly different from that used in Kandep and Kuk. The mound was covered with a layer of soil between two layers of "kunai" grass. No lantern was used as a heat source. Shelters were provided for mounds at Keravat and for six out of the eight mounds at Kuk. No shelters were used at Kandep.

Results from this method showed that sweet potato can be stored successfully for up to 50 days at Kandep, 40 days at Kuk (Table 2), and 30 days at Keravat.

Differences between successful storage periods may have been caused by differences in ambient temperatures . The ambient temperature of Kandep

72

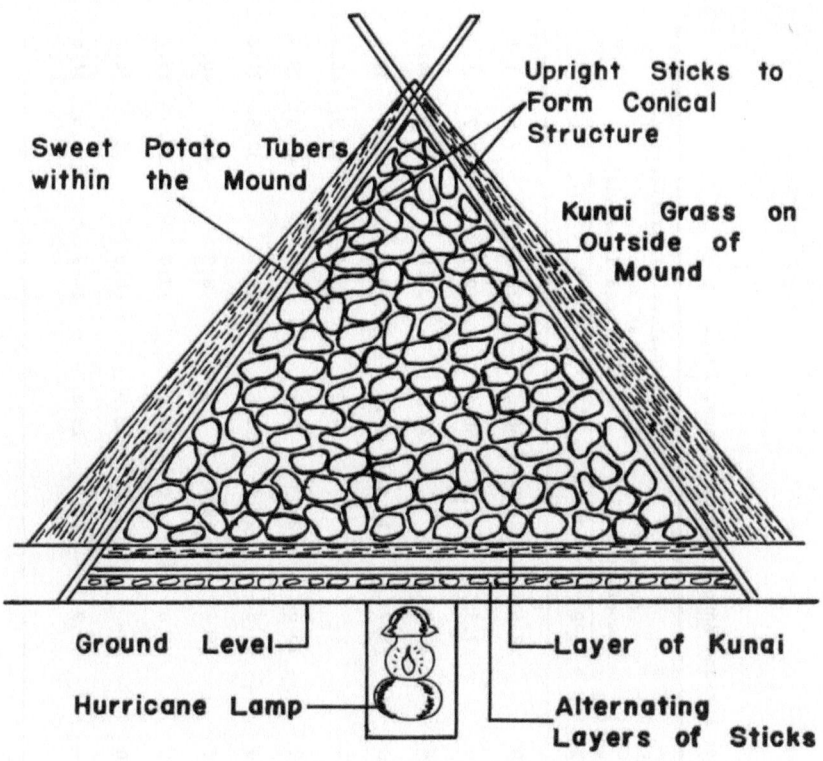

Sweet Potato Tubers within the Mound

Upright Sticks to Form Conical Structure

Kunai Grass on Outside of Mound

Ground Level

Hurricane Lamp

Layer of Kunai

Alternating Layers of Sticks

Figure 2. Diagrammatic sketch of mound construction at Kuk and Kandep (after Aldous, 1976).

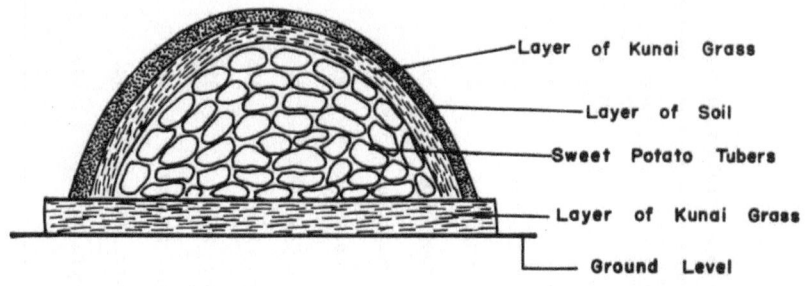

Layer of Kunai Grass

Layer of Soil

Sweet Potato Tubers

Layer of Kunai Grass

Ground Level

Figure 3. Diagrammatic sketch of mound construction at Keravat (Bourke, personal communication, 1978).

73

TABLE 2
Results from Storage Trials with Sweet Potato in Mounds at Kuk and Kandep (after Aldous, 1976)

Treatment	Weight of Tubers in Storage (kg)	Time in Storage (days)	Period of Heat Application (days)	Average Temperature (°C) Top	Average Temperature (°C) Bottom	Tuber Quality Classification at Removal (% placed in mound) A	B	C	Weight Loss (%)
Mound 1 - Kunai roof, Kuk	239	30	10	28.6	27.3	60.0	23.1	7.9	9.1
Mound 2 - Kunai roof, Kuk	271	42	10	27.5	26.6	50.4	12.1	12.2	25.3
Mound 3 - Kunai roof, Kuk	301	66	-	20.8	-	24.4	34.8	6.0	34.8
Mound 4 - Kunai roof, Kuk	272	73	20	25.7	24.6	40.5	13.5	15.0	30.8
Mound 5 - Kunai roof, Kuk	301	107	-	-	-	4.2	11.4	43.9	40.4
Mound 6 - Kunai roof, Kuk	324	108	20	25.6	24.3	3.8	11.0	45.5	39.6
Mound 7 - No roof, Kuk	278	66	10	27.3	27.1	11.6	14.2	52.8	21.4
Mound 8 - No roof, Kuk	322	108	10	28.4	28.4	3.5	8.3	47.5	40.5
Mound 9 - No roof, Kandep	398	49	10	25.0	24.0	58.0	13.0	13.6	15.2
Mound 10 - No roof, Kandep	426	85	10	-	-	3.9	8.0	53.3	34.7
Mound 11 - No roof, Kandep	183	85	-	15.9	14.2	29.8	14.9	18.6	36.6

is close to optimum storage temperature for sweet potato (13 to 15°C) (Keleny, 1965). Storage in mounds in the lowlands (from the Keravat experience) could be considered satisfactory for a period up to one month. However, sweet potato can be stored successfully in a house for almost this long with much less effort. Hence storing in mounds (clamps) is not recommended for the lowlands. However, further work on this method is required using uniform design for all places, and incorporating the suggested improvements may give better results.

Aldous (1976) suggested improvements for the mounds at Kandep and Kuk as follows:

1. The layer of grass covering the mound should be thicker (20 to 25 cm).
2. The stick shell at the apex of the cone should not be completely closed; it must allow adequate ventilation.
3. Mounds should be sheltered to minimize fluctuating temperature in the open area.

Method 2: Storage in a House

This method of storage was studied at Kuk. A total of ten compartments was constructed inside a small circular bush material house which had a diameter of 3 m. The compartments were located around the inside perimeter of the house. Sweet potato tubers were placed within the compartments at different times so that different curing periods could be obtained. A fire was kept burning in the center of the building as a source of heat, and wet bags were hung on the rafters and kept moistened six times a day to keep the relative humidity high. At varying times, tubers were removed and their quality assessed, as outlined under Method 1.

This method of storage was not successful beyond 2 to 3 weeks (Table 3).

Method 3: Pit Storage

At Kuk a hole about 1.2 m square and 1.5 m deep was dug. A trench was dug at the bottom of the hole to assist in drainage. The hole was lined with sticks and grass. Sweet potato tubers were cured in a house (Method 2) for ten days. Tubers were placed in the pit, which was then covered with a layer of sticks. a plastic sheet, and finally soil, to a depth of about 15 cm.

TABLE 3
Results from Storage Trials with Sweet Potato in a House at Kuk (after Aldous, 1976)

Compartment	Weight of Tubers in Storage (kg)	Time in Storage (days)	Period of Heat Applied (days)	Tuber Quality Classification at Removal (% placed in storage)			Weight Loss (%)
				A	B	C	
1 and 2	429	16	14	70.5	12.7	5.5	11.2
3	195	113	21	7.7	20.6	28.5	43.2
4	138	113	21	5.2	27.8	36.0	30.8
5	275	112	21	7.4	11.6	28.4	52.8
6	113	112	21	23.6	4.4	11.6	30.4
7	163	112	21	16.9	24.7	8.6	49.7
8	136	98	7	0.7	24.0	30.7	44.7
9	136	98	7	0.7	5.3	54.3	39.6
10	136	98	7	–	17.3	14.3	68.3

Tubers were left in the pit and when removed were assessed as in Method 1.

Results suggested that the method could be useful for temporary storage for up to four weeks. Long-term storage in this manner requires curing the tubers in a different structure first, before moving them to the pit. Difficulty would occur when the cured tubers are moved from the curing house to the pit. Any damage done to tubers at this stage would mean that the tubers would not store well.

PROCESSING OF ROOT CROPS

Food processing in Papua New Guinea, particularly for root crops, is an area which has received little attention. Apart from sago, all major crops grown in Papua New Guinea for subsistence are consumed without processing.

To answer some of the problems in food processing, the New Zealand Government Aid Scheme made available Mr. H.A.L. Morris, a food technology consultant, who studied local food production facilities and food crop materials. He recommended possible ways to develop the food processing industry in Papua New Guinea. He also gave a demonstration on a range of practical processed foods and processes from adapted technologies possible for Papua New Guinean conditions. Immediately after this report and demonstration, the Government began looking into a program of food industry development, starting with the prototype projects which were recommended in the report.

Two items being processed on an experimental basis are dried sweet potato chips ("kaukau rice") and fried sweet potato chips ("kaukau chips"). "Kaukau" is a Papua New Guinean pidgin word for sweet potato.

Dried "Kaukau Rice"

The Food Marketing Corporation (FMC) at Goroka started processing sweet potato into dried "kaukau rice" on a trial basis, based on Morris's suggestion. "Kaukau rice" is a product based on roller-crumbled shredded sweet potato, with similar cooking nd serving procedures as for rice (Morris, 1975).

The FMC followed up the initial work carried out by Morris on this product to determine processing methods, yields, production costs and equipment for a processing plant operating on 5 tons of raw material daily. The following data are from Haydon, Hepworth, and Hepworth (1975).

The stages in the processing method are as follows:

1. Product reception.
2. Sorting of quality, size grading.
3. Peeling by immersion in 20 percent lye for three minutes, followed by washing with water under pressure.
4. Hard trimming to remove tops, tails, and damaged parts.
5. Shredding.
6. Blanching in water at 100°C for 45 seconds.
7. Sulphite dip in 1.5 percent potassium metabisulphite solution for 45 seconds.
8. Spreading on trays, dewatering and inspection.
9. Loading trays in drier.
10. Hot-air drying under conditions to give a drying time of 6 to 8 hours.
11. Unloading trays and screening out fines.
12. Packing--5 kg of finished product in polyethylene bags, and 4 bags per cardboard carton.

Dried "kaukau rice" has an overall drying ratio of 5:1, and the final product has an attractive appearance of fairly low bulk density. Production costs were calculated for a plant operating with 5 tons of raw material per day and 18 laborers (Table 4).

The cost of raw material represents about 58 percent of the total cost of 56.9 t/kg/ It was suggested that reduction of raw material cost will bring the finished product to a more competitive price.

The equipment required for a production unit of this scale is as follows: sorting table; lye peeler; shredders; sulphide baths; dewatering, inspection, and loading tables; hot or tunnel drier and trolleys; screen; filter; sealers. Most of this equipment could be built in Papua New Guinea with the exception of shredders, heater, and tunnel drier and sealers.

Although processing method, yields, production costs and equipment requirements have been determined, the future of the product is uncertain. There is still need for more extensive taste panel

TABLE 4
Costs of Processing "Kaukau Rice" (after Haydon, Hepworth, and Hepworth, 1975).

Factor	Cost (t/kg) of Finished Product[a]
Raw material 6.5 t/kg	32.5
Processing NaOH	2.0
Sulphite	1.5
Machinery depreciation	5.0
Overheads	5.0
Packaging	2.7
Labor	8.2
TOTAL:	56.9

[a] The PNG currency is a decimal one, with the kina (K) and toea (t) as the basic denominations. 100 toea = 1 kina; K1 - $ Aust. 1.20 ($ U.S. 1.00 = $ Aust. .88).

and acceptability trials of the product before considering commercial production.

Sweet Potato Chips

This product is also being produced in Goroka by the Food Marketing Corporation on an experimental basis. The method of processing is as follows:

1. Wash and peel sweet potato tubers.
2. Soak in 2 percent common salt for 0.5 to 1 hour.
3. Slice 0.5 to 1 mm thick.
4. Drain and fry in cooking oil at 138 to 148°C until brown and crisp.
5. Drain and pack in moistureproof bags.

The product is good, except that it is too oily for widespread consumer acceptance. The oil seeps out of the product, making the package unattractive. Research is still going on at Goroka Food Marketing Corporation into ways of solving this problem.

Sun-Drying and Dehydration

This method has been used as a means of storage in many countries for hundreds of years. In

certain areas of Papua New Guinea it is used on
sweet potato for pig and poultry feed. The process
involves cutting the tubers into chips, drying them
in the sun, and then storing them in bags.

Aldous (1975) evaluated dehydration of sweet
potato in his study. He considered that it could
be a successful method of storing tubers. It is a
simple mode of storage on a small scale, and the
chips store well over extended periods of time.
The only problem would be reconstitution of the
dried product for human consumption. Reconstitution
is not required for livestock food.

SUMMARY AND CONCLUSIONS

Papua New Guinea is largely a subsistence
community which depends mostly on root crops as the
staples in its diet. However, the situation is
gradually changing with rapid development in social
and economic areas. With increasing urban migration, significant advances are required in cultivation, processing, and marketing of traditional food
crops to keep abreast of food problems.

Under the conditions of reasonable rain distribution and with growing seasons extended throughout most of the year, the question of storage has
not arisen. People have also developed cultural
techniques which enable them to obtain a constant
supply of fresh food. Such techniques include year-
round planting, progressive harvesting of individual
plants or a crop, and using crops that store food
in the plants.

The main crop stored for extended periods is
yam. Other crops such as sweet potato and taro
are stored in houses and on platforms for short
periods.

Studies conducted on storage methods have not
come up with any one method that may be considered
satisfactory. Mound storage, however, would seem
to offer the most effective means of storing whole
sweet potato tubers for a limited period of time.

With processing, sun-drying or dehydration
as a storage technique would seem to be acceptable
for storing stock feed. As for dried "kaukau rice"
and "kaukau chips," studies are continuing, but
their future is uncertain.

ACKNOWLEDGEMENTS

Acknowledgements go to those who helped me collect information, especially those listed below, and many Papua New Guineans whose names I have not included. Special acknowledgements go to Michael Bourke,who commented on an earlier draft, Mrs. Bilha Smith, who redrew the figures, and Mrs. Jean Bourke, typist.

The following were consulted during the literature survey and information gathering trips: Gabrielle Lanceley (Food Marketing Corporation); Ken Newton, Fred Fahmy, Peter Byrne, Arthur Charles, Michael Bourke, Allan Kimber, and the librarian (Department of Primary Industry); and Mave O'Collins, Wes Rooney, Bryant Allan, Lance Hill, Rod Lacey, Alfred Bala, Kesi Kesavan, and the librarian (University of Papua New Guinea).

NOTES

1. R.M. Bourke, personal communication, 1978.
2. Awang, personal communication, 1978.
3. Topo-Go, personal communication, 1978.
4. Bourke, personal communication, 1978.
5. Topo-Go, personal communication, 1978.
6. Bourke, personal communication, 1978.
7. Ibid.

REFERENCES

Aldous, T. 1976. Storage of sweet potato tubers. Proceedings 1975 Papua New Guinea Food Crops Conference, 29 April-2 May, 1975, Lae, Papua New Guinea. Department of Primary Industry, Port Moresby, p. 229.

Bourke, R.M. 1978. How farmers can reduce the effects of drought on food production in the lowlands. Harvest 4:122.

Bureau of Statistics. 1963. Survey of Indigenous Agriculture and Ancillary Surveys, 1961-1962. Government of Papua New Guinea, Port Moresby.

Ford, E. 1974. Papua New Guinea Resource Atlas. Jacaranda Press.

Gideon, R. 1975. Preservation of yam. In: Papua New Guinea Agriculture and Resource Technology, Vudal Agricultural College.

Haydon, P., Hepworth, A., and Hepworth, J.H. 1975. Summary Report on Production of Dried Kaukau "Rice." Unpublished report, Department of Primary Industry, Port Moresby.

Jamieson, G.I. 1968. Observation in time of maturity of sweet potato (Ipomoea batatas (Lam.)). Papua New Guinea Agr. J. 20:15.

Keleny, G.P. 1965. Sweet potato storage. Papua New Guinea Agr. J. 17:102.

Kimber, A.J. 1972a. Annual Report for the Department of Primary Industry, 1969-1972. Unpublished. Department of Primary Industry, Port Moresby.

_____. 1972b. The sweet potato in subsistence agriculture. Papua New Guinea Agr. J. 23:80.

_____. 1975. The future development and utilization of root crops in Papua New Guinea. Proceedings Papua New Guinea Stockfeeds Conference, 1-3 October 1974, Department of Primary Industry, Port Moresby.

Morris, H.A.L. 1975. The Development of Food Industries in Papua New Guinea. Unpublished report, New Zealand Government Aid Programme.

Rose, C.J. In press. Yields of sweet potato (Ipomoea batatas (L.) Lam.) from progressive harvesting. Papua New Guinea Agr. J.

Ward, R.G. and Lea, D.A.M. 1970. An Atlas of Papua and New Guinea. University of Papua New Guinea and Collins-Longman.

Beka Fuawe Siki: Highlands Agricultural Experiment Station, Aiyura, Kainantu, Papua New Guinea

4
Processing and Storage of Sweet Potato and Aroids in the Philippines

M. R. Villanueva

ABSTRACT

Root crops, once considered the traditional subsistence crops of low-income families in the Philippines, have also become a prominent food crop. The sweet potato leads the root crops in terms of area and production. It can be grown under extreme conditions in many kinds of soils. It plays an important dietary role in the Philippines, particularly in rural areas. It is generally used fresh, in processed food products, and as animal feed. Practically all the parts are used as food. Taro is used similarly. Only the corms of the other aroids are used as food.

Currently in the Philippines minimal processing and storage of sweet potato and aroids are practiced; the produce is generally used fresh.

New food uses will make sweet potato and aroids more competitive and take marketing and utilization outside the vicinity of the production areas. However, simple, inexpensive storage methods must be designed for use by small farmers to make the industry economically viable.

INTRODUCTION

Root crops have always been referred to as traditional subsistence crops of low-income families in the Philippines. In some ways they still are

83

subsistence crops in certain areas of the country, but in many other ways they have also achieved the status of a prominent food crop like rice or corn. In metropolitan areas, they may even be considered a luxury food because they cost more than the traditional food crops. They are especially cherished for special food preparations served for snacks and desserts.

Of the popular root crops grown in the Philippines, sweet potato has undergone a series of historical events and has even assumed a heroic role. During the World War II Japanese occupation of the Philippines, many people depended on sweet potato for survival while taking shelter in the mountains. The crop was easy to grow, and the shoots were used as a source of nutrition.

Despite the good qualities of sweet potato, including the fact that it can be grown in many kinds of soils under extreme conditions, its production and importance have been taken for granted and the crop has been neglected because some people associate the consumption of sweet potato with poverty. To them, the growing of sweet potato does not offer any challenge, and no special care is needed or taken.

The status of the aroids, particularly Colocasia and Xanthosoma, has always been different from that of the sweet potato. Aroids have always been a respected food commodity. Unlike sweet potato, aroids are generally prepared as vegetables--both tubers and leaves--and as a snack food or dessert. They have always cost more than sweet potato and cassava.

Today, sweet potato and aroids still serve as important food items in the lives of the Filipinos. Particularly in the rural areas, at least a few plants of Colocasia and/or Xanthosoma and a few square meters of sweet potato are found in the backyard of every home.

PRODUCTION

Sweet potato leads all root crops in the Philippines in terms of area and production. Colocasia follows cassava, while Xanthosoma and the minor aroids are even further down the line. Production statistics for the various root crops grown in the Philippines are presented in Table 1.

TABLE 1
Area and Production for the Major Root Crops Grown in the
Philippines, 1975 (Bureau of Agricultural Economics)

Crop	Area (ha)	Production (1,000 mt)	Average Yield (tons/ha)
Sweet potato	195,730	986.0	5.0
Cassava	119,310	679.3	5.7
Taro and yautia (Xanthosoma)	25,970	92.7	3.6
Yam	6,200	25.1	4.1
Potato	4,750	20.5	4.3

UTILIZATION

Generally, root crops are utilized as follows: fresh food for home consumption; processed food products; animal feed; and sources of industrial starch for food and textile industries, for the manufacture of glucose and adhesives, and for the preparation of laundry starches.

Average per capita consumption of sweet potato and taro in the Philippines is as follows:

	Low	High
Sweet potato	5.72	42.52 g/capita/day
Taro	1.21	25.01 g/capita/day

As indicated earlier, sweet potato, among the root crops, is used most extensively for food. In the rural areas, root crops serve as an important source of calories and as ready substitutes and/or supplements when supplies of traditional staples are low. This is evidenced by high consumption of root crops in areas where rice consumption is low.

Practically all parts of the sweet potato and the aroids, particularly taro, are consumed as human food. Tubers and tops of sweet potato are traditionally consumed as food. Tubers are eaten in boiled, baked, or fried form, or are processed into products like candies and chips. Young and tender shoots, about 30 cm long, are steamed and consumed as vegetables.

Similarly, taro corms are prepared and consumed in various ways, primarily in boiled form and as an ingredient in a vegetable dish called "guinatan." Leaves of certain varieties are widely consumed as vegetables, but are prepared differently from sweet potato tops. Young taro rhizomes are a very popular vegetable item.

Among the other members of the aroids like Xanthosoma, Cyrtosperma, and Alocasia, only the corms are used for food. Xanthosoma is as widely used for food as taro in certain regions. Cyrtosperma and Alocasia are consumed only in certain regions. A cottage industry has been firmly established using Alocasia in the preparation of a special confectionary item known as "binagol."

HARVESTING

Harvesting of root crops in the Philippines is done in many ways, but is generally done by hand with the aid of small farm tools ranging from small pointed sticks to a moldboard plow drawn by water buffalo. Harvesting may be done by individual plants or hills without removing the vine or cutting the above-ground parts. In certain instances this is done purposely for sweet potato, so that the field will serve as a source of sweet potato tops for vegetable.

Sweet Potato

Some farmers cut the vines first and roll them to the side of the field. Tubers may be harvested manually with the aid of small tools on an individual plant basis, or a moldboard plow drawn by a water buffalo may be passed two or three times along the rows to expose the tubers. In the latter method, the farmer then gathers the scattered tubers and places them in piles.

Families which raise sweet potato for home consumption may not harvest the entire field at once. They do start harvesting at an earlier stage of the crop, but they harvest by priming; i.e., they select only enough tubers of suitable size for the family's needs at that time.

Taro

Taro and other aroids are generally harvested in the same manner as sweet potato, although harvesting by hand is more general. Depending on the condition of the soil at harvest, the plants may be pulled by hand individually by holding the plant near the base; or this may be done with the aid of small tools like sticks or machetes to uproot the tubers. Again, harvesting of the crop may not be done all at once, depending on the demand. The period of harvesting is more flexible among the aroids than for sweet potato, which must be harvested sooner because of weevil problems. In the case of aroids, too much delay in harvesting results in rotting, deterioration, or regeneration of the corms.

When taro leaves are intended for vegetable use, harvesting is done by priming. Tender leaves, except the youngest two or three leaves, are cut near the base of the petiole. Rhizomes are harvested at any time by severing the tender portions from the mother plants.

Other Aroids

Yautia or tania (<u>Xanthosoma</u>) is harvested in a similar manner as upland taro. Because of the general preference for cormels, plants are usually left in the field for extended periods until the cormels reach marketable size. Mother corms are not normally consumed for food, but are saved for animal feed. Some farmers harvest by priming by scratching the soil around each plant and selecting only the marketable cormels.

<u>Alocasia</u>, because its harvestable portion is above ground, is easily harvested by chopping off the desired portion with a knife or machete. On the other hand, <u>Cyrtosperma</u> requires more work at harvest because of the size of the plant. A shovel is necessary to expose the large corm. The operation may require two or more individuals depending on the size of the plant at harvest.

PROCESSING AND STORAGE

In general, there is very minimal processing and storage of sweet potato and the aroids. Most commonly the fresh tubers or corms are used

87

immediately for food. There is a very short time span from harvest to consumption, implying minimal use of storage. This is the reason for emphasizing the detailed harvesting procedures presented above.

Processing

Many farmers sell their produce directly to consumers at the farm to do away with processing and storing and to eliminate risk of spoilage. Sweet potato and aroids may not be marketed immediately after harvest, inasmuch as they can be stored for a few days before they spoil.

After uprooting, the tubers are cleaned by hand. Non-marketable tubers are separated and discarded in the field. Tubers are then placed in jute sacks or large bamboo baskets. Washing the tubers is optional, and depends on the soil condition at harvest. When it is dry at harvest time, soil particles are shaken from the tubers without difficulty. Taro, Alocasia, and Cyrtosperma are usually washed after harvest. Taro grown in upland or dryland conditions may not be washed. Washing may be done simultaneously at harvest for lowland taro, Alocasia, and Cyrtosperma by dipping the corms in standing water in the field.

Generally, grading is not commonly practiced among sweet potato and aroid growers. Except for the separation of the extremely small and the damaged corms, no sorting is done at any stage between harvest and utilization.

The limited conversion of root crops into processed foods is probably due to the limited market for these products, variability in quality, and inconsistent supply of root crops. Apart from these possible explanations, a major cause of the lack of processing is very likely the limited technology available for small-scale and large-scale operations.

There is no doubt that processing can play an important stabilizing role in promotion of the root crop industry. The conversion of fresh root crops into selected processed products can improve utility and shelf life, and consequently promote their wider marketing and distribution.

Storage

Tubers and corms may be placed in baskets, jute sacks or any containers--depending on the amount harvested--and transported by small vehicles, water

buffalo, drawn sleds, or simply by hand. If not
consumed immediately, corms and tubers may be stored
in the same container used to transport them from
the field, or they may be left in one corner of a
shed until they are consumed. Sweet potato and
aroids may be kept in this manner for two to four
weeks, but spoilage becomes serious with longer
periods of storage.

Taro corms are left with about 30 cm of the
basal petiole, tied in bundles, and hung in the
shade. Some farmers have reported that corms
treated in this way can be stored for as long as two
months. Some farmers bury the corms in the ground
in the shade or even under the house. In most
cases, corms are left in piles in the shade until
they are marketed or consumed.

SUMMARY AND CONCLUSION

Sweet potato and the aroids are grown and consumed as food quite extensively in the Philippines.
They are eaten primarily as boiled or fried tubers,
as ingredients in certain foods, or as processed
simple food preparations like cakes or other confections. In general, however, minimal processing is
done to improve the utility of sweet potato and
aroids outside the home or farm where they are produced. Likewise, minimal storage is done between
harvest, marketing, and utilization because sweet
potato and aroids deteriorate rapidly. There is
also a lack of cheap storage methods that can be
practiced by small farmers.

As a consequence of the above factors, marketing and utilization of sweet potato and aroids in
the Philippines are restricted and limited to the
vicinity of the production area. Furthermore,
because utilization of these crops has concentrated
more on use of fresh tubers, which are bulky and
contain large amounts of water, transportation and
hence mobility become limiting factors in their
market flexibility.

New food uses of sweet potato and aroids must
be developed to make them more competitive. Likewise, simple and inexpensive storage methods must be
designed that will fit the needs of most small
farmers.

M. R. Villanueva: Philippine Root Crop Research and Training
Center, Baybay, Leyte, Philippines

5
Storage and Processing of Some Nigerian Root Crops

A. O. Olorunda

ABSTRACT

Root crops are major staple food crops in many parts of Nigeria; however, storage and processing methods are still farm from satisfactory. Traditional methods of storage and processing of yams (Dioscorea spp.), sweet potato (Ipomea batatas) and cocoyam (Colocasia and Xanthosoma spp.) have been reviewed. Ways of improving the efficiency of these traditional methods by the adaptation of modern and improved technologies have been proposed, and their socioeconomic implications discussed.

INTRODUCTION

Root crops are one of the major sources of dietary carbohydrates in Nigeria and provide a significant part of the total food supply, particularly in southern Nigeria where they account for over 50 percent of the caloric intake (Olorunda, 1973).

The major root crops in Nigeria, and indeed most of the West Africa region, are cassava (Manihot esculenta), yams (Disocorea spp.), sweet potatoes (Ipomea batatas), and cocoyam (Colocasia and Xanthosoma spp.)

Post-harvest losses are high in these commodities, and until very recently relatively little attention was paid to these losses in comparison with the less perishable grains such as cereals and legumes (Coursey and Booth, 1972).

Socioeconomic and agricultural developments in many West African countries including Nigeria over the past decade have generated new interest in the storage and processing of root crops. It is now realized that viable root crop processing industries employing a reasonable proportion of the population could be developed if storage losses could be reduced.

STORAGE OF TROPICAL ROOT CROPS

The dietary importance of root crops has led people to devise various means of storing the crop so that culinary qualities are preserved during storage. Basic requirements for the storage of root crops differ from one root crop to another and from one region to another, depending on several factors. Even for the same root crop, storage methods can differ depending on the reason for storage and the ultimate use (Olorunda, 1973).

Most of the fundamental principles and practices involved in root crop storage have been reviewed (Coursey, 1967). In general traditional methods of storing root crops in Nigeria are based on these principles, although present level of technology and sophistication coupled with relative cost of these commodities very often make most of the methods look inadequate or sometimes primitive.

Storage methods could be classified as follows:

1. Conventional storage methods,
2. Cold storage,
3. Controlled atmosphere storage (C.A.),
4. Chemical storage.

Conventional storage methods cover most of the traditional ways of storage like leaving the crop in the ground where it was grown and harvesting whenever required. Clamp storage is carried out by using loose straw or bales as protection materials for the roots. Barn storage is only used for yams. Other methods include leaving the roots on exposed heaps, and burial in pits.

Leaving the edible root or tuber in the field until needed is perhaps the cheapest, simplest and most primitive method practiced in Nigeria today for yams, sweet potatoes and cocoyam. Although inexpensive, it exposes the tuber unnecessarily to attack

by pre-harvest pests such as yam beetles or termites. Harvesting and collection can at times be
difficult since the ground often becomes hard-baked
during the dry season, or flooded during heavy
rains. The latter factors could also lead to tuber
damage and decay during subsequent storage. Pilfer-
age by man and animal should also not be overlooked.

Another conventional method is barn storage.
As far as is known this method is only used for yams
in Nigeria and is the most popular method of storage
in most of the yam-growing regions. Typical barns
have been described by Irvine (1963). Barns vary
considerably in details of design and construction
between different parts of the yam-growing regions
where they are used, but all consist of a vertical
or nearly vertical wooden framework. To this framework the yam tubers are fastened individually by
means of strings, or more commonly by means of local
cordage material such as raffia. In most cases the
barns are covered with thatched roofs to protect the
tubers from the heat of the sun. This type of
storage structure appears in practice to be highly
satisfactory since the tubers receive adequate ventilation and, being above the ground, they can be
protected from termite attack or from danger of
flooding. The cost of materials used is also relatively low. However, barn construction and yam
tying are laborious and time-consuming, and are not
likely to be satisfactory for large-scale storage
work. Also, appreciable losses can occur; in some
cases up to 40 or even 50 percent of the stored yams
decay within a storage period of 4 to 6 months
(Waitt,1961, 1963; Okafor, 1966; Coursey, 1967; and
Olorunda and Adesuyi, 1970).

Another traditional method of storage is the
heaping of the tubers on the floor of barns or
houses or the laying of the tubers in a layer on a
rock or shelf. In these methods, particularly for
yams, fires are lit once or twice weekly to fumigate
the tubers. Cocoyam and sweet potatoes are often
covered with ashes which also have fungistatic properties. The storage area must always be well ventilated and dry. Although simple, moisture content
and dry matter loss in this method can be considerable. Losses of up to 95 percent have been reported
with sweet potato (Olorunda, 1977).

Experiments have been conducted to compare barn
storage of yams with clamps similar to those used
for storing potatoes in temperate countries (Waitt,
1961, 1963; Olorunda and Adesuyi, 1970). Results

show that traditional barn storage was by far better and cheaper for yams.

Modern storage methods like the use of chemical sprout inhibitors, ionizing radiation, controlled atmosphere storage, and cold storage have been used experimentally with yam (Olorunda et al., 1974; Olorunda and Macklon, 1976; Walker, 1961; Adesuyi, 1976). Some of this work has clearly demonstrated that conventional barns, sprout inhibitors, and controlled atmosphere hold little promise of increasing storage life of yams. Some work, however, indicates that the solution to providing year-round yam supply lies in the direction of cool storage or ionizing radiation.

Efforts are now being made to move from the test tube and laboratory stages to commercialization. Similar research efforts are being initiated with sweet potatoes where the technology of storage is well known (Kushman, 1969). Research currently underway at the International Institute of Tropical Agriculture (IITA) at Ibadan, and the Federal Root Crop Research Institute at Umudike is aimed at extending this technology to Nigerian conditions. The more popular cocoyam is Xanthosoma, which is generally grown as a subsistence crop. As yet, science has had little impact on this crop. However, tubers in storage are subject to rots caused by numerous organisms such as Fusarium solani, Botryodiplodia theobromae, and Sclerotium rolfsii. Storage of tubers in dry places tends to reduce rot incidence. Some research at the Federal Root Crop Research Institute at Umudike is also aimed at improving the agronomy and storage of the crop.

PROCESSING OF TROPICAL ROOTS

Xanthosoma is the main cocoyam used as food in Nigeria. However, Xanthosoma is believed to be inferior to yams and is generally used as a substitute for yams, particularly at some periods of the year when some people cannot afford to buy yams. Hence, it should be quite easy to apply modern techniques of food technology to yam and cocoyam alike to produce the same or similar products.

Yams and cocoyam are prepared as food in many ways. They may be boiled--usually in slices, or in the case of small tubers, especially cocoyam, boiled whole. Boiled yam may be used with meat, fish, palm oil, etc., or may be mashed together with other constituents to form a kind of pottage. Alternatively, the raw material can be cut or grated into stew or soup during preparation and cooked with other ingredients. Tubers or pieces of tubers may be roasted in an oven or over a slow fire, or baked in the ashes of a fire. Small pieces of raw tubers are also fried, usually in palm oil, in a manner similar to potato chips. The most important culinary product made from yam or cocoyam (<u>Xanthosoma</u>) is "fufu." Methods of preparation and consumption of fufu have been described (Coursey, 1965). Dehydrated fufu or "instant yam," similar to instant mashed potato, has been developed and is now being produced commercially. A flow diagram of the process is shown in Figure 1. Unfortunately, this product is still very expensive when compared with instant potato and end product quality still leaves much to be desired.

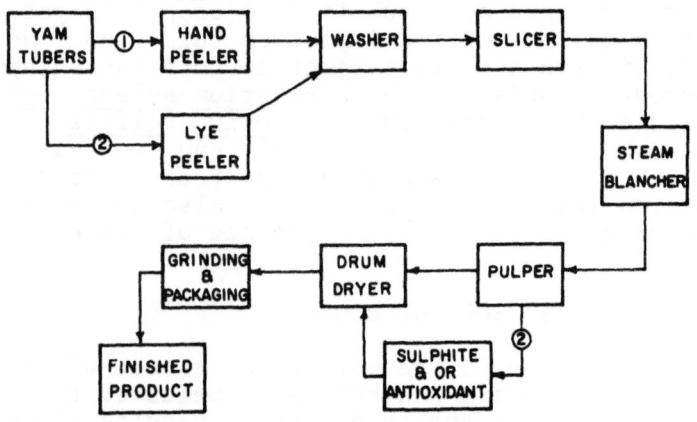

Figure 1.   Flow diagram for instant yam (fufu).
1 - Current technology
2 - Proposed innovations

A proportion of the yam crop is also processed into flour, especially in the yam-growing
areas of West Africa including Nigeria. A flow
diagram of yam flour production is shown in Figure
2. The method of preparation has been reviewed by
Coursey (1965). Yam flour can be stored for a
considerable period and can be reconstituted with
hot water to form a paste or dough, which is considered to be inferior to fufu because of its different color and texture. The color of the paste
or dough is light to dark brown compared with fufu,
which is white. Yam flour, because of ease of
reconstitution and the fact that it can be stored
for months, is gradually gaining popularity in
many yam-growing areas, but the poor color still
hampers complete acceptance among people from areas
where it has not been a traditional food.

Sweet potato is prepared for food by boiling
or frying in palm oil, just as are yams or cocoyam.
Recently there has been a great deal of interest in
chipping sweet potatoes for the local market.
Nigerians generally prefer starchy, non-sweet,
light-colored chips. Most sweet potato cultivars
currently available have a high sugar content which
results in brownish chips. Means of overcoming this
phenomenon by conditioning at elevated temperatures

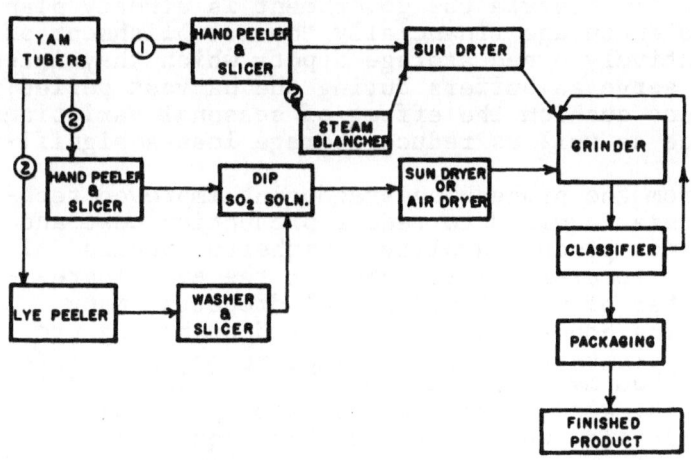

Figure 2. Flow diagram for yam flour.
1 - Indigenous technology
2 - Proposed technology

and vacuum frying at 100 to 110°C have been investigated (Olorunda and Kitson, 1977).

RECOMMENDATIONS AND CONCLUSIONS

It appears that one solution to the problems of providing a year-round supply of starchy root crops such as yam, sweet potato, and cocoyam (Xanthosoma) is cool storage or the introduction of other modern storage methods. However, in making any recommendations to reduce storage losses in tropical roots, it should be pointed out that these commodities are of low unit value and, under the present system of production, are not likely to be able to bear the cost of refrigerated storage or other modern methods. Additionally, technical and organizational constraints as well as a paucity of research on these crops are problems which must be faced. One practical approach might be to encourage cooperative storage. This would involve an organization where the root crops are brought by growers to local collecting centers where they could be graded, cured, stored for a considerable period of time, and then released for sale through marketing channels (Figure 3). An organization similar to the Potato Marketing Board in the U.K. or British Columbia Tree Fruits in Canada might also be appropriate. In Nigeria the government is already planning to encourage financially the establishment of cooperatively owned storage depots which they intend should serve as buffers during the harvest period, and hence cushion the effect of seasonal variations in price as well as reduce storage losses significantly.

From the processing standpoint improved techniques are required to reduce production cost and to enhance product quality. Hitherto, because of the need to create more jobs for the ever increasing number of unemployed school dropouts, many developing countries have insisted on having processing plants that are designed to be labor intensive. From my experience of the food industry in Ghana, the slow development of food processing industries is due in part to this requirement. In these factories, workers are no longer prepared to accept low wages. This, coupled with the problems of excessive handling during processing operations, leads to the manufacture of products which are much more expensive, and inferior to, similar

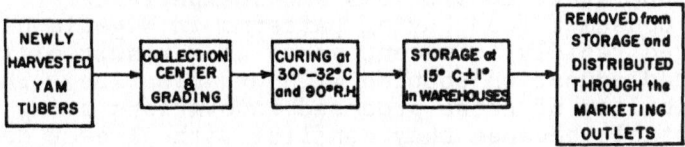

Figure 3.   Proposed flow diagram for yam
and sweet potato distribution.

products produced in developing countries using
highly automated or capital intensive processes.
Hence, what has been witnessed over the past one or
two decades of agro-industrial development in many
developing countries is a sad situation where factories close down soon after their start.   In cases
where such factories are government financed, the
government has always been called upon to intervene
in order that plant operation may continue.   This
usually results in financial hardship for the government concerned.

It is with this in mind that the proposed flow
diagrams for instant yam (Figure 1) and yam flour
(Figure 2) have been proposed.   Lye peeling has been
substituted for hand peeling in the processing operation in order to increase yield and save time and
cost.   Because yams are irregular in size and shape
they cannot be peeled satisfactorily with abrasive
peelers.   If plant breeders come up with a more
uniform root shape, abrasive peeling might be suitable for small-scale processing operations.   In
order to improve color and flavor stability of the
product, sulphiting and use of other antioxidants
have been suggested as process improvements.

It is important to remember that a highly
automated operation is not necessarily large, and
chances of success are probably greater with a
larger number of smaller plants.   Smaller plants
located closer to growing areas avoid transportation
problems and provide an opportunity for developing
managerial and technical skills, while avoiding the
high costs involved in making mistakes in a large
factory.   Many food processing operations in developing countries have failed because markets have

been flooded with a poor quality new product resulting from start-up problems and inexperience.

Consumer food buying patterns in Nigeria have changed rapidly in recent years as a result of increased education coupled with urbanization. The possibility of these proposed innovations being unacceptable because they conflict with already developed food habits is slight.

In conclusion, it should be emphasized that what is needed to improve significantly the supply of tropical root crops in many regions of the world and Nigeria in particular is multidisciplinary research on their agronomy, storage, and processing. Such a research program should have as its objective the application of scientific findings to development of improved stored or processed products based on existing indigenous technology. Storage or processing research should only be regarded as completed when the findings have been incorporated into the building of a pilot storage or processing plant, and feasibility studies of scale-up to commercial operations have been completed.

REFERENCES

Adesuyi, S.A. 1976. The use of gamma radiation for control of sprouting in yams (Dioscorea rotundata) during storage. Nigerian J. Prot. 2:34.

Coursey, D.G. 1965. The role of yams in West African food economies. World Crops 17:74-82.

_____. 1967. Yams. Tropical Agriculture Series. London: Longman Green and Company, Ltd.

_____ and Booth, R.H. 1972. The post-harvest phytopathology of perishable tropical produce. Rev. Plant Path. 51:751-65.

Irvine, F.R. 1963. A Textbook of West Africa Agriculture. Oxford University Press.

Kushman, L.J. 1969. Inhibition of sprouting in sweet potatoes by treatment with CIPC. Hort. Sci. 4:61-63.

Okafor, N. 1966. Microbial rotting of stored yam, Dioscorea sp., in Nigeria. Expt. Agric. 2:179-82.

Olorunda, A.O. 1973. Storage and post-harvest physiology of root crops. Ph.D. dissertation, University of Aberdeen.

_____ and Aboaba, F.O.   In Press.   Preservation by ionising radiation in Nigeria, present
and future status.   IAEA-FAO Symposium on Food
Radiation, Wagenigen, Netherlands, 21-25 November
1977.

_____ and Adesuyi, S.A.   1970.   A Comparison
Between the Barn and Clamp Methods of Storing
Yams.   Nigerian Stored Product Research Institute
Technical Report No. 757-59.

_____ and Kitson, J.A.   1977.   Producing
light coloured sweet potato chips by controlling
storage and processing conditions.   Food Product
Development 11:324.

_____ and Macklon, A.E.S.   1976.   Effect of
storage at chilling temperature on ion absorption, salt retention capacity and respiratory
pattern in yam tubers.   J. Sci. Fd. Agric. 27:
405.

_____, McKelvie, A.D., and Macklon, A.E.S.
1974.   Effect of temperature and chloropropham on
the storage of the yam.   J. Sci. Fd. Agric. 25:
1233.

Waitt, A.W.   1961.   Review of Yam Research in
Nigeria from 1920-1961.   Nigerian Federal Department of Agriculture Research Memo 31.

_____.   1963.   Yam; Dioscorea species.   Field
Crop Abstracts 16:145.

Walker, H.M.   1961.   Sealed Storage of Yam.   West
Africa Stored Product Research Unit Technical
Report No. 18, Lagos, Nigeria.

*A. O. Olorunda: Department of Food Technology, Faculty of
Technology, University of Ibadan, Ibadan, Nigeria*

6

The Technical and Social Problems of Taro Processing and Storage in Nigeria

I. E. Nwana
B. E. Onochie

ABSTRACT

In Nigeria taro crop yields seldom exceed 6 to 7 ton/ha from a potential of 36 ton/ha. Even then post-harvest losses of 30 to 50 percent are common due to Sclerotium rolfsii and Botryodiplodia theobromae.

Processing into "achicha" and flour reduces fungal rot losses but often predisposes the products to damage by stored product pests belonging to the general Tribolium, Tenebrio, Sitophilus, Carpophilus, and Callosobruchus. Potentials for use of taro in medicine are under-exploited.

The possibilities of commercial processing in the light of food needs, economic requirements and social or acceptance considerations are also discussed.

INTRODUCTION

In Nigeria, edible aroids in the general Colocasia and Xanthosoma are referred to as cocoyams. The Polynesian name, taro, which refers to the genus Colocasia, is uncommon and limited to scientific writings. There are two cultivated species, Colocasia esculenta (L.) Schott and Xanthosoma sagittifolium Schott. Several cultivars (varieties) of these species occur, and interest in cultivar taxonomy is

100

increasing. Xanthosoma is a recent introduction and may have been brought in by the eary European explorers, as is suggested by its various local names, e.g., "Iyokho-akra" (Edo), or "Ede Beke" (Ibo), which means the white man's cocoyam. With Colocasia the story is different. Although published works (De Candolle, 1855; Chang, 1958; Plucknett, de la Pena, and Obrero, 1970) put its origin at south-central Asia, perhaps India, there are oral stories of Nigerian traditions--feasts and folk-lores--associated with the discovery of cocoyams as an item of human food, suggesting that the crop is indigenous to Nigeria.

In order of importance, cocoyams are second only to yams (Dioscorea spp.). They used to be cropped mainly by women in the traditional farming systems. Relatively small quantities of cocoyams are used for thickening certain types of native soup for eating "fufu." Cocoyams have their greatest value during the annual famine (May to August) when almost all the yams in the storage barns have been exhausted and the new plantings are still growing in the field.

Cocoyams are usually planted in May or June and harvested between December and February of the following year. Thus there is a period of 4 to 5 months storage between harvest and utilization. During this period of storage cocoyams deteriorate and post-harvest losses of 30 to 50 percent are common (Okeke, 1976). To reduce these losses, efficient storage and processing systems have to be developed.

Cocoyam yields of 6 to 7 ton/ha are common (Enyi, 1973). However, a cultivar (Okorokoro) in our experimental farms has shown a potential yield capacity of 36 ton/ha in an 18- to 24-month crop period.

LOCAL STORAGE SYSTEMS

Surveys conducted by the Cocoyam/Sweet Potato Program, National Root Crops Research Institute, Umudike, Umuahia, Nigeria, revealed three main traditional storage systems for cocoyams in Nigeria. First, cocoyams are stored in heaps under shade of forest or plantation trees. The heaps are loosely covered with leaf or grass litter. In some other localities cocoyams are stored in compound barns constructed with palm fronds in such a way as to allow free air circulation. More permanent compound

barns built with perforated mud walls are usual with
large producers in Awgu and Nsukka areas of Anambra
State. The third method is storage in cylindroid
pits about 1.5 m diameter at ground level, 4 m deep,
and with a basal chamber of about 2.5 m diameter.
The corms and cormels are placed in this basal chamber, allowing an air space of at least 2.5 m above
the heaps. The pits are covered with dried banana
leaves placed on cross bars and then sealed with a
mixture of mashed banana stem and clay until the
stored corms are to be used. Indications are that
post-harvest damage due to rot organisms is least
in this third method. Unfortunately, due to rising
labor costs in Nigeria the pit storage system has
been almost entirely abandoned.

Okeke (1976) isolated Sclerotium rolfsii
and Botryodiplodia theobromae from rotting corms and
showed that these organisms account for almost all
the microbial damage to stored corms, often resulting in 30 to 50 percent loss.

PRODUCTS FROM LOCAL PROCESSING METHODS

Post-harvest changes in stored cocoyams may be
due to primary causes such as enzymolysis and other
biochemical reactions, or secondary causes such as
insect pest and microbial damage. Local processors
have realized this, albeit naively, and developed
methods to arrest their effects. Two main products
are produced, "achicha" and cocoyam flour. Figure
1 shows a flow chart of the processing into achicha.
Raw corms or cormels of C. esculenta are cleaned
and washed. They are boiled for about 3 hours until
they soften and change color from white to flesh,
releasing a pleasant aroma instead of the previous
acrid smell associated with improperly boiled corms.
The skin is then peeled off and the flesh cut into
half-moon slices about 1 cm thick, and sun-dried or
smoked until they break easily between the fingers.
These are then packed in clay pots shielded with dry
leaves, and sealed with a mixture of mashed banana
stems and clay. They are stored in warm dry places
until required for eating. Experience has shown
that, if not properly boiled before chipping,
achicha often grows moldy in storage and develops a
musty taste. On the other hand, improper drying,
especially if achicha is not stored in insect-proof
compartments, predisposes the product to insect
damage. We have identified insects belonging to

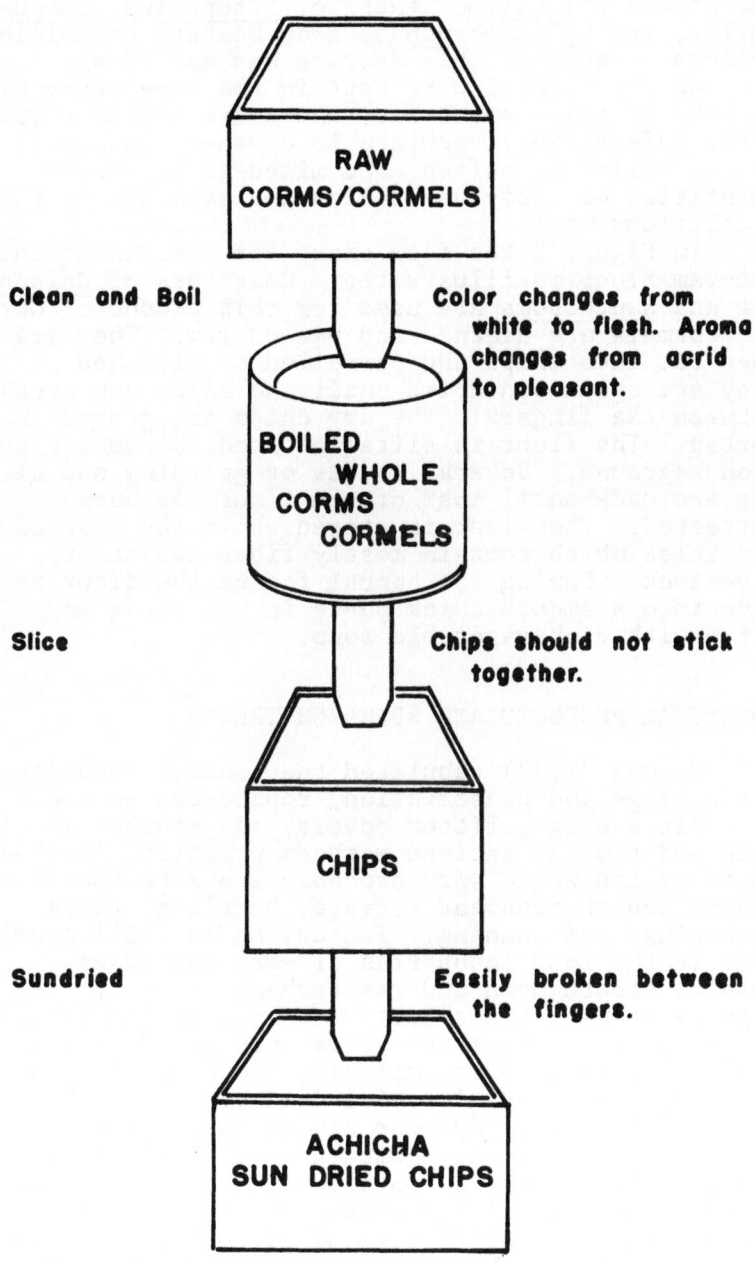

RAW CORMS/CORMELS

Clean and Boil — Color changes from white to flesh. Aroma changes from acrid to pleasant.

BOILED WHOLE CORMS CORMELS

Slice — Chips should not stick together.

CHIPS

Sundried — Easily broken between the fingers.

ACHICHA SUN DRIED CHIPS

Figure 1. Flow chart for processing taro into "achicha."

the genera Tribolium, Tenebrio, Sitophilus, Carpophilus, and Callosobruchus, from baskets containing
achicha. However, this storage was not ideal
because the baskets were kept in the same store with
baskets of maize and beans, and there may have been
cross-infestation. Achicha is crushed into small
grits, boiled to soften, and mixed with copious
quantities of leafy vegetables and palm oil to form
a delicious meal.

In Figure 2 the flow chart for processing into
cocoyam flour is illustrated. Cultivars of Colocasia and Xanthosoma are used for this product. Corms
and cormels are cleaned and peeled raw. They are
then cut into chips and parboiled or blanched.
They are again sun-dried until the chips can break
between the fingers. The dry chips are ground in a
mortar. The flour is sifted out and the coarse portion reground. Several rounds of grinding and sifting are made until most of the flour has been
extracted. The flour is stored while the unground
particles which contain mostly fiber are fed to
livestock. During the annual famine the flour is
made into a smooth thick paste in hot water and
eaten with rich vegetable soup.

POTENTIAL PRODUCTS AND RESEARCH TRENDS

Ngoddy (1977) tabulated the general techniques
for storage and preservation, reproduced in Table 1.

Pit storage, litter covers, and storage in clay
pots and similar ancient methods practiced in other
parts of the world were probably the antecedents of
controlled environment storage, bottling, glass
packaging, and canning. Sun-drying is still prominent in the food industries of many countries
despite modern oven and gas exchange driers. Be
that as it may, the objective of processing is still
to treat and convert raw forms into products that
are not subject to enyzmolysis, biochemical changes,
or insect pest and microbial damage.

Dr. George S. Ayernor of the International
Institute of Tropical Agriculture (IITA), Ibadan,
Nigeria, has pioneered in the study of processing
of cocoyams into flakes which can be made into paste
"fufu," and the dehydration of piece-form cocoyam
similar to the work done by Ngoddy and Wurdemann
(1977) on yams at the University of Ife, Nigeria.

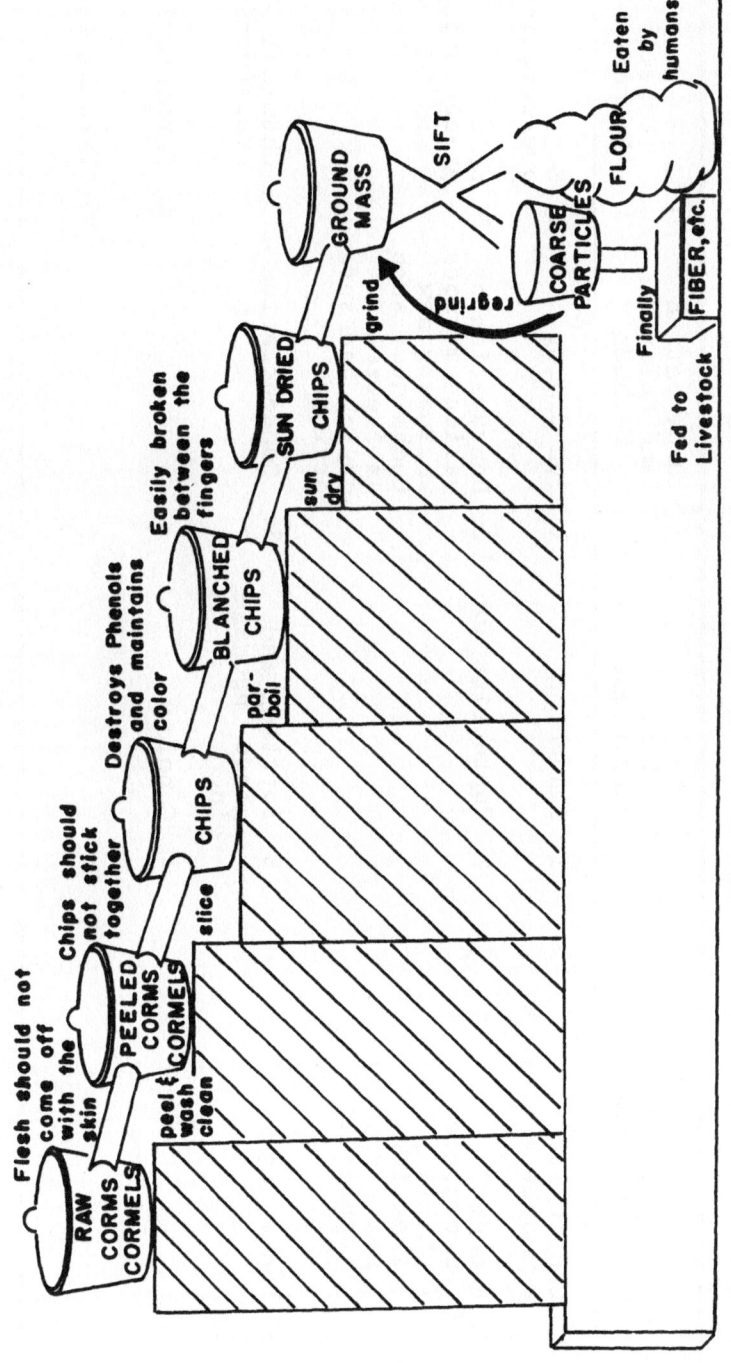

Figure 2.   Flow chart for processing cocoyams into flour.

TABLE 1
General Techniques for Storage and Preservation (after Ngoddy, 1977)

Controlled Environment	Heat Treatment	Dehydration and Related Methods	Other Methods
1. Controlled atmosphere storage	1. Blanching	1. Sun-drying	1. Sugar preservation
2. Reduced temperature - chilling - freezing - newer cryogenic techniques.	2. Pasteurization	2. Dehydration by various methods including new techniques such as freeze-drying and spray drying	2. Salt preservation
3. Packaging	3. Bottling and glass packaging	3. Concentration	3. Chemical preservation
	4. Canning		4. Fermentation
			5. Irradiation

Potential Uses of Cocoyams in Medicine

The most remarkable quality of taro is its
acceptance by people who are allergic to some other
sources of carbohydrates. Together with Dioscorea
alata L., C. esculenta is gradually becoming the
only source of calories for diabetic patients.
Roth, Worth, and Lichton (1967) have demonstrated
the use of "poi," a popular Hawaiian dish produced
from C. esculenta, for the prevention of allergic
diseases in potentially sensitive infants.
It is also believed that possession of cocoyams protects people against witchcraft and evil
spells. People going to war make certain preparations with the corms which are believed to neutralize the charms of the enemy. Such superstitions,
similar to the believed effect of potato on arthri-
tis, do not lend themselves to conventional
scientific investigations.
However, in traditional medicine, scrapings
from the flesh of C. esculenta are used as an
effective embrocation for sprained and swollen
parts. Indeed, the scrapings both embrocate and
act as plaster to hold the sprained parts together.
This use can be investigated and exploited for
modern medicine.

Social Problems in Cocoyam Development

The greatest obstacle to further development
in research on cocoyam processing is lack of
encouragement and support. This is associated with
the low status of cocoyam as an item of food in the
Nigerian's diet. In the humid tropics of Nigeria,
especially in the eastern parts of the country
where roots and tubers form the major staples,
cocoyams are not normally eaten by "he-men" who
consider that it neutralizes certain metaphysical
powers acquired through African traditions and
customs. Men who eat cocoyams are looked down on
as being lazy, weak, and unfortified. Most men
avoid eating and being associated with cocoyams
in order not to be so regarded. Their production
is therefore left in the hands of women and weak-
lings who are not able to carry on tedious production of the more challenging and royal crop, yam
(Dioscorea spp.). The situation arises from lack
of proper knoweldge of the true value of cocoyams.
Tuber crops must be seen in their primary role as
sources of carbohydrates. But even when other

dietary requirements of man are considered, cocoyams
are next only to certain varieties of yam (Dioscorea
spp.) in their high content of crude and true proteins. Cocoyams are exceptionally rich in ash, low
in fiber, sources of oils and fats, and contain
almost all the essential amino acids as well as vitamins in appreciable quantities (Oyenuga, 1968).

Processing and Cocoyam Production

A two-way grower/processor relationship has
been advocated to communicate the processor's needs
for target products to the grower, who will then
develop or select suitable cultivars to satisfy
these requirements. We hope that when this situation occurs, all parties working on cocoyams will
join hands to make its cultivation, processing, and
utilization a success.

ACKNOWLEDGEMENTS

We are grateful to the staff of the Sweet Pota-
to/Cocoyam Programme, NRCRI, for their various
contributions during the preparation of this paper.
Mr. E.O. Siabah, Higher Superintendent of Press
(Graphic Arts) and Mr. O.A. Eke, Senior Artisan
(Graphic Arts) also have our thanks for preparing
the illustrations.

REFERENCES

Chang, T.K. 1958. Dispersal of the taro in Asia.
Ann. Assoc. Am. Geog. 48:255.
De Candolle, A. 1855. Geographie Botanique Rai-
sonee. Paris 2:817.
Enyi, B.A.C. In press. Growth, development and
yield of some tropical root crops. Proc. Third
Intern. Symp. Trop. Root Crops, IITA, Ibadan,
1973.
Ngoddy, P.O. 1977. Determinants in the development
of technology for the processing of roots and
tubers in Nigeria. Proc. First Natl. Seminar on
Roots and Tubers in Nigeria, National Root Crops
Research Institute, Umuahia, Nigeria.
_____ and Wurdemannn, W. 1977. A universal
experimental pilot cabinet dryer--operational and
design features. First Intern. Congr. on Eng.
and Food, Boston, August 1976.

Okeke, G.C. 1976. Annual Report. National Root
Crops Research Institute, Umudike, Umuahia,
Nigeria.
Oyenuga, V.A. 1968. Nigeria's Foods and Feeding-
Stuffs. Ibadan, Nigeria: Ibadan University
Press.
Plucknett, D.L., de la Pena, R.S., and Obrero, F.
1970. Taro (Colocasia esculenta). Field Crop
Abstracts 23:413.
Roth, A., Worth, R.M., and Lichton, I.J. 1967.
Use of poi in the prevention of allergic disease
in potentially allergic infants. Ann. Allergy
25:501.
Talburt, W. and Smith, D. 1967. Potato Processing.
Westport, Connecticut: A.V.I. Publishing Co.,
Inc. (2nd edition).

*Ifedioramma E. Nwana and B. E. Onochie: National Root Crops
Research Institute, P. M. B. 1006, Umuahia, Nigeria*

7
Processing and Storage of Yam in Nigeria

A. O. Adenuga

ABSTRACT

The traditional processing of yam into chips and flour as a menas of storage is discussed. Few species are amenable to processing into yam flour.

Methods of storage of whole tubers vary according to whether the yam sett is to be used for propagation or consumed later. Techniques of storage also vary according to whether short- or long-term storage is required. Diseases and pests to which the yam is prone during storage are discussed, and advantages and disadvantages of different methods of storage are highlighted.

INTRODUCTION

Yam is very important in the economy of Nigeria, second only to cassava in order of importance among root crops. Apart from its nutritional use, it has value in the social and religious festivities of West African people (Coursey, 1965, 1967; Johnson, 1958). Some species are used in traditional medicine as precursors of certain pharmaceutical steroids and alkaloids (Bevan and Hirst, 1958).

FAO (1966) gave the world production of yams and sweet potatoes as 47 million tons per year. Yam was reported to have accounted for about 17 million tons of this total. Kay (1973) estimated

110

world production for yams at 15 to 25 million tons per annum, excluding the People's Republic of China. Twelve million tons of yams per annum are produced by the eastern part of West Africa between the Ivory Coast and the Cameroons (Coursey, 1967); this figure accounts for about two-thirds of world production. Ingram and Greenwood-Barton (1962) reported that Nigeria is the largest producer of yams in the world, responsible for about half of the world production. Gallet, Baldwin, and Dina (1956) asserted that yams rank next to cassava and contribute more than maize in calorie production in Western Nigeria. The species favored as food, in order of popularity, are: Dioscorea rotundata, D. cayanensis, D. alata, D. dumentorum, and D. esculenta.

PROCESSING FOR STORAGE

A large proportion of yams is marketed as fresh tubers and only a small proportion reaches the market in processed form. About 20 to 25 percent of yams harvested in Oyo and Ogun States is processed into dried yam chips. Figures for Ondo and other states, where preparations made from yam chips are less favored, are probably lower.

The major forms of processed yams are flour, chips, and flakes. Whereas storage in tuber form for consumption should not normally be attempted for more than six months, processed chips and flour are easier to store for much longer periods with very little deterioration. Flour and chips are the traditional methods of processing for storage. Yam flakes are only a recent development resulting from increased urbanization.

Yam Chips

Yam tubers are peeled and cut into small slices. The cutting can be done laterally or lengthwise along the tuber. Slices are soaked in water, brought nearly to a boiling point, left to steep in hot water until cool, and finally sun- or air-dried for a few days to reduce the moisture content for storage. The Marconi moisture meter can be calibrated for quick determination of moisture content of dried yam. No such instrument is available for farmers' wives who are engaged in this rural industry; therefore, intuition is the

111

best guide. Slices are normally cut into different thicknesses depending on the amount of sunshine and the possibility of quick drying. If drying is slow, the chips will mold.

Yam Flakes

This is a new form of processing which is unknown to the rural population but becoming increasingly popular in urban areas. Known as "poundo yam" when reconstituted, it is a food of the elite. Its growing popularity derives from the ease of preparing meals from the processed flakes. It is essentially processed by dehydration.

The tubers are peeled, washed, cut into thin slices, and cooked under pressure for about 30 minutes. They are then mashed and the product is roller dried to produce thin flakes which are packed in polyethylene bags for marketing. Onayemi (1973) and Onayemi and Potter (1974) have described in detail the full process of making improved yam flakes.

Yam flakes offer a high potential for processing and storing yam and have the following advantages:

1. Because most of the water (which constitutes about 70 to 80 percent of the yam tuber) has been removed, it is less bulky in transport.
2. It can be stored more easily than fresh yam, and for much longer periods without deteriorating.
3. Preparation time from flakes to white doughy mass is much shorter than preparing pounded yam from fresh tubers.
4. Unlike pounded yam, no pounding is required; all that is needed is stirring and kneading in hot boiling water.

There are now several companies marketing yam flakes extensively in West Africa. "Poundo-Yam" is likely to remain unpopular with rural populations who have developed a taste for pounded yam, because it is not as viscous and sticky--giving the usual "draw" when pulled into small balls for eating--as the pounded yam. It is conceivable that as more research into the mode of processing is done, or additives that can be used to enhance the consistency and increase the drawing power of

112

reconstituted yam flakes are discovered, "poundo yam" will more closely simulate pounded yam.

Not all the yam species or cultivars are good for making flakes; however, D. rotundata and D. cayanensis are superior to others.

Fried Yam Chips

This is a relatively new form of processed yam. They are made in the same way as potato chips, by frying slices of yam in oil and packing them in small polyethylene bags.

STORAGE OF WHOLE TUBERS

Storage of fresh tubers for consumption should not last longer than six months. During storage changes in weight and quality occur. Such losses increase after the first six months of storage. There are two types of losses, sprouting and rotting.

Sprouting causes reduction in the food reserves by translocating carbohydrates from the tuber into sprouts for metabolic processes. It also increases respiration rate, thereby increasing the rate of dry matter loss, and accelerates moisture loss through the permeable surface of sprouts, leading to wilting of the tuber. Progressively, the yam becomes soft to touch from the bottom towards the head, resulting in rotting. Quality and palatability generally deteriorate as a result of sprouting. Rotting generally causes partial or total loss of the tuber, decreases tuber viability if used for seed, and causes increased respiration rate with attendant loss of dry matter.

Methods of storing yam in Nigeria vary according to the period of storage required--particularly whether long or short term, depending on whether the yam is to be used for propagation or for later consumption as food. For yams to be consumed within six to eight weeks of harvesting, storage methods are elaborate. In some areas such tubers are gathered near the base of a big shade tree. Crevices provided by buttresses in the trunk base are used as the storage structure, and tubers are covered with palm fronds.

Storage of yams for longer periods up to six months is traditionally done in clamp or barn.

Clamp Storage

This involves digging of holes or pits in the ground, burial of tubers in the pits, and covering with soil until required.

Barn Storage

A yam barn is constructed by sticking vertical live wooden or bamboo poles into the ground, arranging them about 60 cm apart. They are then held together by more rigid horizontal wooden sticks (Figure 1). It is sometimes preferable to use vertical logs that will root to reinforce the poles. In this way the framework can survive and be used for several years. It is usual to protect the barn from rats by fixing a barrier 1 m high around the yam barn (Figure 2). Iron sheets are most readily available. These should be stuck well below the ground surface to prevent rats digging from beneath.

Pests and Diseases

Processed yam in storage. The common pests of dry processed yam chips in storage are mainly weevils, especially Callosobrunchus maculatus. These insects bore into the chips in store, lay their eggs, and deposit frass into them, causing degradation and making them unhygienic for consumption.

The remedy is either to fumigate the chips before storage or to use insect-proof containers such as polyethylene bags, or both. The chips can also mold in storage if not properly dried beforehand.

Whole tubers in storage. Nematodes which attack the tubers in the field may continue to devastate them in storage. Worse still, they produce secondary effects by opening up wounds through which pathogens (bacteria and fungi) gain entry, causing rots. Nematodes can be controlled by using "nemagon," a commercial product.

Dry rots, caused by Rosellinia and Sphaerostilbe, and soft rots, caused by Penicillium, Fusarium, Botryodiplodia, Rhizopus, and Lasiodiplodia, are quite common. These can be controlled in stored yams by using fungicides such as benlate, thiabenzadole, or captan applied by dipping tubers in a solution of the chemical to give a deposit of 500-1000 parts per million/yam weight.

Figure 1. Traditional yam barn with ropes.

Figure 2. Yam barn showing metal rodent barrier.

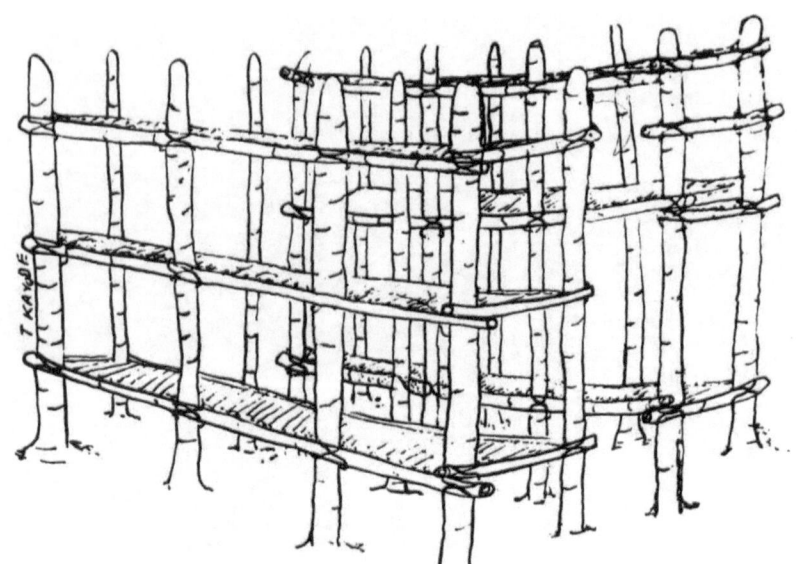

Figure 3a. Shelves for storage of yams.

Figure 3b. Sprouts to be removed by farmer.

116

Ants, rodents, and termites are pests of stored yams both in barns and clamps. Scale insects are also found on stored yams and although they appear to do very little observable damage, they most probably help in spreading some rot organisms. Ants and termites can be controlled by locating their hills, tunnels, and chambers and destroying them with dieldrin solution. The rodents are taken care of by using traps, poison baits, and rodent barriers in barns (Figure 2). The most effective way of dealing with scale insects appears to be removal of affected tubers from the barns.

Advantages and Disadvantages of Both Methods

Olorunda and Adesuyi (1973) have pointed out that both methods of storage result in substantial loss of weight due to rot and sprouting. However, it is easier to inspect and remove the sprouts in barns than in clamps. Therefore yams stored in barns generally have better germination and less weight reduction than those stored in clamps. Food yams can therefore be stored under clamps; however, Olorunda and Adesuyi (1973) maintained that yams stored in barns are more palatable than those stored in clamps.

It is usual practice for peasant farmers to use ropes to tie yams to poles in the barn. This practice is most unnecessary as it can cause damage to yams and lead to rotting. The use of shelves in barns (Figure 3) allows easy removal of any tuber which is deteriorating, and buds as they appear, in order to prevent sprouting in storage.

REFERENCES

Bevan, C.W.L. and Hirst, J. 1958. A convulsant alkaloid of Dioscorea dumentorum. Chem. and Ind. (January 25).
Coursey, D.G. 1965. The role of yam in West Africa food economies. World Crops 17:74.
_____. 1967. Yams. London: Longmans Green.
Gallet, R., Baldwin, K.D.S., and Dina, I.O. 1956. Nigerian Cocoa Farmers. Oxford University Press.
Ingram, J.S. and Greenwood-Barton, L.H. 1962. The cultivation of yam for food. Trop. Sci. 4:82.

Johnson, B.F. 1958. Characteristics of major crops. In: Yams--The Staple Food Economies of Western Tropical Africa. California: Stanford University Press.

Kay, D.E. 1973. Root Crops. Crop and Product Digest Monograph No. 2, Tropical Products Institute, London.

Olorunda, O. and Adesuyi, S.A. 1973. A Comparison Between the Barn and Clamp Methods of Storing Yams. Nigerian Stored Product Research Institute, Technical Report No. 757-59.

Onayemi, O. 1973. Studies of dehydration and storage stability of Dioscorea rotundata Poir (white yam) flake. Master's thesis, Cornell University, Ithaca, N.Y.

Onayemi, O. and Potter, N.N. 1974. Preparation and storage properties of drum dried white yam (Dioscorea rotundata Poir) flakes. J. Food Sci. 39:559.

A. O. Adenuga: Department of Plant Science, University of Ife, Ife-Ife, Nigeria

8
Storage and Processing of Taro in the People's Republic of China

D. L. Plucknett
M. S. White

ABSTRACT

Taro is widely grown in the south and central areas of the People's Republic of China. It is stored for two reasons: as seed for the following year's crop, and for later consumption. Corms can be stored in the field until ready for market. Tops are usually cut, or left to be killed by winter frosts. Corms are then left in the ground for up to two months. In many areas, cormels are stored in piles built up of alternating layers of cormels, straw, and a final layer of soil. At Kweilin, sand is used instead of straw. Storage structures may also be used to prevent freezing. The only form of taro processing was observed in Shanghai, where corms were quick frozen for export.

INTRODUCTION

On a recent trip to the People's Republic of China[2] the senior author had an opportunity to obtain information on storage and processing of taro. The crop is grown widely in south and central China, both as a sole crop and intercrop during the summer, and is especially important in high rainfall areas and in hydromorphic soils.

STORAGE

Taro is stored primarily for two reasons:
(1) to preserve planting materials for next year's
crop, and (2) to preserve corms destined to be used
for food.

Field storage is used widely to keep corms in
the ground in the field until ready for market, and
is used in the north as well as in the tropical
south. In the north the corms are allowed to
remain in the field after winter frosts kill the
top growth. In the lower Yangtze and in the south,
the tops are usually cut in order to prepare the
crop for cold weather storage. For example, at
Hangchow, in Chekiang Province, the tops are cut in
October and corms are stored in the field until
early January. Often other crops, especially short
duration vegetables such as Chinese cabbage, are
grown in the soil above the corms. Two months
after cutting, the corms are harvested. In Canton,
a form of field storage is used in that the mother
corms are harvested first, and sucker corms are
left in the soil to mature and for later harvest.

Piled storage is used in many places. In
Nanking in Kiangsu Province, cormels are stored for
seed as well as for food. Cormels are selected on
or before October 23, checked for signs of insect
or disease attack and to insure that corms are
mature, and are then sun-dried for 2 to 3 hours.

At Nanking, seed cormels about 3 cm in diameter
are stored over the winter in piles. About 5 cm
of straw is placed on the ground before placing a
layer of corms on the pile, then a layer of straw,
then a layer of cormels, and so on. The completed
pile is covered with a straw mat and then a final
layer of soil. Such piles (Figure 1) can be used
to store seed corms from October until April when
the cormels will begin to sprout. Winter temperatures at Nanking do not go much below 0°C. Temperatures in the piles should not go above 25°C. At
Nanking, they plant from 52,500 to 60,000 plants
per hectare.

A modification of the piled storage system is
used at Kweilin in Kwangsi Autonomous Region. In
this system layers of 5-12 cm or so of sand are
alternated with layers of corms. It was said here
that if taro is not stored it will lose quality and
rot in one month.

Storage structures are used in the lower
Yangtze area where wetland taro is stored in early

Figure 1. Pit storage of taro, Hangchow, Chekiang
Province. Photograph courtesy of Professor
Li Shu-Hsien, Chekiang Academy of Agricul-
tural Science, Hangchow.

November while temperatures are still above 10°C,
in order to prevent freezing of the corms.
        Storage rooms are used in Kiangsu Province.
Farmers select a north-facing corner in a room, and
then use clay bricks to build a 2 to 2.5 m square
enclosure 0.5 to 0.6 m high. The floor of the
enclosure is covered with 7.5 to 12 cm of fine
dried soil, selected corms are laid in a 7.5 to 12
cm layer on the soil, and the corms are covered
with 7.5 to 12 cm of soil. This process is repeated
until the top of the square enclosure is reached,
and the pile is covered with 12 cm of fine dried
soil. Such an enclosure can store about 350 to 400
kg of corms. During cold weather, when temperatures
drop below 5°C, grass is placed on top of the enclosure to prevent cold damage. Temperatures in the
room should be held at 8 to 15°C. The fine dry soil

121

Figure 2. Above: Left--quick frozen taro corms; center--
frozen whole tomatoes; right--green peppers.
Vegetable Dehydration and Freezing Factory,
Shanghai.
Below: Frozen taro packed in plastic bags for
export.

helps to keep humidity low. This system is effective, and corms can be stored successfully until about April 20.

   Pit storage structures are built by selecting slightly elevated areas where the ground water table is 1.5 m or more below the soil surface. A rectangular pit 1 m wide, 2 m long, and 1 m deep is dug. The bottom of the pit is covered with a layer of fine soil (ground, if necessary) and a layer of hay, and a ventilation pipe is placed and supported by a pole. Taro planting materials about 0.3 m deep are placed in the pit and covered with a layer of fine soil. Alternating layers of taro corms and soil are added until the pit is full, then the pit is covered with layers of grass and soil. Finally, a 0.15 to 0.3 m layer of soil is added and compacted. The ventilation pipe is opened only on sunny mornings and is closed at night.

   In December, when temperatures drop below 5°C, more soil is added (to make the soil layer 0.3 to 0.4 m thick) to the top of the pit and compacted to prevent water from seeping in. At this time the ventilation pipe is sealed. Storage damage to corms in this system is about 5 to 15 percent.

PROCESSING

   The only type of taro processing seen was at Shanghai, where quick freezing of small corms harvested in August and September was being practiced. A local variety is used.

   The basic processing system used is as follows. Corms are:
   1. Selected and divided into size grades;
   2. Boiled in hot (100°C) water for 3 minutes;
   3. Machine peeled;
   4. Selected by hand and packed in small aluminum trays or plastic bags; and
   5. Quick frozen for 50 minutes.
Figure 2 shows the frozen corms. This product is exported.

Donald L. Plucknett and Margaret S. White: Department of Agronomy and Soils, College of Tropical Agriculture, University of Hawaii, Honolulu, Hawaii

Section III
Post-Harvest Handling and Storage

1
Storage Problems in Aroids and Sweet Potato in India

N. Hrishi
C. Balagopal

ABSTRACT

Sweet potato (Ipomoea batatas) is grown in most parts of India. Presently about 80,000 ha are under cultivation. However, consumption is less than 3 kg per capita per annum, a negligible amount when compared with other countries. Proper curing and storage might help in assuring a good market. Sweet potatoes are usually marketed soon after harvest and are generally stored in jute bags. Main storage problems are the sweet potato weevil, diseases such as soft rot, and physiological changes (reduction of starch and increase in sugars and dextrins).

There is no organized cultivation of edible aroids, although they are widely grown throughout India as intercrops. Corms are processed into chips and flour. During storage, soft rot can occur, but the damage is negligible. There is a general reduction in starch and an increase in total sugars during storage.

Improvement in storage techniques will prevent damage as well as deterioration in quality.

SWEET POTATO

Sweet potato (Ipomoea batatas) is grown in most parts of India, the largest crop areas being in the following states, in descending order: Bihar, Uttar Pradesh, Orissa, Assam, and Madhya

Pradesh. In other states it is cultivated in small
areas. The present crop area is about 80,000 ha,
which is less than 0.2 percent of the area under
white potato. The yield per hectare is only about
7 tons, and is much lower than that reported in
other countries.

Although it is not known authentically, sweet
potato may have been introduced into India by the
Portuguese as early as 1500. There is no doubt
that sweet potato has been in cultivation in India
from very ancient times as it is regarded by the
people as indigenous. This is evident from the fact
that it is used in religious ceremonies, particularly the "Shradh" ceremony, when only indigenous
vegetables are permitted to be cooked.

Based on the total production in the country,
human consumption of sweet potato is less than 3 kg
per capita per annum,which is negligible compared to
the consumption rate in other countries. Sweet
potato is generally prepared in India by simply
boiling.

Storage Problems

Storage of sweet potato tubers has not been
given enough attention, though such methods would
undoubtedly benefit both the producer and consumer
by making them available throughout the year. Proper
curing and storage would help in assuring good market, as judged from good appearance, desirable
culinary qualities, and freedom from other defects.
However, in India the products after harvest are
marketed soon for want of storage facilities and
techniques, and are generally kept in jute bags.
The main problem during storage is due to the existence of the sweet potato weevil.

Another common disease known as soft rot
(caused by Rhizopus spp.) is found during storage.
Similarly, some physiological changes do occur in
the roots after harvest, the major change being the
conversion of starch into sugars and dextrins. There
is scope to minimize such changes, as evaluations of
germplasm indicate that clones differ in the degree
of conversion of starch to sugars and dextrins
during storage and cooking. Decreased ascorbic
acid content and increased carotene content have
also been observed during storage.

Well designed and managed storage facilities
could serve to maintain the quality of sweet potato
after harvest with minimum loss. Efforts are being

made at the Central Tuber Crops Research Institute (CTCRI) to develop such technologies for curing and storage, and which can easily be adopted by farmers and marketing centers. Urgent action is called for to minimize post-harvest losses during storage, especially in view of the food situation in the country and rapid increase in population.

EDIBLE AROIDS AND THEIR STORAGE PROBLEMS

Edible aroids grown in India are Colocasia, Alocasia, Xanthosoma and Amorphophallus campanulatus. It is difficult to give the estimates of area and production of aroids in the country, since there is no organized cultivation. Further, these crops are generally grown throughout the country wherever sufficient moisture is available. Aroids are widely grown mostly as intercrops in homestead gardens in certain regions in West Bengal, Orissa, Andhra Pradesh, Tamil Nadu, Gujarat and Assam.

Aroid corms are used in the form of chips and flour. Sometimes corms are peeled, cooked, and taken with condiments and adjuncts. Colocasia flour is prepared from raw or pre-cooked tubers, and the flour obtained from pre-cooked corms is considered better. Taro malt prepared from flour is considered a good infant and invalid food. During storage, a microbial soft rot can occur, making tubers unfit for consumption. However, the percentage of damage due to this type of pathogenic rot is negligible. A gradual reduction in starch and increase in total sugars are also observed during storage, with appreciable reduction in moisture.

Corms of Amorphophallus campanulatus are used as vegetables or for preparing chips. It is the local practice to store Amorphophallus after dipping in cow dung slurry and ash. Treated corms are preserved for months with no damage. Keeping quality of Amorphophallus tubers is not affected much during storage. Incidence of mealybugs and saprophytic occurrence of sclerotium have been observed during storage of Amorphophallus corms. However, prevalence of high humidity and presence of mechanical injury on such tubers may lead to corm decay.

N. Hrishi and C. Balagopal: Central Tuber Crops Research Institute, Trivandrum, India

2
The Use of Fungicides against Post-Harvest Decay in Stored Taro in the Solomon Islands

G. V. H. Jackson
D. E. Gollifer
J. A. Pinegar
F. J. Newhook

ABSTRACT

Treatment with some of the fungicides tested initially reduced corm rots caused by <u>Phytophthora colocasiae</u> and <u>Pythium splendens</u> in open storage, but infection by <u>Botryodiplodia theobromae</u> from about day 10 led to complete loss by day 20. Wrapping in paper increased the effectiveness of chemical treatment to about 20 days, probably by protecting developing shrinkage cracks against infection by <u>B</u>. theobromae. Without chemical treatment, corms survived well for up to 30 days if kept in polyethylene bags, retaining acceptable taste and texture despite development of roots, cormels, and occasional leaves. Reduced susceptibility to rot appeared to be related to this increased physiological activity. Best results were obtained when tops (petioles and corm apex) were not removed. Dipping in 1 percent sodium hypochlorite before storage in polyethylene bags neutralized the effects of

augmented inoculum of P. colocasiae, analogous to
field contamination. Such a treatment would provide additional protection that might be practicable
and useful in village storage or when corms are
transported to distant markets.

INTRODUCTION

In the Solomon Islands studies on the storage
of corms of taro (Colocasia esculenta (L.) Schott)
have shown that decay is caused by a complex of
fungal organisms. Gollifer and Booth (1973) identified rots caused by Botryodiplodia theobromae Pat.,
Fusarium solani (Mart.) Sacc., and Sclerotium
rolfsii Sacc. In later work, Phytophthora colocasiae Rac. and Pythium splendens Braun were found to
be additional pathogens (Jackson and Gollifer,
1975). Where epiphytotics of P. colocasiae leaf
blight occur in the field, this fungus can account
for 80 percent of the rots formed during the first
5 to 8 days of storage. Later, B. theobromae
rapidly colonizes the corms to complete the decay.
Rots can be substantially reduced by placing
undamaged corms in shallow leaf-lined soil-pits,
where they store without deterioration and maintain
an acceptable taste and texture for at least 4
weeks (Jackson and Gollifer, 1975). This method
does not assist, however, in preventing decay in
corms transported to urban centers. For this,
post-harvest treatments with chemicals is one solution. To be successful, fungicides must have broad
spectrum activity since the decay of stored corms
is caused by representatives of all major taxonomic
groups.
There are few studies reported on the use of
chemicals to control storage rots in taro. In
preliminary tests the use of compounds selected for
their known activity against F. solani and B. theobromae (mainly those degrading to the methyl ester
of benzimidozol-2-ylcarbamic acid, MBC) failed to
prevent decay under Solomon Island conditions
(Jackson and Gollifer, 1975). This was because they
were ineffective against the phycomycetous fungi,
the dominant rot-causing organisms during the early
stages of storage. Additionally, rots due to B.
theobromae were not entirely eliminated. In Jamai-
ca, by contrast, where storage decay is mostly
caused by B. theobromae, dipping in benomyl before
storage reduced the level of rotting and increased

the marketability of corms (Been, Marriott, and Perkins, 1975). Burton (1970) used sodium hypochlorite, socium orthophenylphenate, 2,6-dichloro-4-nitroaniline, hot water, and hot wax treatments on corms reaching the Chicago market from Florida and Puerto Rico. Because they were applied at the terminal market none proved satisfactory, although chlorine and hot wax together reduced decay caused by Botrytis, Rhizopus, Penicillium and Erwinia species.

This communication reports on the results from screening chemicals selected for their broad fungicidal activity and from tests on storing corms in polyethylene bags. This method has been used with success in Fiji to prevent rots caused by P. splendens in taro corms shipped to New Zealand,[1] and in trials in Trinidad where B. theobromae is the most important pathogen in corms of dasheen (Been, Marriott, and Perkins, 1975).

MATERIALS AND METHODS

Preparation of Corms for Storage Treatments

Corms of cv. Akalomamale were used in experiments carried out at Dala Research Station, Malaita. On Guadalcanal, cv. Kumabu was used. Unless stated otherwise, corms were obtained from plants infected by the leaf blight fungus, P. colocasiae. After harvest and depending on the nature of subsequent treatment, shoots (tops) were either left attached or removed by cutting just below the shoot apex. Corms were then gently scraped to remove roots and moribund leaf and petiole tissue, treated with fungicide or placed in polyethylene bags, and laid on paper spread over the concrete floor of a storage building at Dala Research Station (Gollifer and Booth, 1973) and in a similar situation at Dodo Creek Research Station on Guadalcanal.

Evaluation of Fungicides to Control Rots in Stored Corms

Lots of ten corms were dipped (10 min) in 10-liter suspensions (500 ppm a.i.) of each fungicide, allowed to dry, and stored. Other corms were treated with the same fungicides formulated as dusts (20 percent a.i.) prepared by adding Fuller's Earth as a diluent. The preparations were dispersed over the corms by gently shaking 4 g dust onto

8 corms for 3 min in polyethylene bags (80 x 45 cm) to ensure even distribution. The materials evaluated were: captan((N-trichloromethylthio) cyclohex-4-ene-1, 2-dicarboxyimide)50 percent a.i., w.p. (Imperial Chemical Industries (ICI) Ltd.); mancozeb (zinc and manganese bisdithiocarbamate) 80 percent a.i., w.p. (ICI Ltd.); thiram (tetramethyl thiuram disulphide) 80 percent a.i., w.p. (Amalgamated Chemicals Ltd.); dichloran (2,6-dichloro-4-nitroaniline) 50 percent a.i., w.p. (Boots Pure Drug Co. Ltd.); quintozene (pentachloronitro benzene) 75 percent a.i., w.p. (Olin Chemical Corporation (OCC) Ltd.); Terrazole (5-ethoxy-3-trichloromethyl-1,2,4-thiadiazole) 35 percent a.i., w.p. (OCC Ltd.); Terrachlor Super-X (pentachloronitrobenzene and 5-ethoxy-3-trichloromethyl-1,2,4-thiadiazole) 10 percent and 25 percent a.i., w.p. respectively (OCC Ltd.); captafol (N-(1,1,2,2,-tetrachloroethylthio) cyclohex-4-ene-1,2-dicarboxyimide) 80 percent a.i., w.p. (ICI Ltd.); salicylanilide (sodium salicylanilide tetrahydrate) 93 percent a.i., w.p. (ICI Ltd.); sodium hypochlorite (commercial household bleach).

In two trials an experimental compound PP 073 (Plant Protection Ltd., Jeallot's Hill Research Station, Berks, England) was evaluated against storage rots, after tests in vitro (see below) indicated that the material was effective against all five rot-causing fungi isolated from taro corms. Formulation JF 4944, 10 percent a.i., was used in the first trial and JF 4387, 20 percent a.i., in a second at 100, 500, and 1000 ppm a.i., with and without 300 ppm Agral "60" (ICI Ltd.) as a wetting agent. Corms were dipped in the suspensions for 4 minutes.

Evaluation of Fungicides "in Vitro"

Four fungicides were tested at the Tropical Products Institute, London, against the following rot-causing organisms: P. splendens (IMI 176600), F. solani (IMI 176601), B. theobromae (IMI 176602), S. rolfsii (IMI 176604) and P. colocasiae (IMI 176614). The fungicides used were benomyl (methyl-1) (butylcarbamoyl)-2-benzimidazole carbamate) 50 percent a.i., w.p.; bavistin (2-(methoxy-carbamoyl)-benzimidazole); two experimental compounds, PP 073 20 percent a.i., w.p., and DAM 18654 (1,5-cyan-pentylcarbamoyl (methoxy-carbamylamine)-

133

benzimidazole) 50 percent a.i., w.p. (Bayer (Florida) Ltd.).

The minimum inhibitory concentration (MIC) of
each fungicide was determined using the method of
serial dilution in liquid medium. Standard suspensions of each fungicide at 1000 ppm a.i. were
prepared in sterile distilled water and then further
diluted in Sabouraud's liquid medium (Oxoid) to
provide the following final concentrations: 250,
125, 62.5, 31.2, 15.6, 7.8, 3.9, 2.0, and 1.0 ppm.
The inoculum for the assays was prepared by washing
a 4-day-old culture (growing on a slope of Sabouraud's Dextrose Agar) with 5 ml Sabouraud's liquid
medium. The slope was agitated with sterile glass
beads. The spore mycelium suspension was transferred to a sterile universal tube and diluted to
the density of Brown's opacity tube No. 2 (obtained
from Welcome Laboratories, Beckenham, Kent,
England) with Sabouraud's liquid medium. Each tube
in the fungicide series was inoculated with one drop
of standardized inoculum from a sterile Pasteur
pipette calibrated to deliver 30 drops/ml. The
presence or absence of growth was recorded after
incubation at 27°C for 7 days. The MIC was
expressed as the lowest concentration of fungicide
which completely inhibited growth of the fungus.
As P. colocasiae did not grow well in liquid medium,
the effectiveness of the fungicides against this
organism was tested by a plate assay technique. A
series of potato dextrose agar (PDA, Oxoid) plates
containing decreasing concentrations of fungicides
(100, 50, 25, 10, and 5 ppm) was prepared from
aqueous stock suspensions. The plates were inoculated by placing a 7 mm agar disc, cut from the
margin of a P. colocasiae colony growing on PDA,
mycelium downwards in the center of the plate.
There were three replicate plates for each fungicide concentration. After incubation at 27°C for
7 days, the diameter was measured and compared with
the growth on a fungicide-free medium to calculate
the percentage inhibition of hyphal growth. The
percentage of growth inhibition was plotted on a
probability scale against the log of the concentration of fungicide and the ED_{50} determined.

Storage of Taro Corms in Polyethylene Bags

There were three trials: the first two were
carried out at Dala Research Station, Malaita, the
third at Dodo Creek Research Station, Guadalcanal.

Trial 1. Corms were stored in large (80 x 45 cm) polyethylene bags either with (treatment 1) or without (treatment 2) tops attached. Corms were unscraped but most soil was removed by vigorous shaking. Only the largest (3-4) cormels were removed. Ten corms were placed in each bag and the bags were sealed by folding over the open end and fastening with wire staples. In the same trial the polyethylene bag method of storage was compared to one in which corms were dipped for 4 min in a solution containing PP 073 at 500 ppm a.i. (JF 4387). Shoots were detached, corms scraped, dipped in the fungicide, drained of excess liquid and air dried. Nine corms were either stored wrapped singly in brown paper (treatment 3) or unwrapped (treatment 4). Two control treatments were included: tops were detached and corms scraped and either stored dry (treatment 5) or stored after dipping in tap water for 4 min (treatment 6).

Trial 2. This trial, which was essentially a repeat of Trial 1, except for omission of PP 073, further investigated the effects of storing corms in polyethylene bags with or without tops attached. Corms were taken from plants heavily infected by P. colocasiae before harvest. A control treatment was included in which corms with tops attached were stored dry. Three lots of ten corms were used for each treatment.

Trial 3. The concentration of P. colocasiae propagules on the surface of 192 corms was enhanced by spraying with a dense spore suspension collected by washing lesions on 40 naturally infected leaves with distilled water. The corms were divided into two groups: in one tops were removed, in the other tops remained attached. Within each group, corms were redivided into four lots of twenty-four. Treatments were as follows: two lots were dipped in water and two in 1 percent sodium hypochlorite for two minutes. One lot from each of the different dip treatments was stored dry, the other in polyethylene bags. Eight corms were placed in a single bag. Corms were inspected for rots at 15 and 30 days.

Determination of Fresh Weight Changes in Corms Stored in Polyethylene Bags

Changes in the fresh weight of two lots of 20 corms, one stored dry, the other placed in polyethylene bags, were recorded at intervals of 10 and 30 days.

Isolation of Rot-Causing Organisms

In all trials, unless stated otherwise, corms were sampled after 5, 10, and 20 days storage. Corms were peeled, rots located, and isolations made from rot margins onto PDA and 3-PCMA (corn meal agar containing penicillin, 50 ppm, pimaricin, 10 ppm, and pentachloronitrobenzene, 100 ppm). A disease index was calculated according to the method of Gollifer and Booth (1973).

RESULTS

Evaluation of Fungicides for Control of Storage Rots

At 5 days rots caused by P. colocasiae, which were the first to develop in stored corms, were controlled by most fungicides tested. Best results were shown by captan, copper oxychloride, captafol, mancozeb, Terrazole, and sodium hypochlorite (Table 1).

At 10 days, at which time B. theobromae became the dominant rot-causing pathogen, only copper oxychloride, captafol, and mancozeb-treated corms were without serious decay. Even with these fungicides, however, infection by B. theobromae was only delayed, not prevented, and by 20 days all corms were totally destroyed by this fungus.

Essentially similar results were achieved when some of the fungicides were used as dusts (Table 2). Captan, Terrazole, and Terrachlor Super-X prevented infections by P. colocasiae but not those caused by B. theobromae. Negligible decay in this trial was caused by P. splendens. When trials were repeated using only captan, Terrazole, and Terrachlor Super-X as suspensions at the previous rate and dusts at an increased rate of 4 g per five corms, results were the same as in previous trials: the incidence of rots caused by P. colocasiae decreased but they were not totally eliminated and there was no control of B. theobromae infection.

Screening Fungicides "in Vitro"

Of the four fungicides tested, only PP 073 gave control of all fungal isolates grown in liquid

136

TABLE 1

Frequency of Isolation of Rot-Producing Fungi from Taro Corms Stored after Treatment with Fungicides as Dips

Fungicide	Number of Corms (max. 10) Found Infected with Rot-Producing Pathogens at 5 and 10 days			Number of Undecayed Corms (max. 10)			Disease Index[a]		
	Phytophthora colocasiae	Pythium splendens	Botryodiplodia theobromae	5 days	10 days	20 days	5 days	10 days	20 days
Captan	0 1	0 1	0 6	10	4	0	0	26	98
Copper oxychloride	1 0	0 0	0 1	9	9	0	4	6	94
Captafol	2 2	0 0	0 2	8	7	0	4	16	94
Mancozeb	2 2	0 0	0 2	8	8	0	4	12	82
Thiram	5 4	0 0	4 4	5	4	0	12	34	96
Salicylanilide	3 5	0 1	0 6	0	4	0	6	34	92
Dichloran	6 6	1 1	1 6	3	3	0	18	46	88
Terrazole	0 1	0 0	0 4	10	6	0	0	28	82
Quintozene	6 2	0 1	0 3	4	6	0	14	32	64
Sodium hypochlorite	0 1	2 0	3 8	6	2	10	12	26	100
Control	8 8	1 2	2 8	2	0	0	20	66	94
Total Number of Rots	35 32	4 5	10 50						

[a] Disease index = total rating on a 0 to 4 severity scale x 100/max. possible score (see Gollifer and Booth, 1973).

TABLE 2
Incidence of Isolation of Rot-Producing Fungi from Taro Corms Stored after Treatment with Fungicides as Dusts

Fungicide	Number of Corms (max. 8) Infected with Rot-Producing Pathogens at 5 and 10 days						Number of Undecayed Corms (max. 8)			Disease Index[a]		
	Phytophthora colocasiae		Pythium splendens		Botryodiplodia theobromae		5 days	10 days	20 days	5 days	10 days	20 days
Captan	0	2	0	1	0	5	8	1	0	0	60	93
Thiram	3	0	1	1	3	3	4	4	0	18	20	97
Terrazole	0	0	0	1	3	3	5	0	0	8	67	90
Terrachlor Super-X	0	0	0	0	0	4	8	3	0	0	23	97
Quintozene	1	3	3	0	1	4	4	3	0	15	45	97
Non-treated Control	8	7	0	1	1	6	2	2	0	22	64	94
Total Number of Rots	13	12	4	4	8	25						

[a] Disease index = total rating on a 0 to 4 severity scale x 100/max. possible score (see Gollifer and Booth, 1973).

medium (Table 3). Although less effective in suppressing mycelial growth of B. theobromae compared to the other three, it was the only fungicide tested that was effective against P. splendens.

The results from the agar plate test showed that fungicide PP 073 also effectively reduced hyphal growth of P. colocasiae (Table 4).

Storage Trials with PP 073

In the first storage trial, control of P. colocasiae with PP 073 (formulation JF 4944) was poor except at 1000 ppm a.i. (Table 5). Control of B. theobromae, however, was not achieved even at the highest concentration.

TABLE 3
Effectiveness of Fungicides Against Growth "In Vitro" of Fungi Causing Rot in Taro Corms

Fungicide	Minimum Inhibitory Concentration (ppm)			
	Botryodiplodia theobromae	Sclerotium rolfsii	Pythium splendens	Fusarium solani
PP 073	7.8	0.5	1.0	3.9
DAM 18604	0.06	> 250	> 250	2.0
Bavistin	0.02	> 250	> 250	2.0
Benomyl	0.02	7.8	> 250	2.0

TABLE 4
Concentration (ppm) of Four Fungicides Required to Effect a 50% Inhibition (ED_{50}) in Hyphal Growth of Phytophthora colocasiae in an Agar Plate Test

Fungicide	ED_{50} (ppm)
PP 073	54
DAM 18654	Not active
Bavistin	Not active
Benomyl	Not active

TABLE 5
Incidence of Isolation of Rot-Producing Fungi from Taro Corms Stored after Treatment with Fungicide PP 073 (JF 4944) at 100, 500, and 1000 ppm a.i. (plus Agral '60', 300 ppm)

Treatment	Number of Corms Found Infected with Rot-Producing Pathogens at 5 (max. 10) and 10 (max. 5) days						Undecayed Corms (%)			Disease Index[a]		
	Phytophthora colocasiae		Pythium splendens		Botryodiplodia theobromae		5 days	10 days	20 days	5 days	10 days	20 days
PP 073: 100 ppm	5	0	1	0	3	4	10	20	0	55	70	100
500 ppm	4	0	0	0	0	4	60	20	0	23	60	100
1000 ppm	1	0	0	0	0	3	90	20	0	3	65	100
Control (dry)	9	0	0	0	0	5	10	0	0	40	100	100
Control (wet)	8	0	1	0	1	4	0	20	0	73	80	100

[a] Disease index - total rating on a 0 to 4 severity scale x 100/max. possible score (see Gollifer and Booth, 1973).

When the other formulation (JF 4387) was used,
PP 073 at 500 and 1000 ppm, but not at 100 ppm, sub-
stantially reduced infections of P. colocasiae.
Again, the fungicide was ineffective in controlling
B. theobromae infections, and corms were decayed at
10 days (Table 6).

Storage Trials with Polyethylene Bags

Trial 1. Corms in polyethylene bags stored
well. Rots caused by P. colocasiae, P. splendens,
and B. theobromae were present at 5 and 10 days,
but both number and extent were low compared to the
control treatments (Table 7). There was vigorous
cormel and root development on corms in polyethylene bags and, where tops remained attached, leaves
also grew.

When sampled at 20 days corms without tops were
beginning to deteriorate with soft rot bacteria
(Erwinia chrysanthemi Buckholder, McFaddon and
Dimock), and this coincided with a decline in cormel
and root growth. Corms with tops stored well, and
at 20 days eight of ten were without rots. Taste
and texture were acceptable when corms were boiled
in tests comparing stored with freshly harvested
corms.

Fungicide PP 073 included in the trial for comparison with corms stored in polyethylene bags controlled rots up to 5 days, but in later samplings
P. colocasiae was isolated from several rots in
both wrapped and unwrapped treatments (Table 8).
There was a marked reduction in the number of B.
theobromae infections when corms were wrapped, and
this was maintained up to 20 days.

Trial 2. Corms in this trial were taken from
plants where leaf blight infection had been severe
before harvest. After 5 days storage, the control
corms (unscraped with tops detached) were mostly
decayed by multiple (3 to 5) lesions of P. colocasiae. Those stored in polyethylene bags contained
only superficial rots, mostly 2.5 to 3.0 cm diameter
and 0.5 to 1.0 cm deep. Superficial rots from which
P. colocasiae or P. splendens were isolated were
still present at 20 days, when decay of control
corms was complete (Table 8).

In this trial results were similar, irrespective of whether tops remained attached or were
removed.

Trial 3. Corms had been severely bored by the
taro beetle Papuana inermis Prell. prior to harvest,

141

TABLE 6

Incidence of Isolation of Rot-Producing Fungi from Taro Corms Stored after Treatment with Fungicide PP 073 (JF 4387) at 100, 500, and 1000 ppm a.i. with and without wetting agent, Agral '60'

Treatment	Number of Corms (max. 10) Found Infected with Rot-Producing Pathogens at 5 and 10 Days			Undecayed Corms (%)			Disease Index[a]		
	Phytophthora colocasiae	Pythium splendens	Botryodiplodia theobromae	5 days	10 days	20 days	5 days	10 days	20 days
PP 073: 100 ppm	5	2	6	50	20	0	30	45	95
500 ppm	0	0	4	100	60	0	0	15	85
1000 ppm	1	0	6	90	40	0	3	25	75
100 ppm + Agral	6	6	4	40	20	2	25	55	95
500 ppm + Agral	1	0	2	90	80	0	5	5	90
1000 ppm + Agral	1	2	2	90	60	0	3	15	90
Control (dry)	6	4	6	40	20	0	15	65	90
Control (wet)	7	6	6	20	0	0	38	80	100

[a] Disease index = total rating on a 0 to 4 severity scale x 100/max. possible score (see Gollifer and Booth, 1973).

TABLE 7

Incidence of Isolation of Rot-Producing Organisms from Taro Corms Stored after Treatment with Fungicide PP 073 or Stored in Polyethylene Bags

| Treatment | No. Corms | Number of Corms Found Infected with Rot-Producing Pathogens at 5 and 20 Days | | | | | | | | Number of Undecayed Corms | | Disease Index[a] | |
		Phytophthora colocasiae		Pythium splendens		Botryodiplodia theobromae		Erwinia chrysanthemi		5 days	20 days	5 days	20 days
Polyethylene bag:													
Shoots detached	10	3	2	0	0	1	1	0	3	7	0	10	26
Shoots attached	10	0	0	0	0	0	0	0	2	10	8	0	4
PP 073:													
Unwrapped	9	1	0	0	0	1	9	0	0	8	0	4	64
Paper-wrapped	9	0	3	0	1	0	2	0	0	9	3	0	29
Control:													
Dry	9	5	*	3	*	2	9	0	0	1	0	36	73
Wet	9	1	*	1	*	1	8	0	0	6	0	11	58

[a] Disease index = total rating on a 0 to 4 severity scale $\times$ 100/max. possible score (see Gollifer and Booth, 1973).

* Recording impracticable because of severe infection by B. theobromae.

TABLE 8

Frequency of Isolation of Rot-Producing Fungi in Taro Corms Stored in Polyethylene Bags With and Without Shoots Attached

Treatment	Number of Corms (max. 10) Found Infected with Rot-Producing Pathogens at 5 days			Number of Undecayed Corms (max. 10)			Disease Index[a]		
	Phytophthora colocasiae	Pythium splendens	Botryodiplodia theobromae	5 days	10 days	20 days	5 days	10 days	20 days
Polyethylene bags:									
Shoots detached	4	1	3	5	3	3	10	22	26
Shoots attached	2	2	0	6	5	3	10	12	20
Control	7	3	5	1	0	0	46	71	100

[a] Disease index = total rating on a 0 to 4 severity scale x 100/max. possible score (see Gollifer and Booth, 1973).

and their holes harbored sclerotia of S. rolfsii.
After 5 days, mycelium of S. rolfsii developed on
corms in all treatments, but growth was pronounced
only on those stored in polyethylene bags (Table
9). At 8 days under these conditions, several
corms with petioles removed contained large rots.
Infections invariably began at sites where tops or
cormels were removed. No rots caused by P. splendens were located.

Because of the high inoculum of P. colocasiae
there was no advantage in leaving tops attached,
unless corms were treated with sodium hypochlorite
and stored without polyethylene bags. In this
instance rot development was lower (Table 9). For
the same reason, successful storage in polyethylene
bags was achieved only when corms were pretreated
with sodium hypochlorite. Where water was used as
a pretreatment, decay was rapid whether corms were
enclosed in polyethylene bags or stored dry.

Tasting trials on this and subsequent material
showed that corms treated with sodium hypochlorite
and stored in polyethylene bags were acceptable when
tested after 30 days storage but not after longer
periods.

Fresh Weight Changes of Stored Corms

Weight changes in corm, shoot and root tissues
in this experiment were consistent with observations
made in previous trials. Where corms were stored
dry without tops, losses of one-third initial
weight occurred in 30 days, irrespective of whether
corms were scraped or unscraped (Table 10). Comparatively small changes occurred in those corms
maintained in polyethylene bags. Shoot and root
development was greatest when tops were left attached, the effect being particularly noticeable
for tubers in bags and most when unscraped.

DISCUSSION

Our results have shown that rots caused by the
phycomycetous fungi, Phytophthora colocasiae and
Pythium splendens, were substantially reduced in
the early days of storage by the fungicides used,
applied either as suspensions or dusts. However,
by 10 days most corms had become infected by
Botryodiplodia theobromae, and at 20 days they were
completely decayed. The disappointing failure of

TABLE 9

Comparison Between the Incidence of Isolation of Rot-Producing Organisms in Lots of Eight Corms Sprayed with Additional Inoculum of P. colocasiae and Stored Dry or in Polyethylene Bags, (PB), Pretreated in 1% Sodium Hypochlorite (NaOCl) or Water and with Attached or Detached Shoots

Treatment	Number of Corms Found Infected with Rot-Producing Pathogens at 8 and 30 Days (max. 8)				Number of Undecayed Corms (max. 8)		Disease Index[a]	
	Phytophthora colocasiae	Sclerotium rolfsii	Botryodiplodia theobromae	Erwinia chrysanthemi	8 days	30 days	8 days	30 days
Petioles detached; corms dipped:								
Water, stored dry	8 5	1 2	0 7	0 0	0	0	78	95
NaOCl, stored dry	0 0	1 0	2 7	0 1	4	0	10	88
Water, stored PB	6 6	5 0	0 4	0 0	0	1	55	88
NaOCl, stored PB	0 0	4 0	0 0	0 1	4	7	15	5
Petioles attached; corms dipped:								
Water, stored dry	8 5	0 2	0 6	0 3	0	1	68	88
NaOCl, stored dry	0 0	2 2	0 2	0 0	0	4	8	33
Water, stored PB	6 6	2 1	0 0	0 2	1	1	48	85
NaOCl, stored PB	0 0	1 0	0 0	0 1	7	7	5	3

[a] Disease index = total rating on a 0 to 4 severity scale x 100 max. possible score (see Gollifer and Booth, 1973).

TABLE 10
Fresh Weight Changes of Stored Corms (Data are Means of Eight)

Treatment	Initial Wt. (g)	Wt. at 10 days (g)	Wt. Loss (%)	Wt. at 30 days (g)	Wt. Loss (%)	Wt. Roots (g)		Wt. Shoots[a] (g)	
						10 days	30 days	10 days	30 days
Tops detached:									
PB,[b] scraped	605	595	1.65	-	-	0.14	-	0	-
	590	-	-	568	3.81	-	0.04	-	1.39
PB, unscraped	761	753	1.15	-	-	0.63	-	0	-
	669	-	-	620	7.29	-	0.65	-	0
Dry, scraped	600	519	13.54	-	-	0	-	0	-
	580	-	-	344	40.73	-	0	-	0
Dry, unscraped	664	575	13.37	-	-	0	-	0	-
	719	-	-	461	35.83	-	0	-	0
Tops attached:									
PB, scraped	1063[c]	-	-	-	-	5.94	-	8.35	-
	1138	-	-	-	-	-	5.33	-	39.65
PB, unscraped	1125	-	-	-	-	5.92	-	20.52	-
	1040	-	-	-	-	-	21.53	-	62.26
Dry, scraped	1095	-	-	-	-	1.16	-	8.35	-
	1133	-	-	-	-	-	*	-	3.70
Dry, unscraped	1348	-	-	-	-	1.84	-	10.49	-
	1519	-	-	-	-	-	*	-	13.37

[a]Figures include the additional tissue formed during storage.

[b]PB = storage in polyethylene bags

[c]Weights include both corm and shoots.

*Not possible to distinguish roots produced during storage as all roots are decayed.

fungicides to control B. theobromae may have resulted from changes occurring within stored corms, pre-
disposing them to infection. Corms that were devoid
of petioles and shoot material and which were
scraped and stored at ambient conditions rapidly
lost weight, as much as 13.5 percent in 10 days.
This led to development of cracks and fissures and
consequent exposure of inner unprotected tissues
which could have provided entry points for storage
pathogens. That the latter were derived largely
from air-borne inoculum was indicated by the fact
that when corms were treated with the fungicide
PP 073, allowed to dry, and then wrapped in paper,
the incidence of rots caused by B. theobromae was
considerably reduced. Furthermore, the consistency
with which B. theobromae was found to colonize corms
between 5 and 10 days suggests that tissues become
increasingly susceptible as they lose water during
storage. This agrees with previous observations
made on Diplodia tubericolor infection of the
American dasheen: rots developed more rapidly if
corms were kept in the environment of a laboratory
room rather than in a moist chamber (Harter, 1916).
    Maintaining corms in polyethylene bags provided
an alternative to chemicals as a means of prolonging
their storage life. In general, corms survived well
for up to 30 days, retained an acceptable taste and
texture, and suffered little from rots caused by B.
theobromae. Storage life was extended because
undamaged corms were kept physiologically active,
as instanced by development of roots, cormels, and,
in some cases, of leaves. Additionally, conditions
within polyethylene bags, where humidities were
constantly high and temperatures ranged daily from
25 to 32°C, approached those reported necessary for
rapid suberization and periderm formation at
wounded surfaces (Been, Marriott, and Perkins,
1975). Where rots caused by P. colocasiae, P.
splendens, and B. theobromae did occur, they were
notably superficial, even at 20 days, at which time,
by comparison, those corms that were stored dry were
completely decayed.
    While the polyethylene bag method may prove to
be generally effective, there may be occasions on
which B. theobromae or P. colocasiae at unusually
high inoculum levels may be inadequately controlled.
It may be that experience will show that the bag
method should be used in conjunction with a chemical
treatment. Should this be so, sodium hypochlorite
as a corm dip would seem to be a useful choice. In

addition to being effective, it is cheap, is without residual effect, and is safe to use. The one fungus against which it is ineffectual is Sclerotium rolfsii, which was not commonly encountered in our storage trials. In an analogous context, sodium hypochlorite has already proven satisfactory as a pre-storage treatment of potato (Solanum tuberosum L.) tubers contaminated with cyst nematodes and Rhizoctonia solani Kiihn (Wood and Foot, 1975, and in press).

REFERENCES

Been, B.O., Marriott, J., and Perkins, C. 1975. Wound periderm formation in dasheens and its effects on storage. Proceedings of the Caribbean Food Crops Society, University of the West Indies, Trinidad, July 1975.

Burton, C.L. 1970. Diseases of tropical vegetables on the Chicago market. Trop. Agr. Trinidad 47: 303.

Gollifer, D.E. and Booth, R.H. 1973. Storage losses of taro corms in the British Solomon Islands Protectorate. Ann. Appl. Biol. 73:349

Harter, L.L. 1916. Storage-rots of economic aroids. J. Agr. Res. 6:549.

Jackson, G.V.H. and Gollifer, D.E. 1975. Storage rots of taro (Colocasia esculenta) in the British Solomon Islands. Ann. Appl. Biol. 80:217.

Wood, F.H. and Foot, M.A. 1975. Treatment of potato tubers to destroy cysts of potaty cyst nematode: a note. N.Z. J. Expt. Agr. 3:349.

_____ and Foot, M.A. In press. Decontamination of potato tubers grown in soil infested with potato cyst nematodes. N.Z. J. Expt. Agr.

ACKNOWLEDGEMENTS

Technical assistance provided by Kate Theomothe, D. Tua and M. Petakia in the Solomon Islands, and Ms. L. Nicolaides at Tropical Products Institute,

London. We are indebted to the New Zealand
Ministry of Foreign Affairs for supporting the
visits of F.J. Newhook to the Solomon Islands in
July 1976 and April 1978.

*G. V. H. Jackson and D. E. Gollifer: Ministry of Agriculture
and Lands, Honiara, Solomon Islands; J. A. Pinegar: Tropical
Products Institute, 56-62 Gray's Inn Road, London WC1X 8LU,
England; F. J. Newhook: University of Auckland, New Zealand*

3
Importing Root Crops
from the South Pacific Islands
for New Zealand Markets

John Watson

ABSTRACT

The history of importing of taro and other
tropical root crops in New Zealand is one that
parallels the increasing number of Polynesian immigrants from Tonga, Samoa, Fiji, and Rarotonga. The
earliest large shipments of taro began around 1950,
primarily from Fiji. However, flooding in Fiji led
to increased importing of Samoan taro, which has
since grown to greater popularity.

Trial and error led to the use of cool storage
units at 38 to 40°F. A later development was the
use of refrigerated containers which minimized
handling of taro and speeded up the unloading process. Air freight has been used occasionally to
fill market gaps, but has been unsatisfactory.
Fumigation is occasionally necessary, but tends to
cause faster spoilage.

The availability of tropical root crops has
helped Polynesian immigrants to settle more quickly
into their new environment, and has provided a
valuable source of income to the islands of the
South Pacific who have ventured into the export of
such crops.

HISTORY

Although Turners and Growers, Ltd., would like to believe that they were the first to import taro into New Zealand, and probably were the first to import and market taro on a regular basis in the "land of the long white cloud," it is a historical fact that the Maori Polynesian seafarers who set forth on their long voyages of discovery from the heart of Polynesia in the 13th and 14th centuries to eventually settle in Aotearoa did bring with them in their large, ocean-going canoes taro roots as one of the chief items of sustenance during their many days at sea. The descendants of this taro can to this day be found growing wild in parts of North Auckland, but no serious attempt has ever been made to grow it on a commercial basis. The few trials I have conducted with it have brought critical response from the Polynesian people who were given the taro to cook and sample; all complained that it gave them itchy throats. As far as I can ascertain, this early taro was what we call dry land taro and was used by the early Maori settlers as a fill-in food when their preferred sustenance vegetable, the kumara (sweet potato), was in short supply. This would probably be in the months of November and December when their supplies of stored kumaras ran out, and their early planting of kumara was not ready to harvest until January.

Older Maoris in North Auckland still cultivate small patches of taro for their own use, but this is the variety called Taro Hoia which was introduced to New Zealand in 1820 from the Cook Islands and parts of the southern Society Islands. This is what we identify in New Zealand as swamp taro, and it is very similar in appearance and taste to the main variety of taro grown in the Cook Islands.

Taro can be readily observed growing in the backyards of Polynesian homes in and around Auckland, but it is used exclusively for the growing of taro leaf which is used in a wide variety of Polynesian dishes. The variety is predominantly Samoan Pink, a dry land type of taro which, although producing edible leaves, produces corms considered so inferior in eating quality to the fresh imported taro that it is seldom, if ever, eaten.

Demand for taro and other tropical root crops did not appear at the retail level in New Zealand until after World War II.

152

In 1950, at the request of one greengrocer who had inquiries from immigrant Polynesian customers, we brought in 100 cases of "talo" from Fiji on the ship Tofua. With ever-increasing numbers of people from the islands of Tonga, Samoa, Fiji, and Rarotonga coming to New Zealand in search of work, and education for their children, and also seeking to attain a higher standard of living, these shipments became regular fortnightly occasions and have increased in volume to the point where there is a steady demand in New Zealand at viable prices for two thousand 50-lb (22.7 kg) cases of taro per week.

These early shipments of taro from Fiji were transported exclusively in two Union Steamship Company vessels, the Tofua and the Matua. These two vessels maintained a fairly regular fortnightly service. Their main function was carrying manufactured goods and foodstuffs to the islands and to pick up cargos of bananas from Fiji, Tonga, and Samoa for the New Zealand market. On their round trip, Suva was always the last port of call and--being only four days' sailing from Auckland--it was convenient and quite satisfactory to ship taro in holds that predominantly held green bananas at a temperature of 50 to 53°F (10 to 12°C).

This Fijian taro was predominantly a white-fleshed variety, although occasionally some very good pink taro was shipped from the island of Kandavu, approximately 50 miles (83 km) south of Suva. The taro was packed and shipped in uniform wooden slatted cases, which usually held a good 60 lb (27 kg) net weight of taro. It invariably arrived in first class condition. With this fortnightly shipping service, it was proved that holding the consignments of taro in cool storage and spreading the selling over the two-week period between ships gave the best return to the consigner. It also suited the majority of retailers, many of whom did not have suitable cool storage facilities in their stores, to have taro from our cool stores in reasonably fresh condition available for them to purchase at least three days each week. Initially this taro was held in our cool stores at the temperature it was shipped, i.e., 50 to 53°F. As long as we did not have to hold the consignment in cool storage more than 10 days it seemed to be reasonably satisfactory, although the last lot of taro from the cooler was never in as good a condition as the first offerings from the boat.

Eventually, for various reasons beyond our control, such as disrupted shipping schedules through bad weather or industrial trouble on the waterfront, the time came when we tried to hold taro in cool store for more than 10 days, and we found that after 10 days or so at 50-53°F, the quality began to deteriorate quite rapidly.

After many years of trial and error, we have found that 38 to 40°F (3 to 4°C) is the ideal temperature both to ship and to hold taro in cool storage. I have, on odd occasions, seen taro held successfully for six weeks at this temperature and still look fresh and crisp on removal from cool storage. However, after 24 hours or so out of the cooler the quality began to deteriorate quite rapidly. Fortunately, because of reasonably regular ship arrivals we hardly ever had to hold taro consignments for more than two weeks, and the shop life of taro held in a cooler for this period appears to be quite good.

Care must be taken, however, especially in cool storage units relying on cold forced air cycles, to prevent corm dehydration. Taro held in our cool storage rooms is usually slightly wetted with a hose twice a week. This seems to stop dehydration and also makes the taro appear fresh and crisp.

Fiji was practically the sole supplier of taro to the New Zealand market until 1963, when, due to serious flooding in Fiji, the government put an embargo on all taro export for an indefinite period in its concern to preserve the badly depleted food supplies for the population.

By this time the Polynesian population in New Zealand, and particularly Auckland, had grown rapidly to become the largest community of South Pacific islanders outside of the islands themselves. Probably close to 50,000 people had to have taro, so shipments were brought in from Western Samoa. This Samoan taro was dryland taro, pink-fleshed and commonly called Niue Pink, because it was reputed to have originally been propagated from planting material brought into Samoa from Niue Island. It is hard to say whether this was because the Polynesian population in New Zealand was predominantly Samoan, or whether it was definitely a superior eating variety. Not being a taro eater, I would be loathe to promote an opinion, but it is a fact that Samoan taro rapidly became the top selling variety in New Zealand and has remained so ever since, with ever increasing volume, whereas shipments of Fijian taro have gradually declined in quantity.

Use of Refrigerated Containers

In 1968 the Matua and in 1973 the Tofua were taken off the Island run and partially replaced by a container vessel called the Union South Pacific. This was a faster vessel that carried both refrigerated and non-refrigerated containers. With its own unloading gantry, it was designed to unload and load its containers in very quick time. Usually one day would see it unloaded, reloaded, and on its way to the Islands again. The Tofua and Matua, with their cargoes in deep holds, usually took at least a week to unload and reload.

Another unsatisfactory feature of shipping taro in such conventional vessels was that with their mixed cargoes of taro, bananas, watermelons and copra, the taro usually ended up in the bottom and wings of holds. Thus, although loaded first, taro was invariably the last item to be unloaded and inspected by the Port Agricultural Officers. It was also plainly apparent that any deterioration in the cargo stowed on top of the taro adversely affected its condition, appearance, and, consequently, the returns to the consignors.

The Union South Pacific and its containers solved many of these problems, and although she had a short career studded with industrial disputes, mechanical breakdowns and near catastrophes, she demonstrated quite clearly that for shipping produce from the Islands to New Zealand, refrigerated containers were ideal.

If I were a shipper and refrigerated container shipping were available on a regular fortnightly service, I would not consider any other method of shipping even if it were at a less costly freight rate. The minimal handling of the produce, the speed of unloading and placing the taro immediately on the market or in cool storage, the perfect condition of the taro on presentation to buyers and consequent higher returns to growers strongly justify the use of refrigerated containers.

A classic example of the efficiency of shipping in refrigerated containers occurred on the final voyage of the Union South Pacific in December, 1977. The vessel arrived in Auckland on 14 December with 4,031 cases of Samoan taro destined for the Christmas trade. Because of industrial trouble on the Auckland waterfront she was not unloaded until 23 December, which meant that the taro had been locked up in containers for at least 16 days. Despite this long delay the taro was in perfect condition and sold at attractive prices. Had this shipment of taro been in a conventional vessel, I am sure it would not have been unloaded in time to catch the Christmas demand, and would not have been in as good condition.

With the exodus of the Union South Pacific, the Union Steamship Company introduced a new roll-on roll-off container ship called the Marama. This vessel is larger and faster than her predecessor and so far has given an excellent fortnightly shipping service to and from Fiji, Samoa, and Tonga.

Airfreight

Over the past five years, we have airfreighted taro to New Zealand from Fiji and Samoa. This is a costly operation and such shipments have usually been arranged and slotted into gaps on the market, when a short supply situation has occurred through lack of suitable shipping schedules from the Islands. Usually such shipments arrive in virtually fresh-dug condition and fetch high enough prices on the market to offset the high costs involved. Air-freighted taro is usually packed in perforated sheet polyethylene or woven polypropylene bags which hold either 45 or 50 lb of taro. Although taro air-freighted in this way is only a matter of 12 hours or so on the journey, it often arrives in Auckland in quite hot condition. Our usual practice, there-fore, is to bring it straight from the airport and into our cool stores for a few hours to take out the accumulated heat before placing it on the market.

All taro imported into New Zealand, irrespec-tive of the source of supply, must be accompanied by an Agricultural Health Certificate issued by the Agricultural Department of the country of origin. It must be packed in new containers. Secondhand cases or sacks are prohibited.

PLANT DISEASES AND INSECTS

On arrival in Auckland practically all ship-ments of taro are unloaded and transported to a quarantine station near the markets. Samples of each shipper's line are carefully inspected by Port Agricultural Inspectors. This inspection is usually very thorough, as New Zealand--being relatively free of many plant diseases, and relying so heavily on agricultural and horticultural exports--must necessarily take stringent precautions against importing plant diseases.

Insects often found alive in shipments of taro into Auckland, thereby qualifying such shipments for fumigation, are as follows:

1. Ants
2. Mites, such as <u>Rhizoglyphus</u> <u>minutus</u>
3. Snails
4. Mealybugs such as <u>Dyomicocus</u> family
5. Bugs such as <u>Miridae</u> family
6. Weed seeds
7. Coffee bean weevil <u>Caraecarus</u> <u>fasciculatus</u>
8. Wire worm
9. Larvae of the Click Beetle
10. Termites
11. Nematodes such as <u>Pheidale</u> <u>megacephala</u>
12. Spiders
13. Red Mites such as <u>Caloglyphus</u> <u>krameri</u>
14. Scale insects such as <u>Coccus</u> <u>acuminatus</u>
15. Slugs
16. Mites such as <u>Lasioseus</u> <u>penicillinger</u> and <u>Proctdaelaps</u> <u>pygmaeus</u>
17. Beetles
18. Aphids such as <u>Pentalonia</u>
19. Mites of <u>Pyematidae</u> species
20. Pill bugs and forms of wood lice such as <u>Armadilidlum</u> <u>vulgare</u>
21. Earwigs
22. Cockroaches
23. Millipedes

Live insects or larve found in a shipment are immediately sent to Department of Agriculture entomologists for classification. Their findings determine whether or not a shipment must be fumigated.

If found necessary, fumigation is usually carried out at the point of inspection at the quarantine station. Stacks of taro are completely covered in plastic sheeting and fumigated with methyl bromide gas. The dosage is governed by the cubic content and the temperature of the material to be treated. If the temperature of the material is 21°C, 32 g/m^3 is introduced and left for two hours.

If the material is lower than 21°C it is left in fumigation half an hour or so longer. This dosage and treatment has always been 100 percent effective. Even so, fumigated material is always reinspected on removal from fumigation to ensure that such treatment has given a 100 percent kill.

Not only is fumigation an added cost to the shipper, it also has a detrimental effect on the keeping quality of the taro. After a few days, fumigated taro begins to get sticky, initially around the cutoff portion of the stem and then progressively down the skin of the corm, until rot eventually develops within the corm. At retail levels, such taro becomes difficult to sell and the usual result is a costly drop in price.

I cannot emphasize too strongly how important it is for shippers of taro to take every possible precaution, when packing and preparing taro for export, against entry of insects into the cases, and to ensure that the product is visually and bacterially clean. Not to do so is sheer fiscal folly and is detrimental to the good name of the product.

METHODS OF MARKETING TARO IN NEW ZEALAND

Practically all taro imported into New Zealand is sold at the wholesale market level by auction. The quantity to be sold each day is usually decided by the auctioneer, who has the feel of the pulse of the market and allocates the shipment accordingly. The taro is brought out of the cooler the night before the auction and stacked in shipper's lots. Each lot is auctioned separately, and the retail traders bid for their selected requirements.

The top bidder has the option of taking one case, or as many cases as there are, in that shipper's line. The second, third, and fourth bidders then have the option of taking their requirements of the line being auctioned. The obvious advantages of this method of selling are:

1. The purchaser can see what he is buying.
2. He bids on what taro best suits his requirements.
3. The best lines of taro fetch a definite premium.
4. The demand at retail level becomes fully and openly reflected at the market, so that the distribution of taro onto the market can be adjusted daily to ensure that shippers obtain the maximum price.

After the first two days of selling, shippers are usually cabled, giving a progress report on sales, recommendations as to quantities to be shipped on the next vessel, and an estimate of the current state of the market.

On completion of selling the total shipment, shippers are airmailed a complete account of sales, together with a bank draft for the net proceeds. Deductions from gross proceeds are:

1. Ten percent sales commission. (This is the standard rate charged on all fruits and vegetables sold at the Auction Markets in New Zealand.)
2. Freight (if not paid by shipper at point of dispatch).
3. Wharfage.
4. Cartage from wharf to market.
5. Fumigation fee (if necessary).
6. Department of Agriculture Inspection Levy.
7. F.P.A. marine insurance (optional).

CONSUMER PREFERENCES AND FACTORS AFFECTING DEMAND FOR TARO IN NEW ZEALAND

The demand for taro in New Zealand is confined almost entirely to the Polynesian community and their families. The native Maoris eat very little taro and the European community eats practically none. A very large proportion of Polynesian immigrants over the past 20 years has taken employment in the manufacturing industry or the service areas of three localities in the North Island. The majority live in Auckland. Quite a large community has developed in the timber towns of the lower Waikato area about 130 miles (216 km) south of Auckland. Wellington, capital of New Zealand, also has a large Polynesian community. Taro sales are virtually confined to these three areas.

Factors affecting the demand for taro are:

1. Price. If prices at retail level rise much over 50 cents per lb, then a definite consumer resistance becomes apparent and the Polynesian shopper will seek alternatives. The main alternative, when available, is usually green bananas which usually retail at around 30 cents per lb. Another alternative is sweet potatoes, which are available all year round in New Zealand and usually are less expensive than taro and other Island root crops such as tarotarua, ta'amu, or yams. On odd

159

occasions when both taro and the preferred
alternatives have been in short supply,
taro has realized very high prices, as much
as $1.00 per lb.
For ceremonial occasions--weddings,
traditional Polynesian haircutting of boys
attaining manhood, funerals, welcoming of
chiefs, and the festival seasons of Christmas, New Year, and Easter--taro is a must,
and purchases are made regardless of cost.

2. Quality and Appearance. As far as taro is
concerned, the Polynesian shopper appears
to be far more discerning than Europeans.
Whereas a European shopper will pick up the
first package of potatoes she comes to in
the supermarket, the Polynesian will take
time and care to pick out what she consi-
ders to be the best tubers on display.
Small tubers, tubers disfigured from being
grown in rocky ground, and tubers showing
signs of decay and rot are quickly discarded. Shopkeepers usually end up either
selling these well below cost or dumping
them.
The most popular type of taro appears
to be well rounded in shape, between two to
four pounds in weight, small in circumfer-
ence at the root end, with two to three
inches of the trimmed stem end attached to
the tuber. The tuber must be firm and have
a fresh, crisp sound when flicked with the
fingernail.
Long, cylindrical taro corms are
shunned. They are referred to as "wild
bush" taros and are said to cook poorly and
become overly moist.
When in good condition the Samoan "Niue
pink" taro is the number one seller; but
good white taro from Tonga--which has built
up a high reputation for taro of good
keeping qualities--has an increasing following. Unless Samoan pink and Tongan
white taro are in short supply, Fijian taro
is usually difficult to sell. The Cook
Islands black swamp taro--mainly from the
islands of Rarotonga, Aitutaki, and Man-
gaia--have a very high reputation and are
in great demand when available, which is
seldom. Generally, not enough is planted
in those islands to produce an exportable

160

surplus. Also, shipping services to these islands
are widely spaced and irregular.

OTHER SOUTH PACIFIC ROOT CROPS EXPORTED TO NEW ZEALAND

Tarotarua

This is a type of taro with a cluster of corms
which are smaller and thinner than the true taro.
It is often referred to as "poor man's taro," and
is usually eaten when taro is scarce. Two varieties
are imported into New Zealand, the red-crowned
variety which is pink-fleshed, and the white-fleshed
variety. Unlike the preference in taro, the white-
fleshed variety appears to be more popular. The
main source of supply is Tonga, closely followed by
some of the Cook Islands, mainly Aitutaki, Rarotonga
and Mangaia. The demand for tarotarua in New Zealand is approximately four percent of that for good
taro. Between 150 and 200 50-1b cases per fortnight
will usually realize payable prices on the Auckland
market.

Yams

The main supplier to New Zealand is Tonga, although small consignments are sent from Fiji on
occasion. Unlike taro and tarotarua, which are
available all the year found, yams are seasonal,
with supplies available from May till October. Two
varieties are shipped: the long, purple-skinned
type and the round, white-skinned yam. Of the two,
the round shaped yam is the more popular in New
Zealand. The standard method of packing is in
wooden cases, similar to the one used for shipping
Island bananas to New Zealand. These hold approximately 50 lb of yams. Experience has taught us that
great care must be taken in packing, and each yam
tuber should be individually wrapped in paper, wood
wool, or clean coconut fiber. The skin is very tender, and, where rubbing off or bruising occurs, a
blue mold quickly sets in and ruins the saleability
of the product. Yams are also inclined to be brittle, especially the long, purple-skinned variety,
and if not carefully packed with adequate protection
they break and mold quickly sets in. Although we
have tried on many occasions to hold yams in cool
storage, the experiment has not been successful,

mainly due to occurrence of condensation and resultant bacterial infection in the tubers.

Our best results with storage of yams have been produced by holding them in the cellar of our market in relatively dry, even air temperatures. We have successfully held yams under these conditions for two weeks. If we had access to controlled atmosphere storage I am certain that yams could be held quite successfully for longer periods. Hopefully, with regular fortnightly shipments, we will not be required to hold yams for more than 10 days or so.

Compared with taro, shipments of yams are small, reaching a maximum of between 200 and 300 cases per fortnightly shipment. The reason for this is that they are extremely popular in Tonga and fetch relatively high prices in their home market. They are also popular in New Zealand and shipments of the above-mentioned proportions usually realize enough of a premium over Tongan prices to make shipping to New Zealand worthwhile. Shipments of over 300 cases have usually resulted in lower prices, to the point where they would have realized more on the local Tongan market.

Ta'amu

This is another type of taro (<u>Alocasia macrorrhiza</u>) grown in Samoa which often reaches between three and four feet (1 to 1.3 m) in length and 40 lb (18.2 kg) in weight per corm. It is pink-fleshed and, like tarotarua, is eaten mainly when taro is in short supply. Being so large in size it is rather difficult to ship. Building a wooden container to pack them in would be costly. If shipped loose, they become susceptible to damage, and if broken, the corms become virtually unsaleable. We have found that the best method of packing them for shipment is in large, clean jute or polypropylene sacks which hold three or four corms. The demand for ta'amu in New Zealand is limited, but between 100 and 200 corms per fortnightly shipment usually realize good prices, with large samples often realizing as much as NZ$8 or $9 each. Like taro, they travel well at 40°F and will keep for two weeks or so at that temperature.

Kape

This is the Tongan version of the Samoan ta'amu, the only difference being that kape has a

green top shoot and the flesh is white. The demand
and scope for these is very similar to that of
ta'amu.

Cassava, Manioc, Tapioca, Arrowroot

These are all the same crop and are referred to
by the above four names in different countries.
Irrespective of country of origin, in New Zealand
the trade refers to it as cassava. Because cassava
has proved difficult to ship by sea, even under
refrigeration, small consignments of fresh cassava
root are flown in weekly from both Fiji and Tonga.
This is rather costly, but weekly consignments of
40 to 60 cartons, each containing 15 to 17 lb (6.8
to 7.7 kg) of fresh dug cassava, have realized payable returns to the shipper, e.g., between 30 and
40 cents per pound. When cassava can be presented
at retail level in fresh condition it sells very
readily. Because its shop life is very limited,
within 48 hours the flesh begins to turn blue, and
only minimal quantities are bought. If some means
of stopping this rapid deterioration could be found,
I am sure cassava would enjoy a much wider and expanded market.

Pia is the flour which is manufactured from
cassava root and is known to the trade in New
Zealand as arrowroot flour.[1] The sole source of
supply to New Zealand is the Cook Islands, mainly
Rarotonga and Aitutaki.

Because of the long, involved process employed
in the manufacture of arrowroot flour in these
islands, shipments to New Zealand are limited to
probably no more than 15 tons a year. It is usually
consigned by sea in plastic-lined sacks which hold
approximately 150 lb (68.2 kg) of the flour. It is
very popular in the Cook Islands and sells there at
relatively high prices, so that a high value must
necessarily be placed on it in New Zealand to ensure supplies are shipped. NZ$2 to NZ$3 per kilo
appears to be the ruling rate. At these prices,
pia or arrowroot flour is used in making Polynesian
dishes for ceremonial and traditional occasions, and
would not appear in the daily diet as do taro and
other fresh root crops.

CONCLUSION

The rapid growth in demand for taro and other
tropical root crops in New Zealand, experienced

principally in the decade 1965 to 1975, has slowed down over the past two years. In the immediate future it will probably stabilize at somewhere near present levels. The main reasons for this are:

1. The tighter control and limitation of immigration of South Pacific Polynesians into New Zealand.

2. The downturn in the economy in New Zealand has meant a decrease in pay for many Polynesians which, further aggravated by rising costs, has meant that they just do not have as much money to spend on food as they did prior to 1976. Hence, they have become far more price conscious and selective in their purchasing of taro and other foods.

3. The question arises as to whether the offspring of Polynesian immigrants born and reared in New Zealand will follow the same dietary patterns as their parents, or whether they will gradually incline towards the more Westernized varieties of vegetables favored by the average New Zealander. For example, as a New Zealander of European descent, I found it difficult to comprehend why Polynesian shoppers would pay NZ$6.00 for a 2-1b (0.9 kg) can of bully beef, when they could purchase the choicest cuts of fresh rump steak for less than half the price of the canned meat, or why they would buy taro at 50 cents per pound when they could purchase potatoes at 10 cents per pound.

   The answer, of course, is that we are all inclined to follow the dietary pattern we were reared on, and that if such food is readily available we adhere mainly to that range of food. I would like to think that the new generation of young New Zealanders of Polynesian descent will to some degree continue to eat the taro, yams, and cassava as do their parents.

With today's vastly improved plant technology, and with--at long last--a fast and efficient shipping service available to transport such produce, I am sure the islands of the South Pacific will play an increasingly important role in food production for consumption, not only in New Zealand, but for all the countries of the entire Pacific Basin.

164

NOTES

1. Editor's Note: Pia flour may not be made from
   cassava. Pia is the Polynesian name for Tacca
   leontopetaloides, a common root crop of some
   Pacific Islands. Tacca is commonly made into
   starch and is used in Polynesian foods. In
   Hawaii, a coconut/Tacca pudding is called
   haupia.

John Watson: Turners and Growers, Ltd., Auckland, New Zealand

4
The Potential for
Genetic Improvement in
Storage Quality of Root Crops

Jill E. Wilson

ABSTRACT

Application of traditional methods of breeding involving selection of superior plants from genetically variable populations produced through sexual propagation has the potential for improving the storage quality of tropical root and tuber crops.

A selection program has been initiated at the International Institute of Tropical Agriculture (IITA) to study this. Selection in breeding populations grown from seed could provide disease-resistant clones. For example, IITA research has reported some success with weevil resistance in sweet potato.

It should be possible to breed cultivars with characteristics reducing the effects of factors which contribute to storage losses such as fungal and bacterial rots, sprouting, respiration, and mechanical injury.

Recently, methods to promote flowering and seed set and to grow large populations of seedlings of white yam, *Dioscorea rotundata*, and the aroids, Colocasia and *Xanthosoma*, have been developed (Sadik, 1977; Volin and Zettler, 1976; McDavid and Alanu, 1976). This has made it possible to improve these crops using traditional methods of breeding which involve selection of superior plants from genetically variable populations produced through sexual propagation. The application of such breeding has the potential for improving the storage quality of these crops, if particular attention is paid to those characters which affect storage as well as to more conventional characters such as productivity and eating quality.

In West Africa a number of factors contribute to the extensive losses which occur during storage of yam. These include pre-harvest infestation of nematodes, particularly _Scutellonema bradys_, various fungal and bacterial rots, sprouting during storage, respiration, and mechanical injury during harvest and post-harvest handling. It should be possible to breed cultivars with improved storage life by selecting those characters which reduce the effects of these factors.

For example, selection would be for clones which are resistant to nematodes and tuber-rot pathogens, which have a prolonged dormancy period to delay sprouting, and which produce tubers which are round or oval in shape, regular, and tough-skinned to reduce mechanical injury during harvest and transport. The relationship to storage of other factors which may be genetically determined, such as dry matter and phenol contents, should be investigated.

Research to date suggests that genetic variation is available in breeding populations for at least some of these characters associated with yam storage, and a selection program aimed at improving storability has been initiated at IITA.

Likewise, in _Colocasia_ and _Xanthosoma_, appropriate selection in breeding populations grown from seed could produce clones resistant to _Phytophthora_, _Pythium_, _Botryodiplodia_ and other storage-rot pathogens, and thereby improve storage life.

Methods of genetic improvement in sweet potato, _Ipomoea batatas_, have been well established (Harmon, Hammett and Hernandez, 1970) and could be employed to improve the storage of this crop. Damage caused by the weevils, _Cylas_ spp., is the most important factor limiting the storage of sweet potato in many regions of production. Researchers at IITA have reported progress in selecting clones with less susceptibility to this pest, and considerable effort is being invested in developing more efficient methods of screening for weevil resistance. Such efforts should result in the production of sweet potato cultivars more suitable for storage.

REFERENCES

Harmon, S.A., Hammett, R.L., and Hernandez, T. 1970.
Progress in the breeding and development of new
varieties. In: Thirty Years of Cooperative Sweet
Potato Research 1939-1969. South. Coop. Serv.
Bull. 159:8.

McDavid, C.R. and Alanu, S. 1976. Promotion of
flowering in tannia (Xanthosoma sagittifolium) by
gibberellic acid. Trop. Agr. (Trinidad) 53:
373.

Sadik, S. 1977. A review of sexual propagation for
yam improvement. Proc. Fourth Symposium of the
International Society for Tropical Root Crops.
IDRC, Ottawa, Canada.

Volin, R.B. and Zettler, F.W. 1976. Seed propagation of cocoyam Xanthosoma caracu Koch and
Bouche. Hort. Sci. 11:459.

*Jill E. Wilson: International Institute of Tropical
Agriculture, Ibadan, Nigeria*

5
Effect of Storage on Paddy Taro Production System Efficiency

J.-K. Wang
W. E. Steinke

ABSTRACT

Minimum system sizes in terms of inputs such as land, labor, and machinery for optimum system operation have been determined for tropical paddy taro production systems. The procedure developed can be applied to other continuous agricultural production systems.

Two types of land management schemes were considered. When total land in production was allowed to vary, steady output could be achieved.

In general, a steady output can be achieved only with reduced land utilization efficiency. This reduction is compensated by increased labor and machinery utilization efficiencies.

Production system output indices have been determined. It has been determined that these indices, such as kg/man-hour used, are not affected by production constraints.

INTRODUCTION

Taro (Colocasia esculenta (L.) Schott) is an important crop in Hawaii and elsewhere in the tropics (de la Pena, 1970). In Hawaii, it is grown in flooded paddies on a continuous basis throughout the year (Plucknett, 1970). The paddies are situated in low lying areas in some of the major valleys of

169

the State. Successful paddy taro production requires a large quantity of flowing water, thus
limiting cropping sites to areas where sufficient
water is available (Begley, 1976).[1] When water is
available, two of the major limiting resources
required for production of paddy taro in Hawaii are
mechanical equipment and human labor. Therefore,
the primary purpose of this study was to develop a
procedure to optimize the allocation of resources
necessary for taro production, such as labor,
machinery, and land, to meet specific production
goals.

The specialized machinery necessary for temperate zone agriculture has a high investment cost
and thus a high cost per hour unless its use can be
spread over many hours. However, temperate agriculture is also characterized by high timeliness costs
in terms of reduced yield if the planting date is
delayed. Therefore, most research has focused on
the tradeoff between investment in larger machinery
and reduced yield (Holtman et al., 1973; Parsons and
Holtman, 1976; Morey, Peart, and Deason, 1971;
Parsons et al., 1971; Parsons, 1975, Von Bargen and
Peart, 1973; and Von Bargen, 1970). However, these
studies were conducted on grain farms in the mid-
western United States, where climatic conditions do
not allow year-round crop production.

Taro can be grown continuously throughout the
year in Hawaii and other tropical areas, thus creating a situation where previously developed models
cannot be applied to optimize production management.
Many previous studies on continuous cropping have
focused on determining a pattern of crops to utilize
on a yearly rotation (Pal, Pandy, and Mathur, 1973;
Pandy, Pal and Sinha, 1973; Banta, 1973; Mahapatra
et al., 1973). Huang (1974) used linear programming to determine the optimum sequence of crops to
maximize the net return under conditions of resource
availability restrictions. Johnson and Diaz (1973)
explored the feasibility of a continuous system of
rice production to study the area that could be
maintained by one person if a given area was to be
planted every two weeks. No attempt was made to
optimize the use of machinery or labor.

These approaches do not satisfactorily relate
efficient use of production resources in a continuous cropping system to the system efficiency. In
evaluating the efficiency of agricultural production
systems in the area where year-round production is
possible, several efficiency indicators are of
interest. Broadly, there are two categories of

efficiency indicators: (1) the production system
output efficiency indices, such as yield/hectare/
year, yield/man-hour, yield/machine-hour, yield/lb
of fertilizer, etc., and (2) the economic indicators, such as return on investment, benefit-cost
ratio, and internal rate of return. Before economic
indicators can be calculated, optimal management
strategies must be selected. In taro production, as
in most agricultural production systems, the physical system dictates that certain production resource
inputs can only be adjusted in integer units, such
as one unit of machine. As a result, production
scheduling must be used to maximize resource utilization before the appropriate level of technology
can be determined.

For paddy taro production in Hawaii the basic
considerations in production management are the
limited--although relatively constant--demand, and
the lack of long-term storage technology. The
basic production resource variables to be optimized
in taro production, given sufficient water and
existing agronomic practices, are labor, machinery,
and land.

Effective utilization of all resources is
necessary for optimization, especially in developmental planning. For any given production system
where short-term or part-time production resources
cannot be obtained from outside the system, the
availability of labor and machines must remain constant, and must be incremented in steps, such as
one machine and one-half of a person. Therefore,
it should be obvious that there is a minimum size of
land where utilization of labor and machines can be
maximized. In other words, for a given price structure and set of operational constraints, a production system can be characterized by a set of values
describing the minimum land size, and maximum labor
and machine utilization efficiencies, provided that
these are the only input variables considered. Once
a minimum size S_m is determined, any production
system which consists of multiples of the minimum
size will have identical system output indices.
Production systems with size S, where

$$n \times S_m < S < (n + 1)S_m , n = 0, 1, 2,...$$

will be less efficient than production systems with
size NS_m, where

$$N = 1, 2, 3...$$

Continuous agricultural production systems can
be divided into two categories:

1. Type I, where all land in the system is used all the time. If a fixed rest period is specified in the production schedule, then that period is considered to be an operation which is necessary for production.
2. Type II, where amount of land actually involved in production is subject to adjustment between successive plantings.

In type I systems the total land area in production is a constant. In type II systems the total land area in production is a time-dependent variable, changing as different plantings are being harvested and planted. For type II systems the maximum land area required by the production is considered to be the land resource needed.

TARO PRODUCTION

Table 1 shows the production activities for production of paddy taro.

Four levels of mechanization can be considered in taro production: human power only, animal assistance with certain tasks, 2-wheel tractors, and 4-wheel tractors. To illustrate the modeling approach, only the 4-wheel tractor level is used. However, it should be obvious that the basic optimization procedure is not affected by the level of mechanization chosen.

Both labor and machinery availabilities are chosen to be 39 hours per week. For paddy taro production a total of 66 weeks are required to produce a crop, including a 4-week rest period between crops. For the sake of convenience, the 66-week production cycle is divided into 33 bi-weekly periods. Since land preparation work for paddy rice and paddy taro is identical, standard work rates for paddy land preparation were used (Johnson, 1963). Krishnan (1976), Smith and Shen (1972), and Smith (1975) provided the information regarding mechanical harvesting of taro. Plucknett, de la Pena, and Ezumah (1973) provided useful information on potential mechanization in taro production.[2] Molinyawe (1967), Plucknett (1970), de la Pena (1970), and Warid (1970) were used to establish cultural practices as well as labor requirements for the various production activities for the production of paddy taro.

TABLE 1
Production Activities and Resource Demands

Production Activities	Bi-weekly Periods During Which Task is to be Performed (No. Periods Since Beginning of Cycle For any Particular Crop)	Production Resource Demands	
		Labor[a]	Machine[b]
A. Land preparation	1	27.5	27.5
B. Planting	2	30.6	10.2
C. Fertilization	5,10,15	114.0	0
D. Weeding	6,10	48.0	0
E. Bank Maintenance	8,14,20,26	10.0	10.0
F. Harvest	31	0.0093	.003
G. Rest	32,33	--	--

[a] Labor demands are in man-hours/hectare, except for the harvest task which is in man-hours/kg.

[b] 40-h.p. 4-wheel tractors with appropriate mechanical harvesting aid.

Figure 1 shows the yield function for paddy taro. As is shown, taro yield follows an annual cycle. Taro harvested during any bi-weekly period was assumed to have been planted 30 bi-weekly periods before.

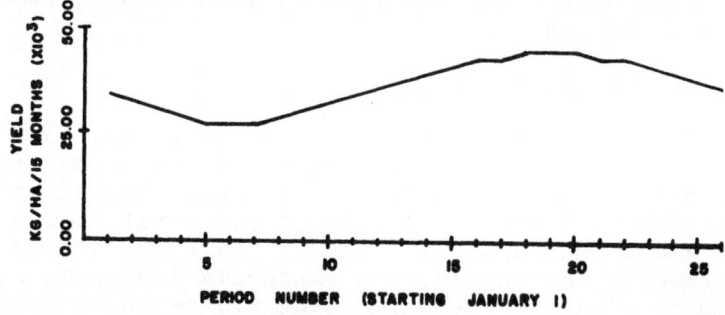

Figure 1. Yield function for wetland taro kg/ha harvested during bi-weekly period shown.

Figure 2 shows the schedule of production activities in paddy taro production.

ISSUES TO BE ADDRESSED

The Problem of Steady Supply

If fresh taro is to remain a staple food for a sizeable segment of the population in the tropics and sub-tropics, a steady supply of the commodity throughout the year is a necessary goal in taro production system planning and management. Since inexpensive long-term storage technology has not yet been developed, the first purpose of this modeling effort is to determine the possibility of a taro production system to deliver steady output through careful regulation of planting and harvesting schedules.

The ability to deliver a steady supply has a number of other important implications for improved taro production in many developing countries. At present, most taro is produced by subsistence farmers for home or local consumption. When intensified farming practices are introduced, the existing simple marketing structure will have to change, because improved farming practices can bring large fluctuations in supply, which in turn can induce undesirable price instability in taro.

Taro can be processed into poi, which also has a short shelf life, and taro flour and related products. Some properties of taro starch make it a highly desirable source for baby food and for processing into biodegradable plastics. Therefore, taro starch can become industrially important in the near future. A steady year-round supply will reduce taro processing costs and make it more competitive in the marketplace.

Rural Income Distribution

It is well recognized that the first goal of agricultural development in the developing countries is to bring benefits to the rural poor. Almost by definition, the great majority of the rural poor are found among the landless farmers. As such, steady employment opportunities are extremely important to the improvement of the well-being of the rural poor. Therefore, in agricultural developmental planning, the ability to maintain a reasonably constant labor

1	2	3	4	5	6	7	8	9	10	11	12	13	14	15	16	17	18	19	20	21	22	23	24	25	26	27	28	29	30	31	32	33
A	B		C	D		E		CD			E	C						E						E					F		

OPERATION

OPERATION CODE KEY

A - PRIMARY TILLAGE AND LEVELING

B - PLANTING

C - FERTILIZATION

D - WEEDING

E - BANK MAINTENANCE

F - HARVEST, SEED PREPARATION AND TRANSPORTATION OF CORMS FROM PADDY

Figure 2. Schedule of activities within each planting of paddy taro.

demand in rural areas should be an important consideration in production system design.

Mechanization is important for the purpose of increasing labor efficiency in agricultural production. However, machines replace labor, thereby reducing employment opportunities for the unskilled. Present planning methodologies tend to create high peak labor demand during the production cycles so that mechanization becomes necessary to meet the schedules of improved agricultural production.

Therefore, the second area to be investigated by this modeling effort is to determine the feasibility of reducing much of the seasonal variation in labor demand generally associated with agricultural production.

Production Efficiency

By careful regulation of the production schedule, desirable solutions can be obtained for the two above problems. However, what are the effects of these management strategies on production efficiencies? A procedure must be developed to deliver a relatively accurate estimation of these effects so that economic analysis can be performed to guide governments, corporations, or individual farmers to make proper decisions.

FORMULATION OF THE PROBLEM

The paddy taro production system was fitted to a linear programming model, and the optimization was carried out by the IBM MPS/360 solution package. A matrix generator and translator developed by Huang (1974) was extensively modified to input the LP model into the IBM MPS/360 package.

The basic time interval used was the bi-weekly period. A total of 33 bi-weekly periods, including two rest periods, are normally recommended for one cycle of paddy taro production in the tropics.

Using the 33 bi-week cropping schedule, and considering that there are 26 bi-weekly periods in a year, a total of 33 years or 858 bi-weekly periods, or 26 cropping cycles is required to completely recycle all the production activities to their original starting point. To do this would have overburdened the computing facilities. After careful evaluation, an optimization horizon of 264 weeks, or 4 production cycles was chosen. The taro

176

yield function follows an annual pattern, and 264
weeks is slightly more than 5 years, thus creating
a slight gap in timing. This slight offset in
timing is expected to introduce a very slight error,
and compensatory corrective actions were taken to
reduce the error.

Figures 3 and 4 show the cyclic production
activity schedules for types I and II production
management strategies, respectively. In type I
operation, a total of 33 crops were scheduled and
the land resources, within given constraints, were
allocated to the 33 crops. Each crop went through
four production cycles, and land, once allocated to
a crop, was not allowed to change between production
cycles. In type II operation, the optimization
procedure was allowed to change land devoted to a
crop between plantings, so that land resources allocated to a crop could be changed from one production
cycle to another. There were 132 crops for the type
II operation, corresponding to the 132 bi-weekly
period in the optimization horizon.

The variables and/or constraints used were:
land, labor, machinery and production limits. Land
was treated as a continuous variable, labor was
incremented in ½-man steps, and machines were
incremented in integer units. The production limit,
in kg/bi-weekly period, was imposed in order to
achieve uniform and steady production.

SUMMARY OF RESULTS

Table 2 gives a summary of results. A harvest
limit of 16,330 kg per period was imposed on the
system, so that harvest would not exceed 16,330 kg
of taro during any 2-week period.

As Figures 5 and 6 show, under the given constraints, using type II management strategy, the
system could deliver a steady production of 16,330
kg of taro per period. The type I strategy with
fixed land resources allocation scheme was unable
to do so. The imposition of harvesting limit merely limited the peak yield of each land unit without
increasing production elsewhere.

Some of the system output indices, for
instance, kg/(m-hr used) and kg/(t-hr used), were
affected by neither the management strategies nor
the imposition of production limit.

Figures 5 through 10 show the yield pattern as
well as labor and machine demand patterns. Figure

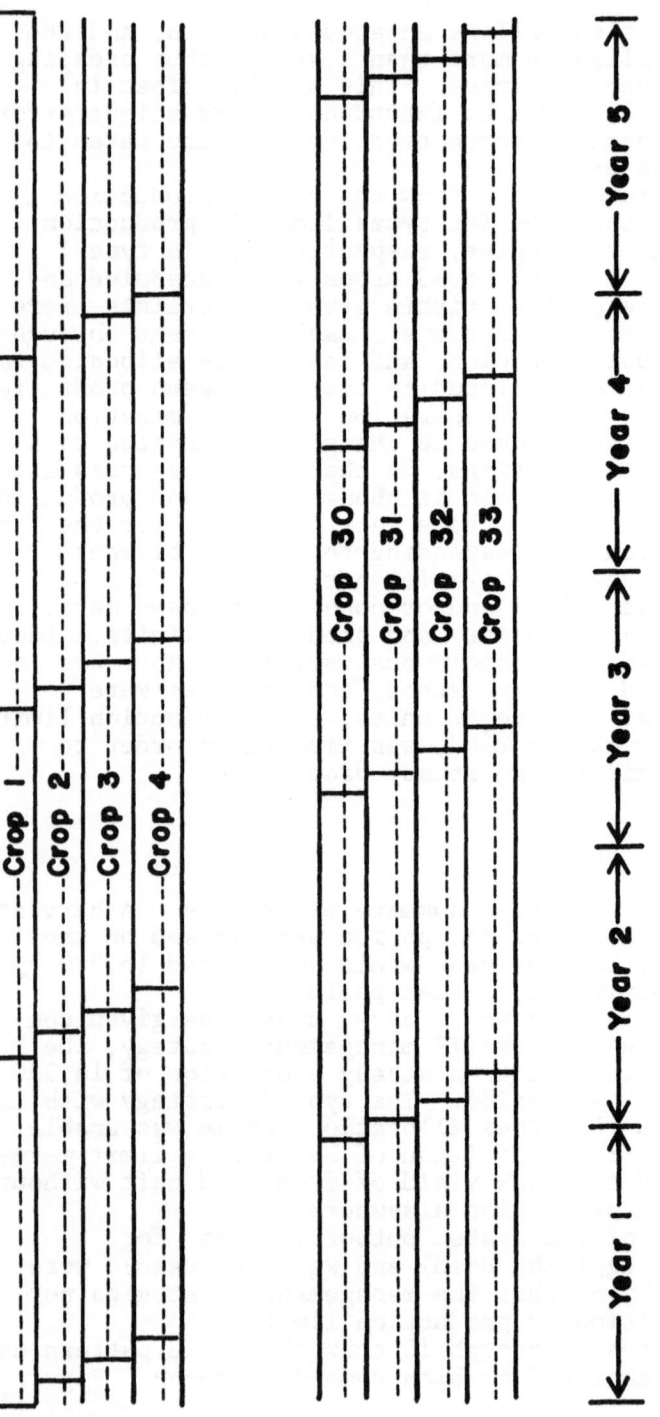

Figure 3. Type I schedule of crops (each crop consists of four successive growth cycles).

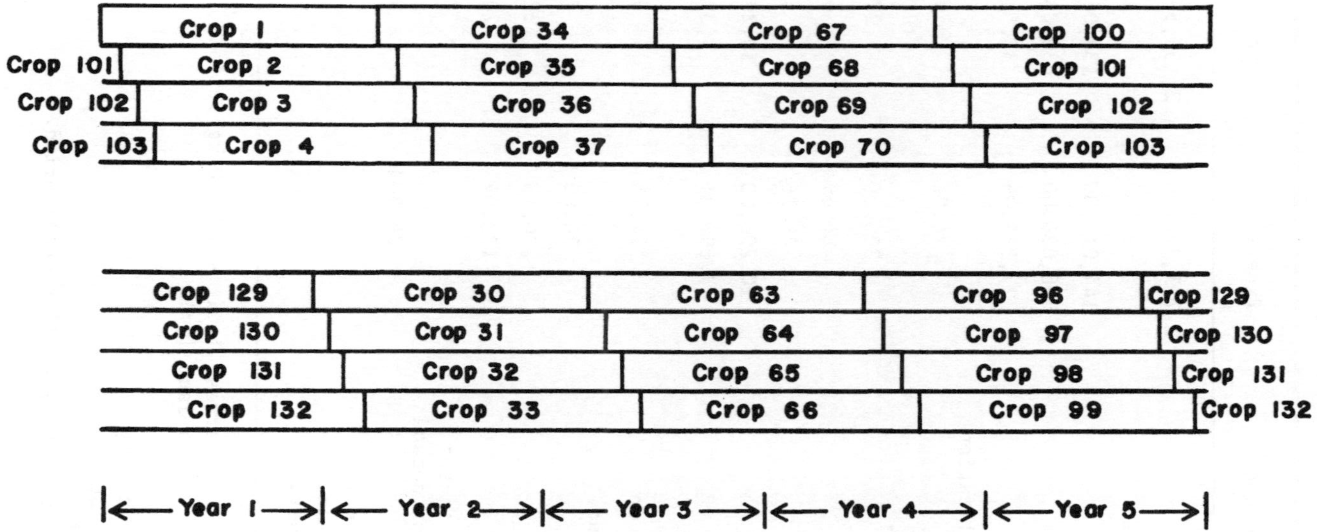

Figure 4. Type II schedule of crops (each crop is one growth cycle).

TABLE 2
Summary of Results

System Out-Put Indices	Type I[a]	Type II[b]
Total yield, kg[c]	1,771,374.00	2,155,560.00
Max. area used, ha	12.30	16.20
Kg/ha	144,014.00	133,059.00
Available m-hr	36,036.00	36,036.00
M-hr used	27,382.06	33,514.40
Overall labor utilization factor	76.00	93.00
Peak labor demand, m-hr/2-week	236.20	270.10
Kg/m-hr used	64.69	64.32
Kg/m-hr available	49.16	59.82
Available t-hr	10,296.00	10,296.00
T-hr used	7,696.31	9,393.61
Overall machine utilization factor	74.75	91.24
Peak machine demand t-hr/2 week	74.80	76.50
Kg/t-hr used	230.16	229.47
Kg/t-hr available	172.04	209.36
Labor utilization factor during peak labor demand periods	86.50%	98.90%
Machine utilization factor during peak machine demand periods	95.90%	98.10%

3.5 laborers   Harvest limit - 16,330 kg/period
1 tractor   No land limit

[a]See Figures 5, 7, 9, 11.
[b]See Figures 6, 8, 10.
[c]Yields are based on a 264-week (4 crops) period.

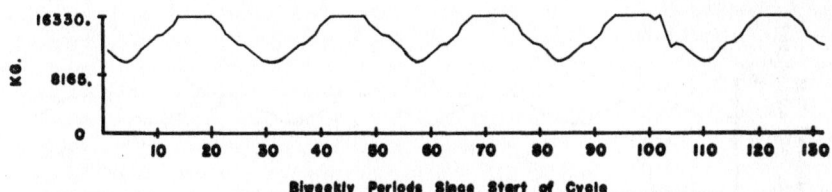

Figure 5. Paddy taro harvest, kg/period. Type I, resource availability as in Table 2.

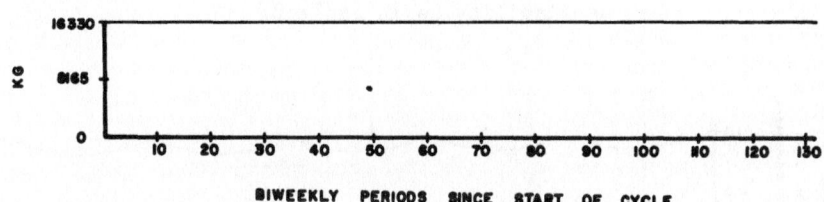

Figure 6. Paddy taro harvest, kg/period. Type II, resource availability as in Table 2.

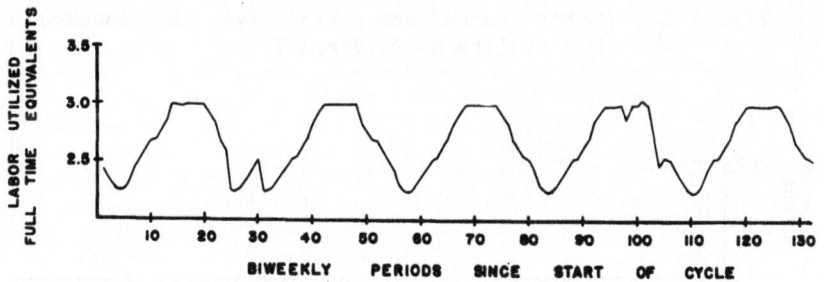

Figure 7. Labor demand per period. Type I, resource availability as in Table 2.

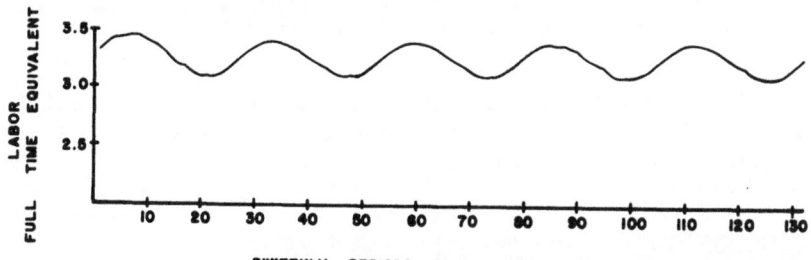

Figure 8. Labor demand per period. Type II, resource availability as in Table 2.

181

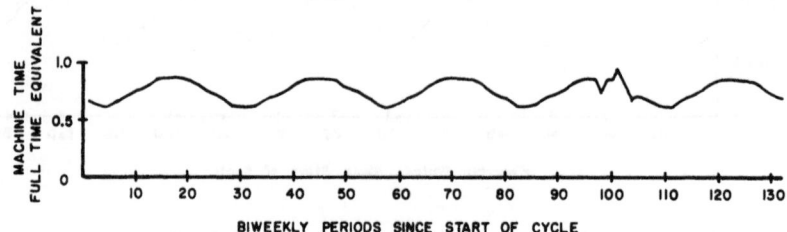

Figure 9.   Machine demand per period. Type I, resource
            availability as in Table 2.

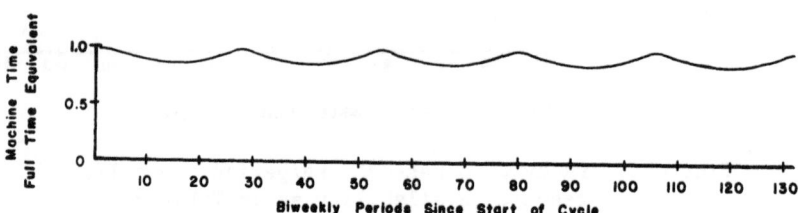

Figure 10.   Machine demand per period. Type II, resource
             availability as in Table 2.

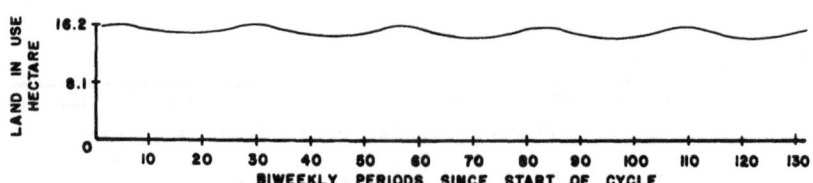

Figure 11.   Land use pattern. Type II, resource
             availability as in Table 2.

11 shows the land use pattern for type II operation.
The land use pattern for type I operation is, by
definition, a constant.

DISCUSSION

It was expected that type I management strategy
would not have been able to provide a steady yield.
But since it is the most efficient management

strategy with respect to land utilization, in that all land resources are used all the time, it does provide a standard of measurement in production system efficiency.

Table 2 shows that type I strategy gave an estimated yield of 144,014 kg/ha per 4-crop production cycle, while type II gave 133,059 kg/ha during the same length of time, a net reduction of 7.61 percent. Whether or not this was too high a price to pay for steady production cannot be answered until an economic analysis of storage, processing and marketing costs is performed. What was important was the fact that a steady production output was achieved, and the cost for achieving it has been estimated.

The fact that kg of taro produced per man-hr and tractor-hr used did not change is significant. Whatever changes are shown in Table 2 are certainly within the probable error of this type of analysis, indicating that the two types of management strategies did not affect the physical indices. In other words, both manpower and machine were used efficiently and the reduction in land productivity was mostly caused by the fact that some of the land under type II operation was forced to be idle at times.

When seasonal variations in the availabilities of labor and machine were not allowed, as was the case under study, the system output was constrained by the peak demands on production resources.

Table 2 shows that the peak demand made by the type I operation on labor was only 86.5 percent of available labor resource (3.03 persons), and the peak machine demand was 95.9 percent (0.959 machine).

This clearly indicated that improper constraints were imposed on the system. In other words, the choice of 16,330 kg per period as the harvest limit forced the production system to be optimized at land sizes less than the minimum size required for efficient operation. However, the minimum size for efficient operation can be calculated from information provided by Table 2.

Since 12.3 hectares were under cultivation and the peak labor demand was 3.5 x 0.865 = 3.0275, each hectare had a peak labor demand of only 0.246 person. In the same fashion, it can be shown that only 0.078 machine would have been required to meet the peak demand made by a hectare of land in production of paddy taro. Since 500 is the smallest number that, when multiplied with 0.246 and 0.078, can make both numbers integers, the minimum size for efficient operation was very close to 500 hectares.

If the harvest limit had been adjusted to allow 500 ha in production with 39 machines and 123 laborers, the optimization results would still have identical kg/(m-hr used), kg/(t-hr used) and kg/ha indices, but both the labor and utilization factors during peak demand periods would have very nearly approached 100 percent.

Following the same procedure, it can be shown that the minimum size for efficient operation for type II operation was calculated to be 5,000, 1,070 persons and 303 machines.

The overall labor and machine utilization efficiencies, when operated under these optimal conditions, will be very close to:

	Type I	Type II
Labor	87.86%	94.03%
Machinery	77.95%	93.01%

The shapes of the curves in Figures 4 through 10 will not change appreciably, but the scales of the coordinates will. It has been shown that to increase the size of operation will bring about increases in resource utilization factors, but the extent to which improvement can be made is limited in type II operations. The improvement in overall labor and machinery utilization factors for type I operation is significant.

It is clear that, in comparison with type II operation, the higher land utilization for type I operation has been obtained at the expense of lowered labor and machine utilization efficiencies, since the basic system output indices such as kg/ha, kg/(m-hr used), and kg/(t-hr used) remain unchanged.

NOTES

1. Bryan Begley, "Taro in Hawaii: Study of a Food System," unpublished research manuscript, Agricultural and Resource Economics Department, University of Hawaii, Honolulu, Hawaii, 1976.

2. Donald L. Plucknett, Ramon S. de la Pena, and H.C. Ezumah, "Mechanization of Taro in Hawaii," unpublished research manuscript, Agronomy and Soil Science Department, University of Hawaii, Honolulu, Hawaii, 1973.

REFERENCES

Banta, Gordon R. 1973. Mechanization, labor and time in multiple cropping. Agr. Mech. Asia 4:27.

de la Pena, Ramon S. 1970. The edible aroids in the Asian-Pacific area. Proc. Second Intern. Symp. Trop. Root and Tuber Crops, College of Tropical Agriculture, University of Hawaii, Honolulu, Hawaii, Vol. 1:136.

Holtman, J. Ben, Pickett, Leroy K., Armstrong, D.L., and Connor, Larry J. 1973. A systematic approach to simulating corn production systems. Trans. Am. Soc. Agr. Engr. 16:19.

Huang, Wen-Yuan. 1974. Optimal use of Hawaii agricultural reservoir system: a multiple cropping approach. Ph.D. dissertation, University of Hawaii, Honolulu, Hawaii.

Johnson, Loyd. 1963. Power requirements in rice production. Presented at the Conference on Agricultural Engineering Aspects of Rice Production, International Rice Research Institute, Los Baños, Laguna, Philippines.

_____ and Diaz, Alfonso. 1973. A continuous rice production system. Agr. Mech. Asia 4:109.

Krishnan, Palaniappa. 1976. Evaluation of auger plow for digging wetland taro. Master's thesis, University of Hawaii, Honolulu, Hawaii.

Mahapatra, I.C., Leeuwirk, D.M., Singh, K.N., and Dayanand. 1973. Green revolution through multiple cropping in India. Agr. Mech. Asia 4:37.

Molinyawe, C.D. 1967. Status of root crop research in the Philippines. Proc. First Intern. Symp. Trop. Root Crops, Trinidad, Vol. 1:69.

Morey, R.V., Peart, R.M., and Deason, D.L. 1971. A corn-growth harvesting and handling simulator. Trans. Am. Soc. Agr. Engr. 14:326.

Pal, Mahendra, Pandy, S.L., and Mathur, B.P. 1973. Cropping patterns in multiple cropping systems. Agr. Mech. Asia 4:31.

Pandy, S.L., Pal, Mahendra, and Sinha, A.K. 1973. Cropping patterns and irrigation problems in multiple cropping. Agr. Mech. Asia 4:22.

Parsons, S.D. 1975. An event-oriented corn production simulation model. Am. Soc. Agr. Engr. Paper No. 76-5539.

_____ and Holtman, J.B. 1976. An event-oriented corn production simulation model. Am. Soc. Agr. Engr. Paper No. 76-5539.

_____, Pickett, L.K., Holtman, J.B., and Fridley, R.B. 1971. Modelling man, machine and crop relationships for corn combine simulation. Am. Soc. Agr. Engr. Paper No. 71-625.

Plucknett, Donald L. 1970. Colocasia, Xanthosoma, Alocasia, Cyrtosperma, and Amorphophallus. Proc. Second Intern. Symp. Trop. Root and Tuber Crops, College of Tropical Agriculture, University of Hawaii, Honolulu, Hawaii, Vol. 1:127.

Smith, M. Ray. 1975. Mechanical harvesting of wetland taro. Am. Soc. Agr. Eng. Paper No 75-1024.

_____ and Shen, H. 1972. Pickup mechanism for harvesting wetland taro. Trans. Am. Soc. Agr. Engr. 15:1005.

Von Bargen, K. 1970. Analysis and simulation of a field machine and transport system for row-crop planting. Ph.D. dissertation, Purdue University, West Lafayette, Indiana.

_____ and Peart, R.M. 1973. Simulation of field machine and transport activities for a row crop planting system. Trans. Am. Soc. Agr. Engr. 16:1010.

Warid, Warid A. 1970. Trends in the production of taro in Egypt (United Arab Republic). Proc. Second Intern. Symp. Trop. Root and Tuber Crops, College of Tropical Agriculture, University of Hawaii, Honolulu, Hawaii, Vol. 1:141.

Jaw-Kai Wang and William E. Steinke: Agricultural Engineering Department, University of Hawaii, Honolulu, Hawaii

Section IV
Processing Techniques
and Products

1
The Processing of Yams

D. G. Coursey
C. E. M. Ferber

ABSTRACT

During the past decade, there has been
increased research and development activity on the
post-harvest technology of the Dioscorea yams.
There has been little departure to date from
traditional yam storage practices in most developing countries. Although the manufacture of various
processed products has received some attention,
and pilot scale operations to produce dehydrated
yam products have been initiated, none is yet in
full commercial operation. There is now, however,
a much greater knowledge than before of the technological parameters related to yam processing.

INTRODUCTION

Until just over a decade ago, the yams, like
most tropical food crops and in particular the root
and tuber crops, were seriously neglected by
agriculturalists and food technologists alike.
Little work had been done then beyond botanical
description of species, and processing, in particular, was very poorly studied. Such earlier work as
had been done was fully reviewed about that time
(Coursey, 1966; 1967) and it is therefore proposed
to review here only those developments which have
taken place since the base line of 1967. In this

last decade there has been a great upsurge of interest in tropical food crops. Specifically in the case of yam, at least half as much published work on the crop appears to have been produced in the last decade as in the whole of previous history. At the same time, there has been a growing appreciation of the great contribution that root and tuber crops in general can make to the feeding of the populations of the tropical world (Coursey and Haynes, 1970).

Yams are major food crops in only three parts of the world: West Africa, the Caribbean islands, and some of the island groups of the Pacific Ocean. The first of these three regions is by far the most important, and Nigeria alone accounts for considerably more than half the world's total production of yam. Research and development activities have therefore tended to concentrate on the two major, closely related, West Africa species, Dioscorea rotundata Poir. and D. cayanensis Lam. (Ayensu and Coursey, 1972). Indeed, food yam production is derived largely from these two species, together with D. alata L. which is grown in all three major yam-growing areas, and D. esculenta (Lour.) Burk. which is largely confined to the Pacific area and Southeast Asia.

Yams have not as yet been utilized to a significant extent on a commercial basis for the manufacture of processed foods, although traditionally, various types of dried flours and similar products are manufactured in various parts of the world (Coursey, 1967). The manufacture of processed products with longer storage lives could help to reduce food losses and to stabilize seasonal fluctuations in supply. Dehydrated products, in particular, could also minimize marketing and transportation costs.

Yams are prepared for consumption in a variety of ways, including boiling, baking, or frying. One of the most important traditional culinary preparations of the yam in West Africa is "fufu" (Coursey, 1966; 1967), in which the peeled, sliced tuber is boiled until soft and then pounded to give a stiff dough. Strong individual and local preferences exist for different species or cultivars, and for different methods of preparation. All need to be taken into account in planning any processing operation.

PROPERTIES OF YAM STARCHES AND THEIR INFLUENCE
UPON PROCESSING

In the development of any food product from
yams, knowledge of their food technological properties, in particular those of the starch which is
the major component, is needed to predict behavior
under given processing conditions. The properties
of Dioscorea starches, particularly their rheological properties, have been the subject of several
recent investigations, some of which relate to the
processing characteristics of yams.

The starches of four major West African
species of yam were studied by Rasper and Coursey
(1967). Rheological examination of their starch-
water pastes indicated that the viscosities
obtained with D. rotundata starches were significantly greater than those of other species, and the
gel strengths were also moderately high. These
properties (Table 1) relate to the fact that this
is the preferred species for the production of
fufu, for which a stiff dough is required. D.
alata exhibited a very high gel strength, but the
starches of the other two species studied (D. esculenta and D. dumetorum (Kunth.) Pax) have low
viscosities and low gel strengths and are not
suitable for making fufu. These rheological parameters were further compared with those of the
starches of several other West African starch
crops. Most of the yam starches gave very viscous
pastes with high gel strengths compared to the
starches of other tropical starch crops (Rasper,
1969a). These rheological characteristics can be
related to the swelling and solubility properties
of the starch (Rasper, 1969b) and also to granule
size (Rasper, 1971). The gelatinization characteristics of yam, together with various other root
and cereal starches, are given in Table 2.

Yam starch is also noted for its heat stability, which is indicated by the absence of a peak
viscosity when the starch is subjected to heating:
the viscosity, rather, increases throughout the
cooking and cooling cycles (Cruz-Cay and Gonzalez,
1974). The swollen granule is thus very resistant
to fragmentation and solubilization upon heating
and stirring (Rosenthal, Pelegrino, and Correa,
1972), which explains why the production of fufu
is such a tedious process requiring up to 60
minutes of pounding (Coursey, 1967).

TABLE 1
Rheological Properties of Various Yam Starches

	Pasting Temp. °C	Viscosity (Brabender Units)		Gel Strength (ml) After		
		On Attaining 95°C	Maximum Reached Before Cooling	24 hours	96 hours	168 hours
D. rotundata cv. Puna	76	450	630	8.8	13.6	14.1
" cv. Labreko	78–79	260	470	4.3	6.2	8.0
" cv. Kplinjo	77	330	490	10.6	12.7	13.3
" cv. Tantanpruka	79	610	650	12.4	17.2	20.5
" cv. Templ	80–82	430	520	7.5	10.5	10.8
D. alata white fleshed	85	25	110	14.8	16.5	17.2
" purple fleshed	81	80	200	14.8	18.5	19.4
D. esculenta	82	25	55	2.5	4.0	4.6
D. dumetorum	82	25	25	-	-	-

After Rasper and Coursey (1967).

TABLE 2
Gelatinization Characteristics of Yam and Other Starches (after Rasper, 1969b)

| Origin of Starch | Pasting Temperature[a] °C | Maximum Viscosity[a] (B.U.) | Period of Increasing Viscosity[a] min | Apparent Rate of Viscosity Increase | At 95°C | |
					Swelling Power	Critical Concentration Value
Root starches						
Potato (Solanum tuberosum)	62.5	1920	8.0	0.4	>100	<1
Cassava (Manihot utilissima)	63.5	690	12.5	1.8	48.7	2.6
Cocoyam (Xanthosoma sagittifolium)	77.0	350	2.5	0.8	30.9	3.2
Sweet potato (Ipomoea batatas)	77.0	590	5.0	0.9	27.2	3.7
Yam (Dioscorca)						
D. rotundata	73.5	980	22.5	2.3	24.9	4.0
D. esculenta	78.5	500	steadily increasing		23.0	4.3
D. cayenensis	75.0	690	16.5	2.4	21.3	4.7
D. alata	77.5	620	steadily increasing		18.3	5.5
D. dumetorum	83.0	185	"	"	13.9	7.2
Colocasia (Colocasia antiquorum)	77.0	260	"	"	16.0	6.3
Alocasia (Alocasia macrorrhiza)	73.0	160	"	"	16.6	6.0
Cereal starches						
Maize	79.5	265	9.0	3.4	16.5	6.1
Other starches						
Plantain (Musa paradisiaca)	75.5	530	15.0	2.8	22.1	4.5

[a]Data obtained from Brabender Viscograph curves.

The desirable glutinous dough-like texture of fufu is a transient characteristic which fades rapidly under normal holding conditions. This decline in quality has been attributed to starch retrogradation (Ayernor, Brennan, and Rolfe, 1974). These authors followed the changes in the rheological properties of the dough after preparation by means of compression tests, and retrogradation was confirmed by microscopy and X-ray diffraction studies. The high degree of retrogradation observed in Dioscorea starches was also noted by Rasper (1969a) and Rosenthal, Pelegrino, and Correa (1972). It is suggested that the high content of linear amylose in the starch is responsible for this property (Cruz-Cay and Gonzalez, 1974). The resulting textural changes, wherein the dough loses its elasticity and becomes hard and brittle due to the growth of crystals in the dough structure upon retrogradation, are temperature dependent (Brennan and Ayernor, 1973), increasing with decreasing storage temperature down to 2°C. The rate of retrogradation in yam starches is several times faster than in wheat starches, the process being virtually complete at 20°C in yam dough after 12 hours, compared with 16 days for wheat starch gels. Further studies of this phenomenon are desirable, as their results could lead to some means of retarding these undesirable changes.

Yam starch has been shown to possess a relatively high gelatinization temperature compared with other starches (especially cassava and potato), and a low susceptibility to exogenous bacterial α-amylase (Park, 1974). This relative resistance of yam starch to digestive enzymes was found to be typical of most tropical root starches, with the notable exception of cassava starch (Rasper, 1969b), and may be of importance in determining the functional properties of yam starches in composite doughs (Rasper, Rasper, and Mabey, 1974).

YAM FLOUR AND ITS USE FOR FUFU AND OTHER PREPARATIONS

Yam flour is produced in West Africa on a small scale by the sun-drying of the peeled, sliced tuber (Coursey, 1967). This product provides a possible method of extending the yam supply through the off-season and of reducing storage losses and marketing and transportation costs. The traditional flour is of poor quality and the introduction of improved production systems such as that outlined by Jarmai and Montford (1968) would be desirable. This suggests the preparation of yam flour for fufu production from precooked yam, by oven-drying in 2-cm layers at 50 to 70°C for 6 to 8 hours. The product, of less than 10 percent final moisture content, was used to make traditionally prepared fufu, which was reported to be indistinguishable from fufu obtained from fresh tubers.

The effect of fortification of yam flour with soy flour on the quality of fufu has been investigated recently (Collins and Falasinnu, 1977). Yams contain only a limited amount of protein, so the successful fortification of yam flour for fufu production would be nutritionally advantageous. Soybeans were suggested as providing a suitable protein source (Abe, 1973). Studies showed that yam flour, produced by freeze-drying of cooked 1.25 cm slices, could be fortified to levels of 5 or even 10 percent protein on a dry weight basis using soy flour without detracting from the quality of the fufu derived from it. Fufu samples containing soy were slightly softer and stickier than those without, but this could be adequately compensated by decreasing the proportions of rehydration water (Table 3).

TABLE 3
Instron Force Required to Extrude Rehydrated Mixture of Yam and Soy Flours (after Collins and Falasinnu, 1977).

Protein Content (%)	Water Held Constant		Water Reduced	
	Parts Water: 1 Part Flour	Mean[b] (kg)	Parts Water: 1 Part Flour	Mean[b,c] (kg)
1.25 (control)	3	13.1 ± 0.7	3	13.1 ± 0.7
5	3	9.0 ± 0.0	2.5	13.4 ± 0.5
10	3	6.6 ± 0.2	2	13.4 ± 0.4

[a]Reprinted from J. Food Sci., Vol. 42, No. 3, pp. 821-823, 1977.
[b]Means of 9 observations with one standard deviation
[c]Means are not significantly different at 0.05 level of probability.

The use of yam flour (D. alata, purple cvs)
for various popular traditional Philippine food
preparations has been shown to give products of
acceptable quality (Afable, 1970). The flour was
produced by oven-drying the precooked, grated yam
at 50°C for 4 hours. No deterioration in the flour
was observed after storage at 30°C for six months.

A comparison of air-drying at 42 ± 0.5°C for
4 to 6 hours, and freeze-drying at 180 mm Hg from an
initial yam temperature of -31°C to a final temperature of 26°C after 6 hours, of peeled 1 to 2 mm
slices of yam (D. batatas Decne, or D. opposita
Thunb) was made (Misawa and Matsubara, 1965). There
was little loss of vitamin C from the original 5 mg
percent during either type of drying, although some
browning was observed with the air-dried samples.
The viscosities of the dried yam slurries were significantly reduced from their original values.

The freeze-drying of various vegetables
including yams was studied by Rai and Jain (1970).
The effect of freeze-drying of yam on the properties
of a Japanese confection made from rice powder,
sugar, and yams was reported by Ishitani, Kojo, and
Kimura (1973).

INSTANT YAM FLAKES AND THEIR POTENTIAL FOOD APPLICATIONS

It may be reasonably anticipated that the
application of techniques currently used for potato
processing should be applicable to yams (Coursey,
1966; Abe, 1973). A considerable amount of research
has been done mainly on the production, by drum-
drying, of "instant" yam flakes, which are capable
of rehydration to yield the desired product without
further preparation.

Control of the alterations in textural characteristics of such a product, associated with the
changes in properties of the starch upon processing,
is one of the main problems involved. The type of
product required will, however, depend upon its
intended use. A fufu-type product for a West
African market must be capable of yielding a sticky,
elastic dough upon rehydration, while in contrast,
a fluffy, mashed-potato-like product would be
required for West Indian markets (Steele and Sammy,
1976b).

The introduction of an instant fufu would be
of considerable value to schools, hospitals, and

other such institutions, as well as to restaurants, or even in the home in Africa, where labor requirements for traditional methods of making fufu are becoming prohibitive. A mechanical aid for the production of pounded yam in Nigeria has been suggested (Makanjuola, 1974), but the design appears to be at the development stage and further modifications to improve its performance are required.

The alternative, instant fufu, has been investigated using the traditional D. rotundata yam Onayemi and Potter, 1974). Dehydrated flakes of less than 10 percent moisture content can be readily produced by drum-drying of precooked, mashed yam such that rehydration gave a product which was very similar to conventionally prepared pounded yam and much superior to the products from traditional yam flour. Starch damage was monitored, but not controlled, throughout the process, since the resulting sticky product is the one desired. Flakes stored at 37.8°C retained their acceptability for at least 90 days, with slight color deterioration in samples of higher moisture content.

The cooking time and mashing conditions in the production of drum-dried flakes from D. rotundata have been related to the characteristics of the fufu prepared from flakes (Ayernor, 1976). Increased cooking time and more intense mashing conditions, although conferring greatest starch damage and less rigidity to the resulting dough, gave greater elasticity, which is a desirable characteristic in pounded yam. The influence of cooking time on the properties of the dried product and the reconstituted dough is summarized in Tables 4 and 5. It was concluded that the commercial development of an instant yam, which, on rehydration, has the traditional rheological characteristics of fufu, would be feasible on this basis. Pilot scale production of such a material, under the trade name of "Poundo-yam," has been carried out in Nigeria for some years, but there has been no expansion from this level to date.

With the Caribbean market in mind, a drum-dried yam (D. alata cv Lisbon) was produced under conditions such that excessive stickiness in the product was avoided (Gooding, 1972). Cooked, mashed yam was drum-dried and the degree of starch damage, as measured by the quantity of free starch, was monitored at all stages of the process. Although cooking was associated with minimal damage, the effect of mashing depended upon procedure. Ricing of the product followed by gentle mashing in a mechanical mixer was recommended. The drying process was found to cause most starch damage and must be closely controlled. The maximum amount of free starch liberated by such damage consistent with an acceptable product was indicated by a Blue Value Index (BVI) of less than 200 for the final product. Addition of glyceral monostearate at 0.5 percent on a dry matter basis also reduced stickiness. After storage for 12 months in cans or polyethylene pouches, the product was still acceptable. These investigations were developed further in collaboration with the Tropical Products Institute, and a semi-commercial plant was operated for some years by the Agricultural Development Corporation in Barbados.

TABLE 4
Influence of Steam Pressure Used to Cook Diced Yam for 10 Minutes on Four Powder Characteristics (after Ayernor, 1976) [a]

Steam Pressure (psig)	Powder Bulk Density (g/cm^3)	Starch Damage BVI	Swelling Power $(cm^3/g$ solids)	Solubility $(g/cm^3$ water)
0	0.31 ± 0.05	70 ± 2	11.7	2.4
5	0.30 ± 0.07	100 ± 3	11.7	2.6
10	0.28 ± 0.03	110 ± 1	13.2	3.1

TABLE 5
Rheological Properties of Yam Dough at 15 Minutes after Reconstitution (after Ayernor, 1976) [a]

Steam Pressure (psig)	Modulus of Elasticity $(x\ 10^5 dyne/cm^2)$	Ultimate (Yield) Deformation	Recoverable Elasticity (%)	Energy Ratio
0	5.2	0.50	66	0.77
5	4.9	0.48	73	0.71
10	4.4	0.45	81	0.55

[a] Reprinted from J. Food Sci. Vol. 41, No. 1, pp. 180–182,1976.

A similar dependence of stickiness on the degree of starch damage, which was higher with increased cooking time of the yam (D. alata cv Florido), was found by Rodriguez-Sosa and Gonzalez (1972). Table 6 shows that the gelatinization temperature of the mashed yam decreased with increased cooking time, indicating partial starch gelatinization during cooking. This gives rise to a greater degree of cell rupture during subsequent dehydration as is indicated by an increased BVI. As the cooking time is lengthened, the starch also exhibits a peak viscosity (Table 6). This is characteristically absent in unmodified yam starch (Rosenthal, Pelegrino, and Correa, 1972), which is very stable to heating.

In a further study on the same cultivar, Rodriguez-Sosa and Gonzalez (1974) found that smaller flakes had a higher degree of starch damage compared to the larger ones, presumably due to greater surface exposure. They consequently exhibit higher viscosity and greater stickiness on rehydration. Similarly, the gelatinization temperatures of yam flake slurries were lower than those for the unmodified starch, and a peak viscosity was again observed. The inclusion in the process of a short but distinct precooking stage prior to the

TABLE 6
Pastiness Characteristics of Mashed Yam: BVI and Total Solids Content of Yam Flakes

	Mashed Yam			Yam Flakes	
Cooking Period (min)	Gelatinization Temperature (C)	Viscosity 90 C (Brabender Units)	Viscosity After 40 Minutes Cooking at 90 C (Brabender Units)	BVI	Total Solids (%)
5	66.4	137	248	200	94.11
10	63.0	237	237	239	93.99
15	60.3	278	197	273	94.15
20	59.5	295	177	244	93.12

After Rodriguez-Sosa and Gonzales (1972).

199

steam-cooking stage appears to lower the amount of starch damage in the final product as shown in Table 7 (Rodriguez-Sosa et al., 1974). Precooking at 71.1°C for up to 25 minutes caused partial starch gelatinization and conferred resistance to damage by further processing. Slurries of flakes obtained from precooked yams showed heat stability similar to unmodified starch slurries in that they also did not exhibit a peak viscosity.

Yam cultivars differ in their processing characteristics and hence in the quality of the final product (Rodriguez-Sosa, Gonzalez, and Martin, 1972). Free starch content appeared to affect the ease of product handling during dehydration but did not appear to be the major determinant of product quality in this study, in which color and flavor of the product emerged as more important quality aspects. Steele and Sammy (1976b) prepared drum-dried flakes from six cultivars of yams and concluded that free starch, as indicated by the BVI, was not the only factor contributing to stickiness in the final product. The flakes stored satisfactorily in glass jars, but after 24 weeks in polyethylene bags, deterioration (browning and rancidity) occurred due to moisture absorption (Table 8) in the case of flakes from D. alata cv Lisbon. The acid value and optical density at 400 mμ of a 60 percent ethanol extract of the product appeared to be good indicators of deterioration in flake quality.

A patent (Fyns Konsum Industri, 1972) for the treatment of starchy fruits and vegetables with sulphur dioxide suggests its specific use for the production of mashed potato or yam intended for subsequent drying, to improve the color and keeping qualities of the dried product. Sulphur dioxide is introduced during the steam cooking stage of the raw material, the sulphur dioxide content of the steam being gradually reduced to zero during the last part of the steam treatment. Residual sulphur dioxide is removed by the subsequent drying stage.

YAM FLOUR USED IN COMPOSITE FLOURS FOR BAKERY PRODUCTS

Although bread is not a traditional dietary item in most developing countries, its consumption is rapidly increasing. Due to climatic restraints, wheat suitable for breadmaking cannot be grown

TABLE 7
Pasting Characteristics of Instant Florida (D. alata) Yam Flake Slurries Prepared from Yam Tubers Pre-Cooked in Water at 71.1 C (modified from Rodriguez-Sosa, Gonzalez, and Parsi-Ros, 1974).

Precooking Time (minutes)	Pasting Temperature (C)	Viscosity Measurement[a]					
		Initial at 30 C	On Reaching 92 C	After 1 Hour at 92 C	On Reaching 50 C	After 1 Hour at 50 C	BVI
5	68	157	83	112	330	337	227
10	74	150	62	80	265	270	206
15	74	115	65	95	275	282	227
20	75	107	45	80	225	238	114
25	88	53	3	59	160	163	98

[a] Viscosity was measured in Brabender units.

TABLE 8

Changes in Properties of Stored (28°C) Flakes from Lisbon Yam (after Steele and Sammy, 1976b)

Maturity	Storage Period (Week)		Package	Moisture Dry Wt %	pH	Acidity mg NaOH/g	Optical Density of 60% EtOH Extract at 400 mμ	Organoleptic Properties				
	Tubers	Flakes						Appearance (10)[a]	Odor (10)	Texture (10)	Taste (10)	Total (40)
Mature	0	0	Clear glass jar	10.0	5.6	3.3	0.115	9	10	8	10	37
	0	12	"	10.1	5.5	3.8	0.165	7	9	8	9	33
	0	24	"	10.6	5.4	4.6	0.240	6	8	8	8	30
	20	0	"	10.2	5.6	4.2	0.125	9	10	8	10	37
	20	12	"	10.4	5.6	4.4	0.145	8	9	8	9	34
	20	24	"	10.5	5.6	4.8	0.170	7	8	7	8	30
	0	0	Amber glass jar	10.8	5.6	3.3	0.115	9	10	8	10	37
	0	12	"	10.8	5.5	4.0	0.150	9	9	8	9	35
	0	24	"	11.0	5.5	4.2	0.215	8	8	8	9	33
Over mature	0	0	Clear glass jar	8.0	5.6	2.8	0.065	9	10	8	10	37
	0	12	"	8.6	5.6	3.0	0.070	8	9	8	9	35
	0	24	"	8.7	5.5	3.2	0.110	7	8	8	9	32
	20	12	"	10.0	5.6	3.2	0.080	9	10	8	10	37
	20	12	"	10.5	5.6	3.6	0.145	7	8	7	8	30
	20	24	"	11.0	5.5	3.8	0.195	6	7	7	7	27
	0	0	Poly bags	8.0	5.6	2.8	0.065	9	10	8	10	37
	0	12	"	11.2	5.5	3.2	0.085	7	7	8	7	29
	0	24	"	11.9	5.6	3.8	0.125	6	6	7	5	24
	0	0	Clear glass jar under N₂	9.0	5.6	3.2	0.080	9	10	8	10	37
	0	12	"	9.4	5.6	3.6	0.130	9	9	8	10	35
	0	24	"	10.7	5.7	3.8	0.160	8	8	8	8	32

[a] Maximum acceptability value.

satisfactorily in many of these countries, and utilization of indigenous sources of starch such as yams could lead to reductions in importation of wheat or wheaten flour.

In an attempt at total substitution of wheat flour by root crop flours in bread, the effect of various starch binders which must be incorporated to maintain loaf volume in the absence of gluten, and of protein additives to yam flour, were investigated (Kim and de Ruiter, 1968). The cake-like dough and resulting poor loaf volume indicates that total substitution is likely to be more successful with baked goods other than bread.

Substitution of wheat flour by yam flour at the 20 percent level in breads has been shown by Martin and Ruberté (1975) to give a satisfactory product, while with other bakery products such as pancakes and cupcakes even higher levels (50 and even 100 percent) can be acceptable depending upon the yam variety. D. alata cvs. Forastero and Farm Lisbon were reported to yield superior baking flours. It was, however, stressed that sophisticated methods for producing drum-dried flakes were likely to prove too expensive for an economically viable flour. Yams suitable for machine processing are required, perhaps necessitating the development of new cultivars. A study with various composite flours, including yam flour, by Dendy, Clarke, and James (1970) also concluded that in general up to 20 percent of non-wheat meaterial may be used, but that it may be possible to use higher levels if mechanical dough development instead of bulk fermentation is employed.

The functions of components in composite doughs including non-wheat flours other than starch may be more significant than the differences in the starches themselves (Rasper, Rasper, and Mabey, 1974). The rheological quality of the bread doughs and the loaf volume of the final product obtained using 15 percent yam flour appeared to be better, especially with D. rotundata, than those obtained with other root crop flours, despite the high gelatinization temperature and low amylolytic susceptibility of yam starch which, on its own, gave rise to a less desirable product. The results of the baking tests using composite flours based upon 15 percent non-wheat flour are shown in Table 9. Further work on D. rotundata and D. alata showed that the non-starchy polysaccharide components of yam might indeed contribute to the more favorable

rheological characteristics obtained with yam flours
(Hahn and Rasper, 1974). Their effect was mainly
attributed to changes in the water-binding capacity
of the dough, although in the case of D. rotundata
there was an indication that another, as yet
unidentified, mechanism was operating which gave
rise to the pronounced strengthening of the rheological quality of wheat dough observed on addition
of the non-starchy components extracted from this
cultivar (Sefa-Dedeh, MacDonald, and Rasper, 1977).
     The economics of the utilization of flours
derived from tropical starch crops, including yams,
for breadmaking in developing countries are discussed in a report by Lehmann (1971) together with
some possible baking techniques outlined subsequently (Lehmann and Munzberg, 1972). Yam is normally
expensive to produce, compared to cassava, on
account of high labor requirements and other inputs
in its production. Also, yam requires relatively
fertile soil.

FRIED YAM PRODUCTS

     The development of yam products similar to
existing potato products such as chips (known as

TABLE 9
Baking Tests with Composite Flours; 15% HRS Wheat Flour Replaced by Non-wheat Flour
(after Rasper et al., 1974)

Flour Mixture	Moisture Absorption on 14% m.b. %	Specific Loaf Volume ml/100g flour 14% m.b.	Crumb Compressibility 1/10 mm		
			1 day	2 days	3 days
HRS wheat flour only	64.2	827	107	74	51
" + cassava flour	61.2	688	72	60	41
" + yam flour (D. rotundata)	60.3	740	77	56	53
" + yam flour (D. alata)	61.2	740	72	54	47
" + yam flour (D. cayenensis)	60.0	680	56	46	38
" + plantain flour	59.5	710	61	46	38
" + kokonte	60.5	705	62	48	42
" + gari flour	59.9	615	57	40	35
" + sorghum flour	61.5	680	75	44	37
" + millet flour	61.8	710	86	50	36

crisps in the U.K.), French fries (known as chips in the U.K.), or other snack items, may prove to be successful for more sophisticated markets.

An investigation of the suitability of 25 varieties of yams for the production of chips and French fries (Martin and Ruberté, 1972) led to the identification of several suitable varieties. A white, dense-fleshed yam such as D. alata--particularly Forastero, Farm Lisbon, and Feo cultivars--of suitable size and shape for processing was recommended. The performance of all the varieties tested, with an assessment of their overall suitability for chips and French fries, is summarized in Table 10. The quality of the product was dependent upon the frying medium, but the inclusion of a suitable antioxidant was thought to be one of the most important aspects for quality. Storage studies on yam (D. alata cv Farm Lisbon) chips (Rodriguez-Sosa et al., 1973) indicated that vacuum packing in tins was superior to cellophane packing, but by using lard as a frying medium, products in cellophane packing were still acceptable after 9 weeks. By contrast, products fried in corn oil deteriorated rapidly due to moisture absorption.

CANNING

Canning trials with D. alata, D. rotundata, and D. cayenensis yams indicated that the best product was obtained using blanched (5 minutes in boiling water) ½-inch slices (1.27 cm) in 3 percent brine at 60 percent fill, followed by a 60-minute processing at 240°F for an A2 can (Crowther and Kramer, 1974). The products were acceptable after 18 months storage at room temperature (20°C). The major problem was the possible gelation of free starch during processing, which reduces heat penetration and increases processing time, and also gives a product of less attractive appearance.

A similar procedure for ¼-inch (8 mm) slices or chips (Needham, 1971) utilized a thorough water washing after blanching to remove excess starch, and a maximum fill of 70 percent (preferably 50 to 60 percent) was recommended to minimize starch gelation. A Ghanaian standard (1970) for canned yams covers packing liquids, other ingredients, quality criteria, tolerances, labelling, and analytical procedures for the product.

205

TABLE 10
Advantages and Disadvantages, Rating of Kitchen Convenience, and Final Judgment of Yam (Dioscorea spp.) Varieties as French Fries and Fried Chips (after Martin and Ruberté, 1972)

Variety	Advantages	Disadvantages	Kitchen Convenience[a]	Rating
Morado	–	Stickiness	Fair	Acceptable
Ashmore	–	Poor color	Good	Poor
Vino Purple	–	Purple color	Fair	Not acceptable
Purple Lisbon	–	Purple color	Fair	Not acceptable
Macoris	–	Poor color, bitter	Fair	Not acceptable
Florido	–	–	Excellent	Acceptable
Yellow Lisbon	Good flavor	Poor shape	Good	Excellent
Bottleneck Lisbon	–	Very poor shape	Poor	Excellent
Oriental	–	Poor flavor	Good	Not acceptable
Seal Top	–	–	Excellent	Acceptable
Smooth Statia	–	Poor color	Excellent	Not acceptable
Farm Lisbon	Excellent flavor	Poor shape	Poor	Excellent
Feo	Good color and flavor	Poor shape	Poor	Excellent
Gordito	–		Fair	Acceptable
Hawaii Branched	–	Poor color	Fair	Poor
Forastero	Excellent color, flavor, crispness	–	Good	Excellent
Hunte	Good flavor		Fair	Very good
Vino Blanco	–	Poor shape	Fair	Poor
Guinea Blanco	Good color and crispness	Bitterness	Good	Not acceptable
Guinea Puledo	Good color and crispness	Bitterness	Good	Poor
Gunda (sharp angled)	–	Acrid taste	Fair	Not acceptable
Gunda (round)	–	Acrid taste	Fair	Not acceptable
Papa	–	Bitterness	Excellent	Not acceptable
Pana	Good color	Bitterness	Excellent	Not acceptable
Mapuey Largo	–	Stickiness	Fair	Acceptable

[a]Kitchen convenience is an estimate closely related to ease of peeling, low waste, and moderate size of tuber.

LYE PEELING

The economic success of any future commercial development of yam products would depend upon the adaptability of each processing stage to mechanization. Most products require peeled yams as a starting material, and the efficiency of the peeling process affects both the quality of the finished product and the amount of waste. Hand peeling is labor-intensive and often wasteful, while mechanical peeling is unlikely to be applicable due to the irregularity of size and shape of yam tubers. Lye peeling which utilizes chemical and thermal action on the tuber surface may therefore be applicable.

Rivera-Ortiz and Gonzalez (1972) studied the relationship between lye concentration and exposure time in boiling solutions required for satisfactory peeling of D. alata cv Florida. Several sets of peeling conditions were identified, all of which gave substantially lower waste (between 12.2 and 14.4 percent losses for good peeling) than for hand peeling, which gave 22.2 percent losses on the average. Lye peeling did not result in any change in sensory quality, acidity or hardness of the peeled yam, but was less efficient with tubers stored for more than 28 days. In a study on lye peeling of the effect of varietal differences between yams, a 10 percent lye solution at 104°C was generally recommended with varying immersion depending upon cultivar (Steele and Sammy, 1976a). Flesh losses were again reduced compared to hand peeling and there was insignificant free lye retention after peeling. The proximal end of the tuber consisting of older tissue showed greater resistance to peeling than the distal end of the fresh tuber, and, while storage again reduced peeling efficiency, more uniform peeling was obtained due to maturing of the distal end in storage. Overall, it was concluded that yams could be successfully lye-peeled on a commercial basis, although conditions for each batch may have to be individually assessed.

POTENTIAL USES OF STARCHES AND FLOURS FROM INEDIBLE YAMS

Certain species of yams are not significantly utilized as a food source, except in times of shortage, due to their toxicity attributed to the presence of the alkaloid dioscorine or related compounds (Coursey, 1967). These species may be detoxicated by aqueous soaking. Sulit (1967) suggested the detoxification of D. hispida Dennst. either by brine or alcohol extraction to yield edible starches or flours for use in the preparation of liquid starch pastes or baked goods. A similar use for D. cinnamonifolia as a source of liquid starch paste for both food and non-food preparations has been suggested (Rosenthal et al., 1972). The unsuitability of D. dumetorum for fufu manufacture and its small starch granule size (Rasper and Coursey, 1967) may indicate a potential as a substitute for rice starch in printing, laundry, and cosmetic industries (Osisiogu and Uzo, 1973).

NOTES

1. P.C. Crowther and E.C. Kramer, personal communication, 1974, Industrial Development Department, Tropical Products Institute, England.

REFERENCES

Abe, M.O. 1973. Adaptability of potato drying to yam processing. J. Milk Food Technol. 36:456.

Afable, L.A. 1970. The preparation of ubi powder. Philippine J. Plant Ind. 35:19.

Ayensu, E.S. and Coursey, D.G. 1972. Guinea yams; the botany, ethnobotany, use and possible future of yams in West Africa. Econ. Bot. 26:301.

Ayernor, G.S. 1976. Particulate properties and rheology of pregelled yam (Dioscorea rotundata) products. J. Food Sci. 41:180.

Ayernor, G.S., Brennan, J.G., and Rolfe, E.J. 1974. Rheology as a tool in product development. Fourth Intern. Congr. Food Sci. Technol., Madrid.

Brennan, J.G. and Ayernor, G.S. 1973. A study of the kinetics of retrogradation in a starch-based dough made from dehydrated yam (Dioscorea rotundata (L.) Poir). Die Starke 25:276.

Collins, J.L. and Falasinnu, G.A. 1977. Yam (Dioscorea spp.) flour fortification with soy flour. J. Food Sci. 42:821.

Coursey, D.G. 1966. Food technology and the yam in West Africa. Trop. Sci. 8:152.

_____. 1967. Yams. Tropical Agriculture Series. London: Longmans.

_____ and Haynes, P.H. 1970. Root crops and their potential as food in the tropics. World Crops 22:261.

Cruz-Cay, J.R. and Gonzalez, M.A. 1974. Properties of starch from Florido yam (Dioscorea alata L.). J. Agr. Univ. Puerto Rico 58:312.

Dendy, D.A.V., Clarke, P.A., and James, A.W. 1970. The use of blends of wheat and non-wheat flours in breadmaking. Trop. Sci. 12:131.

Fyns Konsum Industri A.S. 1972. Method and apparatus for preservation treatment of starchy fruits and vegetables. British Patent No. 1 261 416.

Ghana National Standards Board. 1970. Canned yams, Ghanaian Standard G.S. F 1 0. (In: Food Sci. Technol. Abstr. 1972, 4:U452.)

Gooding, E.G.B. 1972. The production of instant yam in Barbados. Part 1: process development. Trop. Sci. 14:323.

Hahn, P.P. and Rasper, V. 1974. The effect of non-starchy polysaccharides from yam, sorghum and millet flours on the rheological behavior of wheat doughs. Cereal Chem. 51:734.

Ishitani, T., Kojo, T., and Kimura, S. 1973. Studies on the Freeze Drying of Foods. 9. Freeze Drying of Yam and its Utilization for Karukan Manufacture (in Japanese). Report of the Food Research Institute (Japan) 28:88. (In: Food Sci. Technol. Abstr. 1975, 7:L991.)

Jarmai, S. and Montford, L.C. 1968. Yam flour for the production of fufu. Ghana J. Agr. Sci. 1:161.

Kim, J.C. and de Ruiter, D. 1968. Bread from non-wheat flours. Food Technol. 22:867.

Lehmann, G. 1971. Food and nutrition problems in developing countries. I. The bread situation (in German). Ernahrungs-Umshau 18:505. (In: Food Sci. Technol. Abstr. 1972, 4:M770.)

_____ and Munzberg, F. 1972. Food and nutrition problems in developing countries. II. Flour and bread manufacture (in German). Ernahrungs-Umshau 19:316. (In: Food Sci. Technol. Abstr. 1973, 5:M173.)

Makanjuola, G.A. 1974. A machine for preparing pounded yam and similar foods in Nigeria. Appropriate Technol. 1:9.

Martin, F.W. and Ruberté, R. 1972. Yams (Dioscorea spp.) for the production of chips and French fries. J. Agr. Univ. Puerto Rico 56:228.

_____ and Ruberté, R. 1975. Flours made from yams (Dioscorea spp.) as a substitute for wheat flour. J. Agr. Univ. Puerto Rico 59:255.

Misawa, M. and Matsubara, M. 1965. Studies on the manufacturing of the powder of tuber of the yam. Japan. J. Food Sci. Technol. 12:23.

Needham, W.H. 1971. Canning of Tropical Vegetables. Metal Box Co., Ltd., Food Packaging Information No. 15700.

Onayemi, O. and Potter, N.N. 1974. Preparation and properties of drum dried white yam (Dioscorea rotundata Poir.) flakes. J. Food Sci. 39:559.

Osisiogu, I. and Uzo, J. 1973. Industrial potentialities of some Nigerian yam and cocoyam starches. Trop. Sci. 15:353.

Park, Y.K. 1974. Microscopic observations of cassava and yam starch gelatinization and enzyme susceptibility. Fourth Intern. Congr. Food Sci. Technol., Madrid.

Rai, M.M. and Jain, N.L. 1970. Some studies on the freeze-drying of common vegetables. J. Food Sci. Technol. (Suppl.) 7:22.

Rasper, V. 1969a. Investigations on starches from major starch crops grown in Ghana. I. Hot paste viscosity and gel forming powder. J. Sci. Food Agr. 20:165.

_____. 1969b. Investigations on starches from major starch crops grown in Ghana. II. Swelling and solubility patterns: amyloclastic susceptibility. J. Sci. Food Agr. 20:642.

_____. 1971. Investigations on starches from major starch crops grown in Ghana. III. Particle size and particle size distribution. J. Sci. Food Agr. 22:572.

_____ and Coursey, D.G. 1967. Properties of starches of some West African yams. J. Sci. Food Agr. 18:240.

_____, Rasper, J., and Mabey, G.L. 1974. Functional properties of non-wheat flour substitutes in composite flour. I. The effect of non-wheat starches in composite doughs. Can. Inst. Food Sci. Technol. J. 7:86.

Rivera-Ortiz, J.M. and Gonzalez, M.A. 1972. Lye peeling of fresh yam, Dioscorea alata. J. Agr. Univ. Puerto Rico 56:57.

Rodriguez-Sosa, E.J. and Gonzalez, M.A. 1972. Preparation of yam (Dioscorea alata L.) flakes. J. Agr. Univ. Puerto Rico 56:39.

_____ and Gonzalez, M.A. 1974. Effect of flake size on pasting characteristics of instant Florido yam (Dioscorea alata) flake slurries. J. Agr. Univ. Puerto Rico 58:219.

_____, Gonzalez, M.A., and Martin, F.W. 1972. Evaluation of ten varieties of yam (Dioscorea spp.) for production of instant flakes. J. Agr. Univ. Puerto Rico 56:235.

_____, Cruz-Cay, J.R., Gonzalez, M.A., and Martin, F.W. 1973. Shelf-life study of Farm Lisbon yam (Dioscorea alata) chips. J. Agr. Univ. Puerto Rico 57:196.

_____, Gonzalez, M.A., and Parsi-Ros, O. 1974. Effect of precooking on quality of instant flakes from Florido yam (Dioscorea alata L.) J. Agr. Univ. Puerto Rico 58:317.

Rosenthal, F.R.T., Pelegrino, S.L., and Correa, A.M.N. 1972. Studies on the starches of Dioscorea, Dioscorea alata (edible), and Dioscorea cinnamonifolia (non-edible). Die Starke 24:54.

Sefa-Dedeh, S., MacDonald, B., and Rasper, V.F. 1977. Water-soluble non-starchy polysaccharides of composite flours. II. The effect of polysaccharides from yam (Dioscorea) and cassava flours on the rheological behaviour of wheat dough. Cereal Chem. 54:813.

Steele, W.J.C. and Sammy, G.M. 1976a. The processing potential of yams (Dioscorea spp.). I. Laboratory studies of lye peeling of yams. J. Agr. Univ. Puerto Rico 60:207.

_____ and Sammy, G.M. 1976 b. The processing potential of yams (Dioscorea spp.). II. Precooked drum dried flakes--instant yams. J. Agr. Univ. Puerto Rico 60:215.

Sulit, J.I. 1967. Method of processing and utilization of nami (Dioscorea hispida) tubers. Araneta J. Agr. 14:203.

D. G. Coursey and Carol E. M. Ferber: Tropical Products Institute, 56-62 Gray's Inn Road, London WC1X 8LU, England

2
Processing and Storage of Taro Products

J. H. Moy
W. K. Nip

ABSTRACT

There are various ways of processing and
utilizing taro (<u>Colocasia esculenta</u> (L.) Schott).
Forms taro might take for consumption include poi
(fresh paste, canned, canned-acidified, freeze-
dried), flour, cereal base, beverage powders, chips,
sun-dried slices, and drum-dried flakes. This
paper highlights previous and current work done on
taro processing, with some attention given to storage of certain of these processed products. A
review of processing techniques and products indicates the possibility of developing and marketing a
number of useful products from taro, and from other
root crops as well. Problems to bear in mind in
considering commercial production are economic
feasibility and consumer acceptance.

INTRODUCTION

Taro (<u>Colocasia esculenta</u> (L.) Schott) is
cooked and eaten in much the same manner as potato
and sweet potato in many parts of the tropics and
subtropics. In Hawaii taro is commonly eaten in
the form of poi. A rather extensive study on taro
was conducted in the mid-1930s by Payne, Ley and
Akau (1941). In reviewing other pertinent literature it appears that as a root crop, taro has perhaps been prepared or processed into more consumable

forms than other root crops. These include poi (fresh paste, canned, canned-acidified, and freeze-dried), flour, cereal base, beverage powders, chips, sun-dried slices, and drum-dried flakes.

Obviously, then, various ways of processing and utilizing taro are possible. The two main factors to be considered for commercial production and marketability are economics and end use. This may account for the fact that some of the history of taro products and processing is repeating itself.

In the following sections, attempts will be made to highlight previous work and state-of-the-art of processing taro products. Wherever applicable or available, the storage of the processed products will also be mentioned.

TARO PASTE (POI): FRESH, FERMENTED, CANNED, DEHYDRATED, AND IRRADIATED

Poi is a purplish-gray paste made from taro, most frequently taken from var. Lehua maoli. In Hawaii the product is sold commercially in plastic bags, jars, or cans. Poi is easily digested and is therefore an excellent food for people with specific health problems. It has long been popular as an excellent baby food and is bottled for this purpose and sold in the baby food section of grocery stores. Poi is practically non-allergenic and is one of the foods recommended to patients with food allergies. Patients allergic to certain grains can eat poi with no adverse reaction (Derstine and Rada, 1951; Glaser, Harrison, and Ball, 1967; Roth, Worth, and Lichton, 1967).

The preparation of poi from taro involves pressure cooking of the corms, washing, peeling, trimming, grinding, blending with water to reach 30 percent total solids, and passing of the paste through a series of strainers, the final one having openings of 0.5 mm in diameter.

The processing of taro into poi changes the distribution of the starch components. In the acid fermentation of taro paste the major starch components of taro change from amylose to amylopectin. A significant amount of dextrin and a very small amount of glucose are also produced. No change in the amount of pentosan during the processing of cooked taro into poi has been observed. Reducing sugars decrease rapidly in the early stages of fermentation and remain fairly constant in amount from the stage of 12 hours to 8 days (Allen, 1933; Bilger and Young, 1935; Standal, 1970).

During the acid fermentation, the causal agents include Lactobacillus and Streptococcus species and several species of yeast. Acid production is usually very rapid within the first 24 hours, during which time the acidity commonly increases from a pH of 6.3 (fresh ground taro corms) to a pH of 4.5. Thereafter, the acidity increases gradually until it reaches an average pH of 3.8. From the third to sixth day, a flora of yeast, mycoderms, and oidia become increasingly prevalent in the poi. The fermentation products identified are lactic acid, acetic acid, formic acid, alcohol, acetaldehyde, and carbon dioxide (Allen, 1933; Bilger and Young, 1935).

Canned fresh poi is the unfermented product, less than 4 hours old, containing not less than 18 percent total solids. In a 20-oz (560 g) can (U.S. No. 2), the thermal process requires cooking for 100 min at 116°C.

Canned acidified poi is the unfermented product, less than 4 hours old, to which 1 percent commercial grade lactic acid (50 percent lactic acid) has been added. It contains not less than 18 percent total solids. The product is filled and heated in a steam jacketed kettle to 96°C. The can is closed at a minimum temperature of 93°C, inverted for 2 minutes to sterilize the cover, and cooled in water.

Experimental studies have been made to prepare a dehydrated poi by freeze-drying.[1] The product was quite acceptable but it was considered an expensive process. At present, bottled freeze-dried poi is available in supermarkets in Honolulu. Gamma-radiation was used to extend the shelf life of poi. A minimum of 700 krad was required to increase the shelf life to 7 to 10 days (Moy and Lee Loy, 1967).

TARO FLOUR

Taro flour was manufactured in several ways.

The procedure developed by the Hawaii Agricultural Experiment Station was as follows: raw taro corms → cooking in pressure retort at 104-121°C for 1 hr → mechanical or hand peeling and washing → grinding to "paiai" → refrigeration at 2°C for 36 hr → shredding in mechanical food shredder or slicer → drying in recirculating tray-drying cabinet 5 to 6 hours at 63°C (initial) to 93°C (final)→ grinding in hammer mill → sifting through mechanical sifter with silk or metallic 66-grits gauze → packaging.

Taro flour manufactured by this process had good quality. It could be reconstituted to fresh, unfermented poi and could also substitute up to 20 percent of wheat flour during baking Potgieter, 1940; Payne, Ley, and Akau,1941). The only disadvantage of this process was the use of refrigeration before drying in order to reduce the gumminess of the intermediate product (paiai), which was rather expensive.

The process developed by the Central Food Technological Research Institute, Mysore, India, was as follows: washing of corms in water → peeling by scraping with knives → cutting into slices of ¼" (6 mm) thickness → keeping the peeled tubers and slices in water, and scrubbing in water to remove mucilaginous matter → keeping the slices overnight → washing and scrubbing → immersion in 0.25 percent solution of SO_2 for 3 hr → removing the slices from the SO_2 solution and washing in water → drying on trays or blanching in boiling water for 4 to 5 min, then drying on trays, drying temperature preferably at 57 to 60°C → grinding and sieving → packaging.

The dried product was generally better than the unblanched product. The taro flour produced by this procedure had good quality. It could be used in the preparation of chapatties with a substitution of 25 percent wheat flour and the coarse grainy material left on the sieve formed a by-product resembling that of soji or rawa (Jain, Das, and Lal, 1953). The main difference between this procedure and the others was the soaking of the slices in water, scrubbing, and SO_2 treatment. Taro flour produced by this procedure does not give a gummy or sticky product upon reconstitution.

The Food Processing Laboratory of the Government of Weatern Samoa has a rather direct procedure of manufacturing taro flour. Taro is peeled, sliced into slices or cubes 0.5 to 1.0 cm in thickness, and dried at 66°C for 12 to 15 hr. Dried slices or cubes are then ground to approximately 12 to 15 mesh through a Fitzmill, followed by a further grinding through a pin mill. This produces a flour which is quite stable and could be successfully stored over one year.[2]

The procedure for the manufacturing of instant taro powder suggested by the Food Industry Research and Development Institute of Taiwan, Republic of

China, was as follows: washing of corms → abrasive peeling → trimming → slicing to 1.7 to 2.2 mm thick slices → steam cooking at 100°C for 30 min → pureeing → adjustment of puree to contain 20 percent solid with 5 percent glucose solution → drum-drying at 2 kg internal drum pressure, at 2.5 rpm, and 0.01 in spaces between drums. The reconstituted product was said to have good flavor, texture, and color, but the viscosity was lower than the product made from fresh taro (Hsu and Chiou, 1971). The cost of producing taro flour (powder) by this procedure would be lower than the Hawaiian procedure, but higher than the Indian and Western Samoan procedures.

A more recent version of preparing taro flour at the Hawaii Agricultural Experiment Station was somewhat similar to that used in the Food Processing Laboratory of Western Samoa. The taro was peeled, sliced into 5mm thickness for air-drying (60°C for 9 hr) and freeze-drying (10 hr), and 2.5 mm thick for solar drying. The dried slices were ground to approximately 5 to 10 mesh and then further ground to 60 mesh in a Fitz impact grinder. Both the flour and slices were found to be stable at 38°C for more than one year, with minor changes in acidity, pigment, moisture contents and with moderate changes in catalase activities and flavor (see Moy, Wang, and Nakayama, this volume). It was also found that these products could be used for remanufacturing into usable forms such as rice and noodles.

STARCH

Langworthy and Deuel (1922) compared the digestibility of raw rice, arrowroot, canna, cassava, taro, tree fern and potato starches. The granules of taro root starch were found to be extremely small, in the same range as the smallest of the arrowroot (Zamia floridana) starch granules, both measuring between 1 and 7 microns. It was observed that "raw corn, wheat, cassava, rice and taro root starches are completely digested in amounts as large as 250 g daily...."

Payne, Ley, and Akau (1941) found the average size of starch grain of Kai Uliuli variety was 4.5 microns, and the maximum 9.3 microns.

Through a series of steps including grinding, wet milling, filtration, centrifugation, and drying, Tu, Nip, and Nakayama (in this volume) separated

40 percent of the taro starch from a total of 70 to 80 percent on a dry weight basis. The taro starch granules formed a clear, stringy paste somewhat resistant to enzyme action, indicating that taro starch could be utilized for both human consumption and industrial use.

The residue in the nylon filter bags after drying was ground to flour which contained 60 to 70 percent starch. Noodles, flakes, cookies, muffins, and biscuits were made by mixing this raw flour fraction with wheat flour. Results indicated that taro flour could be substituted for part of the wheat flour required for pastry preparation.

CEREALS, BREAD, AND CAKE

In the 1930s a company called Hawaiian Taro Products, Ltd., was organized in Honolulu after several years of extensive research in the agronomic, nutritional, and processing phases of taro production. A line of products was planned and produced, including flour, cereals, infant and invalid foods, and beverage powders.

The baby formula, called "Taro-Lactin," contained fully pre-cooked, dehydrated taro, skim milk, unrefined cane juice, and 1.5 percent sodium chloride. It was used either as a gruel or in the preparation of a formula.

The taro flour produced by Hawaiian Products, Ltd., was essentially a dehydrated poi made from finely ground meal of the entire cooked taro corm. Besides being marketed as dehydrated poi, it was also promoted for preparing bread, cookies, wafers, breakfast cereals, puddings, and soups.

The following percentages of the taro flour were recommended for various baked goods: bread, 15 percent; cookies, 25 to 50 percent; biscuits, doughnuts, pancakes, waffles, pastry, and layer cakes, 20 to 30 percent.

It was not completely clear exactly when and how the business and product line were discontinued, but this occurred sometime after the second world war.

Commercial pamphlets indicate a similar product line was distributed by the Galen Co., Inc., Berkeley, California, around the late 1930s and 1940s under the brand names Poyolin and Poyo-meal, which contained about 60 percent dehydrated poi on a dry weight basis.

BEVERAGE POWDER

A beverage powder using dehydrated, cooked taro as its base was marketed in Hawaii in the 1930s and 1940s under the name "TA-RO-CO." It was a chocolate-flavored taro beverage with sugar, cocoa, malt, and salt as added ingredients.

Although TA-RO-CO was intended mainly for use in milk beverages, it also found favorable application as a nourishing and tasty ingredient in ice cream, puddings, custards, and other preparations. Again, economic reasons probably caused it to disappear from the market.

TARO FLAKES

Payne, Ley, and Akau prepared taro flour by tray drying, drum drying, and spray drying. They concluded that while tray-dried taro flour was the most economical and practical of the three, drum-dried taro flakes had a pleasing flavor and might be used in puddings, ice cream, and beverages.

Drum-drying has been adapted to dehydrating various root crops such as potato (Cording et al. 1957), sweet potato (Spadaro et al., 1967), pumpkin (Hoover, 1973), yam (Onayemi and Potter, 1974; Steele and Sammy, 1976), tanier (Xanthosoma)(Rodriguez-Sosa and Gonzales, 1977), and taro (Hsu and Chiou, 1971).

In the early 1970s there was interest among some taro growers in Hawaii to increase the cultivation and use of taro. Hing (1974)[3] did some preliminary work on drum drying of poi mixed with some tropical fruit purees. Nip recently studied the drum-drying properties of combined taro and tropical fruit purees as well as their storage stability and acceptability (Nip, 1978; see also this volume).

The tropical fruit purees included guava, papaya, pineapple, and mango. It was found that a fruit-to-taro ratio of 3 to 2 was preferred by the taste panel over that of 1 to 2. In spite of some decreases in ascorbic acid content and the development of browning in some samples during storage, general acceptability was rated by the taste panel between "like slightly" and "like moderately."

TARO CHIPS

Taro can be and has been marketed as chips by deep fat frying, much like the popular potato chips. One difference is the acridity factor which exists in taro but not in potato.

The acridity factor in taro, commonly known as "taro itch," is an area of taro chemistry deserving more in-depth study. The presence of calcium oxalate crystals in the form of raphides in taro corm tissues has been suggested as a cause of the irritation. Osisiogu, Uzo, and Ugochukwu (1974) reported that the irritant principles of taro were water-soluble and volatile, and that they were not destroyed by heat, but by volatilization.

To make an acceptable taro chip product, a variety that will lose the acridity factor during frying should be an important criterion. The alternative approach is to make a taro chip similar to a currently popular potato chip product which is really a flash-fried, reconstituted, dehydrated potato flake.

CONCLUSIONS

A review of various processing techniques and products for taro suggests that it is quite feasible to develop and market a number of useful products from this root crop and from others. Some of the problems requiring further identification and studies would be:

1. Technical and economic feasibility. Should the processing techniques be thoroughly studied to the point of making them compatible with economic feasibility?
2. Marketing and consumer acceptance. Is the market ready for some of the products? Should the objective be defined for the development of each product before it is commercialized?
3. Constraints and limiting factors. What are some of the constraints and limiting factors in increasing the cultivation and utilization of taro and other root crops?
4. New approaches. Are there some new approaches to processing and product development that have not been and should be considered?

NOTES

1. J.H. Moy, 1967, unpublished data.
2. C.J. Pedrana, 1976, personal communication.
3. F.S. Hing, 1974, unpublished data.

REFERENCES

Allen, O.N. 1933. The Manufacture of Poi from Taro in Hawaii, with Special Emphasis upon its Fermentation. Hawaii Agr. Expt. Sta. Bull. 70, Honolulu, Hawaii.
Bilger, L.H. and Young, H.Y. 1935. A chemical investigation of the fermentation occurring in the process of poi manufacturing. J. Agr. Res. 51:45.
Cording, J., Jr., Willard, M.T., Jr., Eskew, R.K., and Sullivan, J.F. 1957. Advances in the dehydration of mashed potatoes by the flake process. Food Technol. 11:236.
Derstine, V. and Rada, E.L. 1951. Some Dietetic Factors Influencing the Market for Poi in Hawaii. Univ. Hawaii Agr. Econ. Bull. 3.
Glaser, J., Lawrence, R.A., Harrison, A., and Ball, M.R. 1967. Poi, its Use as a Food for Normal, Allergic and Potentially Allergic Infants. Univ. Hawaii Coop. Ext. Serv. Misc. Pub. 36.
Hoover, M.W. 1973. A process for producing dehydrated pumpkin flakes. J. Food Sci. 38:96.
Hsu, H.Y. and Chiou, K.M. 1971. Instant taro powder. I. Comparison of different dehydrating methods. J. Chinese Agr. Chem. Soc. 9:164.
Jain, N.L., Das, D.P., and Lal, G. 1953. Arvi (Colocasia) flour. Indian J. Hort. 10:19.
Langworthy, C.P. and Deuel, H.J., Jr. 1922. Digestibility of raw rice, arrowroot, canna, cassava, taro, tree-fern, and potato starches. J. Biol. Chem. 52:251.
Moy, H.J. and Lee Loy, B. 1967. Shelf Life Extension of Poi by Gamma Irradiation. UH-235-P-5-2. Report to USAEC-Isotopes-Industrial Technology.
Nip, W.K. In press. Development and storage stability of drum-dried guava- and papaya-taro flakes. J. Food Sci.
Onayemi, O. and Potter, N.N. 1974. Preparation and storage properties of drum-dried white yam (Dioscorea rotundata Poir) flakes. J. Food Sci. 39: 559.

Osisiogu, I.U.W., Uzo, J.O., and Ugochukwu, E.N. 1974. The irrigant effects of cocoyams. Planta Medica 26:166.

Payne, J.H., Ley, G.J., and Akau, G. 1941. Processing and Chemical Investigation of Taro. Hawaii Agr. Expt. Sta. Bull. 86, Honolulu, Hawaii.

Potgieter, M. 1940. Taro (Colocasia esculenta) as a food. J. Am. Dietetic Assoc. 16:536.

Rodriguez-Sosa, E.J., and Gonzales, M.A. 1977. Preparation of instant tanier (Xanthosoma sagittifolium) flakes. J. Agr. Univ. Puerto Rico 61:26.

Roth, A., Worth, R.M., and Lichton, I.J. 1967. Use of poi in the prevention of allergic disease in potentially allergic infants. Ann. Allergy 25: 501.

Spadaro, J.J., Wadsworth, J.I., Ziegler, G.M., Gallo, A.S., and Koltum, S.P. 1967. Instant sweetpotato flakes--processing modifications necessitated by varietal differences. Food Technol. 21:326.

Standal, B.R. 1970. The nature of poi carbohydrates. Proc. Second International Symp. Trop. Root Crops, Coll. Trop. Agr., Univ. Hawaii, Vol. I: 146.

Steele, W.J.C. and Sammy, G.M. 1976. The processing potential of yams (Dioscorea spp.). II. Precooked drum dried flakes--instant yams. J. Agr. Univ. Puerto Rico 60:215.

James H. Moy and W. K. Nip: Department of Food Science and Technology, University of Hawaii, Honolulu, Hawaii

3
Processing of Taro into Dehydrated, Stable Intermediate Products

J. H. Moy
N. T. S. Wang
T. O. M. Nakayama

ABSTRACT

Taro (Colocasia esculenta L. Schott) is an
underexploited tropical root with promising economic
value and the potential to help ease the world food
problem. It produces more calories per unit area of
land than other tropical root crops. For practical
and efficient utilization, it must be processed into
ration dense, stable forms. A study was made to
process the Lehua Maoli variety into flour and dried
slices which could be stored at 38°C for at least
one year. The eating and storage qualities of the
air dried, freeze dried, and sun dried samples were
evaluated and compared. Results indicate this
approach to be feasible. Further studies are continuing to determine the feasibility of utilizing
these intermediate products for making ready-to-use
products such as rice and noodles.

INTRODUCTION

Taro (Colocasia esculenta L. Schott) is an
underexploited root crop with promising economic
value (Engel, 1975). Plucknett, de la Pena, and
Obrero (1970) suggested that root crops, particularly taro, had an excellent potential in solving the
current shortage in the world food supply.
Taro has been an important food crop in different parts of the world for more than 2,000 years

(Whitney, Bowers, and Takahashi, 1939). Although it is widely grown, only in Egypt, the Philippines, Hawaii, and some Pacific and Caribbean Islands is taro a commercial crop (Engel, 1975). Like potato, taro can be boiled, baked, roasted, or fried. The Hawaiians cooked and pounded the taros into a paste which is called poi, and is still extensively eaten throughout the Hawaiian Islands. It was the "staff of life" of the old Hawaiians, and has been an important source of carbohydrates in many South Pacific Islands such as Cook Islands, Fiji, and New Caledonia (de la Pena, 1970).

One of the goals of our current research was to develop a systematic plan which could potentially increase by 10 percent the supply of calories delivered to consumers in the tropics and subtropics. Besides minimizing post-harvest crop losses, some of the problems were to detect potential processing and storage anomalies and to test consumer acceptance. These types of problems could be exposed earlier and attacked more rapidly if the total concept of food delivery was tested within an interdisciplinary group to allow sufficient feedback for making effective corrections.

The delivery of calories from root crops, especially taro, was chosen because they produce the most calories per unit area of land. In terms of food preservation, the tubers have great potential because of their ability to be dried and stored under less than ideal conditions.

The primary objectives of our study were:

1. To determine the feasibility of converting taro into ration dense forms which (a) will be stable for a year at 38°C devoid of nonfunctional ingredients and water (preservation of nutrients and calories), (b) will require a minimal amount of energy to prepare, and (c) will have sufficient appeal to the consumers.
2. To detect problems in dehydration and processing which would affect the palatability and storage quality of the dried products.

In order to develop taro products to meet the above-mentioned objectives, dehydrated products were considered to be the most logical and practical. Among the various dehydration techniques available, an optimum has to be chosen in terms of product quality, storage stability and processing economics.

EXPERIMENTAL

In attempting to reach the objectives, the following steps were identified and considered:

1. Characterization--analyzing, defining, and comparing the properties and characteristics of different varieties of taro.
2. Stabilization--preparation of corms into dehydrated, stable, intermediate forms for storage studies.
3. Detoxification--removal or deactivation of calcium oxalate and other toxicants to neutralize the acridity of taro.
4. Processing--preparation of taro into usable forms like flour, "rice," or noodles.
5. Storage--studies to follow previous steps in meeting objectives.
6. Evaluation--check quality and acceptability of product.

Wherever practical, energy intensive and labor intensive methods were used in product preparations so that processing costs could later be determined and compared.

Taro of the Lehua maoli variety was purchased from a local supplier, Honolulu Poi Company, and prepared in slices by three methods of dehydration: air, solar, and freeze drying. The tubers were washed, peeled, and sliced into 5 mm thick, non-uniform but somewhat round shapes for freeze drying and air drying. For solar drying, it was found necessary to use 2.5 mm thick slices, otherwise it might take longer than two eight-hour days to dry.

Freeze drying, though generally recognized to be expensive, was used to preserve the best quality possible as a dried, uncooked product so as to compare with the other two dried samples. A Virtis sublimator (Model-15) was used with condenser at -50°C, vacuum at 0.20 ± 0.05 mm Hg, and shelf temperature at 60°C. Air drying was carried out in a Proctor and Schwartz forced-air tray dryer, with air thermostatically controlled at 60°C. Solar drying was conducted outdoors with samples placed on screen trays in a direct dryer with ambient air temperature between 24° and 29°C, wind velocity at 15 to 22 ft/sec.

The dried samples were measured for their weight loss, and residual moisture determined with the vacuum oven method. Samples were packaged in polyethylene bags and stored under three conditions:

21°C, 70 ± 10% R.H.; 38°C, 33 ± 3% R.H.; and 60°C, 10 ± 1% R.H. The first condition simulates United States supermarket conditions, the second represents severe tropical temperature, and the third represents areas of extreme heat.

The following tests were conducted on the three types of dried samples as measurements of acceptance and quality changes.

pH. To measure the accidity change that might have a possible effect on protein and lipids. A 2-g ground taro sample was mixed with 20 ml distilled water, stirred well, let stand for 30 min, and pH measured with a Corning Model-10 pH meter.

Color change. To measure extent of pigment degradation. This is particularly true of anthocyanins, the major pigments found in this variety. Absorbance of an ethanol extract of a sample of dried taro flour was measured at 520 nm and 422 nm with a Bausch & Lomb Spectronic 88 to detect changes in the red pigment (anthocyanin) and brown pigment (degradation of anthocyanin), respectively (Sastry and Tischer, 1952).

Catalase activity. To measure the degree of catalase activity in the sample as an index of quality change. A modified iodometric titration method was used to measure oxygen production at 15-minute intervals of a 6-g taro flour sample in 100 ml phosphate buffer (pH 7) and 1.0% H_2O_2 at 30° C (Hans, 1963).

Moisture. To measure the stability of the dried samples--an index of cooking quality which might also reflect the relationship with flavor change and browning.

Eating quality. To measure the overall quality and acceptability of dried samples. Preparation for taste panel involved two hours of soaking in cold water and one hour of boiling. A freshly boiled sample served as control. A trained panel of twelve from the Food Science and Technology Department was served two separate sets of 4 samples at a sitting, each set consisting of a control and three dried samples stored at the same temperature (21°, 38°, or 60°C). A hedonic scale of 7 was used for the ranking, which ranged from "like very much" (7), "neither like nor dislike" (4), to "dislike very much" (1). Samples from each storage temperature were tested in triplicate every month.

The differences in the overall eating qualities among samples dried by the three methods were evaluated by analysis of variance from the average scores of triplicate runs of the multiple comparison test.

Gelatinization temperature. To determine
whether the dried taro samples in flour form might
have undergone changes in their cooking behavior as
a result of storage. Gelatinization temperature
range was determined with the method used by Schoch
and Mayfield (1956) which measures the change in
polarization crosses in the taro starch granules in
a 0.3 percent suspension being heated at the rate
of 2°C per minute.

Energy value. To compare energy value of
dried taro samples with other cereal products. A
Parr adiabatic oxygen bomb calorimeter was used for
the measurement.

RESULTS AND DISCUSSION

Yield of Peeled Taro Corms

The yield of peeled taro corms increased with
increasing initial weight (Figure 1). This seems
logical, since the smaller the corm, the larger the
percentage of peels.

Drying of Taro Samples

Drying rates of taro slices by three methods of
drying are compared in Figure 2, where sample weight
remaining was plotted as a function of drying time.
Total drying time required for 5 mm slices was 9 to
10 hr by both the air drying and freeze drying
methods, but with different residual moisture (ca.
1.64 percent by freeze drying, versus 4.35 percent
by air drying, wet weight basis).

Solar drying required about two days to complete (4.86 percent residual moisture) because of
fluctuation of available sunshine and wind velocity.
Direct solar drying was not always efficient.
Improved sanitation and technology are needed.

pH Change

Results of pH measurements on various dried
samples during a 12-month storage period showed a
0.2 to 0.6 pH unit change in both slice and flour
samples stored at 21° and 38°C. In those stored at
60°C, a 0.8 to 1.4 pH unit change was observed,
suggesting possible quality change (Tables 1 and 2).

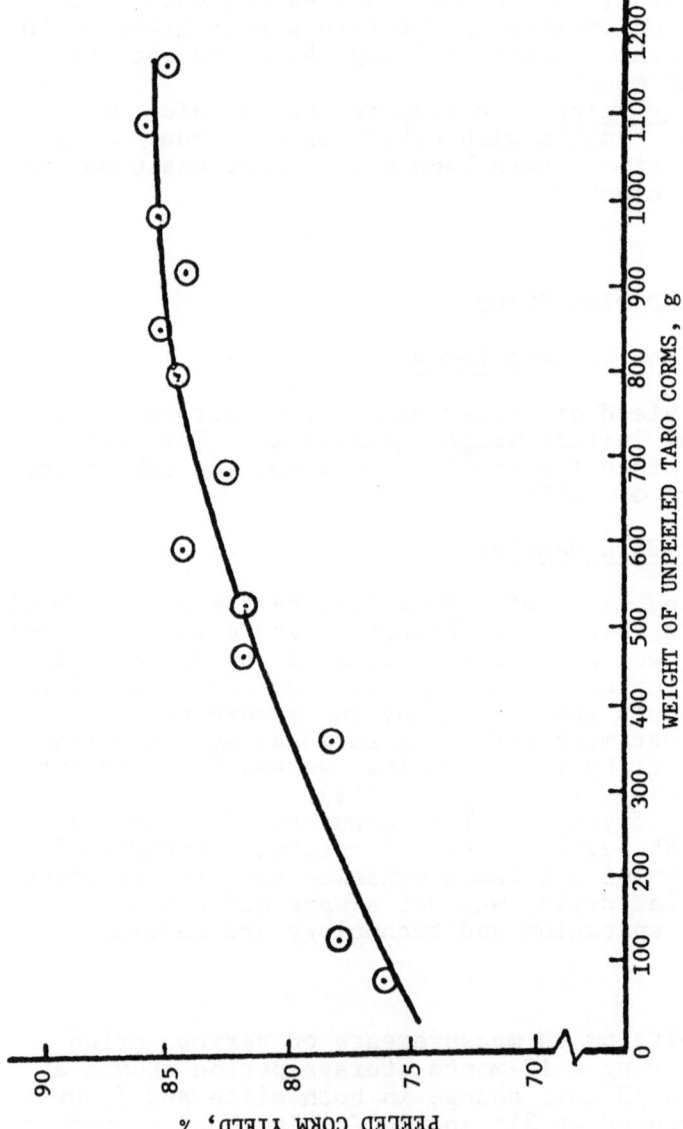

Figure 1. Yield of taro corms after peeling (var. Lehua).

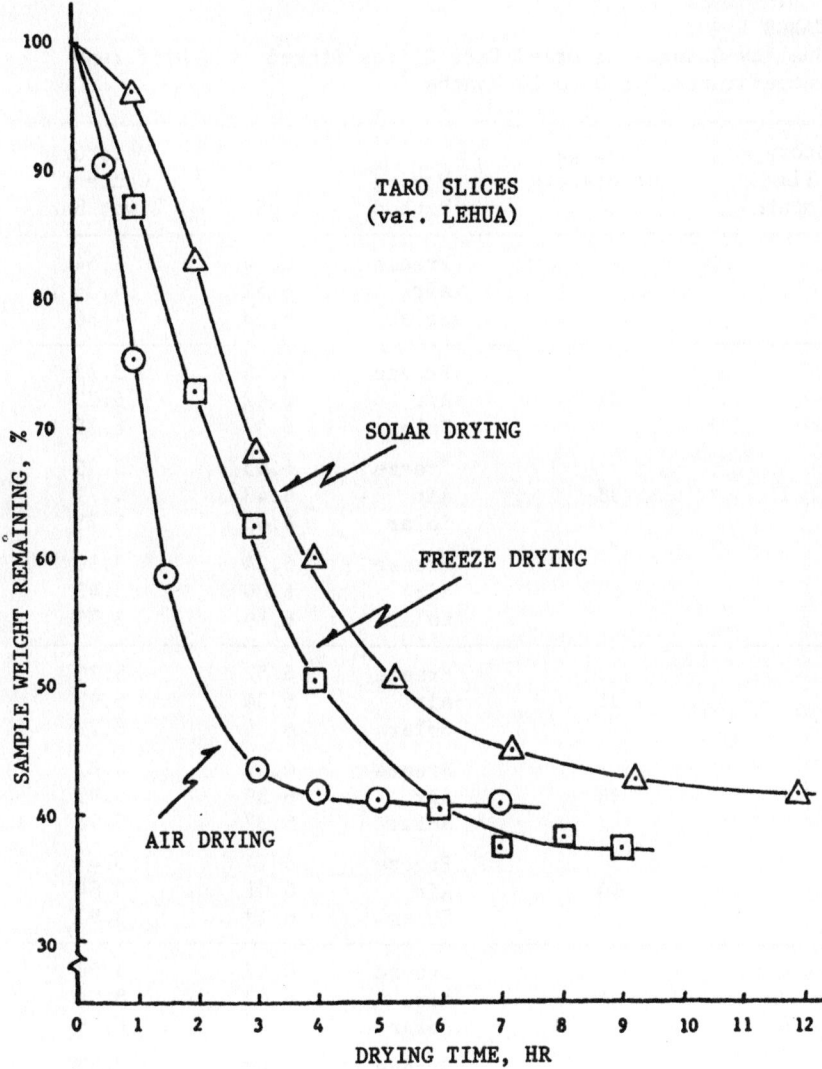

Figure 2. Freeze drying and air drying of taro (var. Lehua) slices of 5 mm thick vs. solar drying of taro slices of 2.5 mm thick. Heated air at 60°C, 15 to 22 ft/sec; shelf temperature in the freeze dryer, 60°C, and vacuum at 0.20 mm Hg; ambient air at 24-29°C, and 70 ± 5% R.H.; loading density 1.5 lb/ft² (0.73 gm/cm²).

229

TABLE 1
Quality Changes of Dried Taro Slices Stored at 3 Different
Temperatures for 0 to 12 Months [a]

Storage Time (month)	Storage Temperature °C	Drying Method	pH	Moisture Content % wet basis
0	--	Freeze	6.44	1.64
	--	Air	6.51	4.35
	--	Solar	6.58	4.86
1	21	Freeze	6.46	3.72
		Air	6.47	6.75
		Solar	6.47	6.75
	38	Freeze	6.45	3.95
		Air	6.43	4.71
		Solar	6.45	5.75
	60	Freeze	6.24	3.10
		Air	6.00	3.42
		Solar	6.19	3.73
2	21	Freeze	6.52	5.20
		Air	6.34	6.92
		Solar	6.34	8.23
	38	Freeze	6.47	4.81
		Air	6.39	5.95
		Solar	6.37	5.93
	60	Freeze	6.14	3.43
		Air	6.01	3.84
		Solar	6.19	3.93
3	21	Freeze	6.47	9.58
		Air	6.47	8.94
		Solar	6.40	10.79
	38	Freeze	6.43	7.29
		Air	6.48	7.35
		Solar	6.37	8.00
	60	Freeze	6.05	5.04
		Air	5.96	4.93
		Solar	6.00	5.00
4	21	Freeze	6.56	9.16
		Air	6.45	11.62
		Solar	6.70	12.49

TABLE 1 (continued)

Storage Time (month)	Storage Temperature °C	Drying Method	pH	Moisture Content % wet basis
4	38	Freeze	6.30	7.94
		Air	6.35	7.95
		Solar	6.40	8.88
	60	Freeze	5.96	4.67
		Air	5.74	5.04
		Solar	5.96	5.23
5	21	Freeze	6.41	9.81
		Air	6.39	11.74
		Solar	6.54	12.65
	38	Freeze	6.18	7.90
		Air	6.38	7.57
		Solar	6.48	8.02
	60	Freeze	5.90	4.00
		Air	5.70	4.12
		Solar	5.50	4.40
6	21	Freeze	6.06	11.33
		Air	6.39	12.50
		Solar	6.35	13.05
	38	Freeze	6.17	7.65
		Air	6.32	7.89
		Solar	6.28	8.39
	60	Freeze	5.85	3.50
		Air	5.50	3.51
		Solar	5.25	3.91
7	21	Freeze	6.20	11.98
		Air	6.50	12.48
		Solar	6.45	13.50
	38	Freeze	6.10	7.90
		Air	6.30	8.00
		Solar	6.25	8.40
	60	Freeze	5.80	4.27
		Air	5.40	4.31
		Solar	5.15	4.58
8	21	Freeze	6.30	12.36
		Air	6.55	12.21
		Solar	6.45	13.14

TABLE 1 (continued)

Storage Time (month)	Storage Temperature °C	Drying Method	pH	Moisture Content % wet basis
8	38	Freeze	6.06	6.86
		Air	6.30	7.53
		Solar	6.23	7.35
	60	Freeze	5.85	3.52
		Air	5.62	3.48
		Solar	5.08	4.10
9	21	Freeze	6.40	13.30
		Air	6.60	13.92
		Solar	6.60	14.26
	38	Freeze	6.06	8.70
		Air	6.28	9.31
		Solar	6.21	8.41
	60	Freeze	5.72	3.49
		Air	5.43	3.60
		Solar	5.10	4.27
10	21	Freeze	6.35	13.10
		Air	6.40	12.75
		Solar	6.38	13.24
	38	Freeze	6.00	6.97
		Air	6.14	7.60
		Solar	6.25	7.28
	60	Freeze	5.40	3.53
		Air	5.30	3.66
		Solar	5.00	4.38
11	21	Freeze	6.15	12.96
		Air	6.30	13.24
		Solar	6.45	13.30
	38	Freeze	6.00	6.68
		Air	6.23	7.60
		Solar	6.30	7.34
	60	Freeze	5.42	3.85
		Air	5.65	3.77
		Solar	5.00	4.10
12	21	Freeze	6.20	13.01
		Air	6.30	13.57
		Solar	6.25	13.30

232

TABLE 1 (continued)

Storage Time (month)	Storage Temperature °C	Drying Method	pH	Moisture Content % wet basis
	38	Freeze	6.10	6.75
		Air	6.25	7.64
		Solar	6.00	7.43
12				
	60	Freeze	5.72	3.58
		Air	5.47	3.86
		Solar	5.10	4.27

[a] Results are average measurements of duplicate samples.

Color Change

The absorbance of ethanol extracts of various dried flour samples at 520 nm and 422 nm were measured, with freeze dried samples in most cases having the highest absorbance. During storage, the absorbance of all samples at 520 nm decreased, indicating loss of anthocyanin, while absorbance at 422 nm also decreased, but slower than those at 520 nm, indicating gradual loss of brown pigments. The ratios of A_{520} nm/A_{422} nm were gradually decreasing from about 2.98 to less than 1.0. These are shown in Table 2.

Catalase Activity

Results of O_2 generation in the modified iodo-metric titration method showed decreasing catalase activities in various samples of dried taro flour. Initially, solar dried samples had higher activities than those in freeze dried samples which in turn were higher than those in air dried samples. This trend continued with storage time. In terms of storage temperature, those stored at 21°C had higher residual catalase activities than those stored at 38°C, while the activities in all three types of dried samples stored at 60°C became too low to be measured even after one month of storage (Table 3).

TABLE 2
Quality Changes of Dried Taro Flour Stored at 3 Temperatures for 0 to 12 Months[a]

Storage Time (month)	Storage Temperature °C	Drying Method	pH	Anthocyanin Degradation A520/A422	Moisture Content % wet basis
0	--	Freeze	6.56	2.98	2.05
	--	Air	6.68	2.80	4.10
	--	Solar	6.76	2.70	4.50
1	21	Freeze	6.66	2.49	4.10
		Air	6.61	2.11	5.67
		Solar	6.10	2.23	6.02
	38	Freeze	6.53	2.39	3.56
		Air	6.58	2.03	7.06
		Solar	6.14	2.20	9.13
	60	Freeze	6.25	1.51	3.68
		Air	6.03	1.55	4.01
		Solar	6.12	1.03	4.23
2	21	Freeze	6.70	2.64	6.67
		Air	6.62	2.14	7.52
		Solar	6.68	2.45	8.72
	38	Freeze	6.56	2.44	6.61
		Air	6.55	2.00	9.60
		Solar	6.69	2.05	9.69

TABLE 2 (continued)

Storage Time (month)	Storage Temperature °C	Drying Method	pH	Anthocyanin Degradation A520/A422	Moisture Content % wet basis
2	60	Freeze	6.27	1.41	5.57
		Air	6.00	1.17	5.12
		Solar	6.06	1.10	5.69
	21	Freeze	6.72	2.53	6.99
		Air	6.58	1.94	8.16
		Solar	6.60	2.10	8.78
3	38	Freeze	6.55	2.35	5.76
		Air	6.53	1.95	7.46
		Solar	6.59	1.95	8.56
	60	Freeze	6.23	1.31	4.48
		Air	5.94	1.11	4.78
		Solar	5.99	1.01	4.67
	21	Freeze	6.53	2.46	7.72
		Air	6.53	2.45	8.23
		Solar	6.32	1.94	8.64
4	38	Freeze	6.33	2.20	6.72
		Air	6.44	1.87	8.07
		Solar	6.31	2.00	8.78
	60	Freeze	5.83	0.94	4.43
		Air	5.82	1.04	4.40
		Solar	5.73	1.37	4.80

235

TABLE 2 (continued)

Storage Time (month)	Storage Temperature °C	Drying Method	pH	Anthocyanin Degradation A520/A422	Moisture Content % wet basis
5	21	Freeze	6.57	2.51	8.75
		Air	6.50	1.75	9.51
		Solar	6.54	2.03	10.21
	38	Freeze	6.36	1.32	7.76
		Air	6.39	1.65	8.75
		Solar	6.44	1.73	8.88
	60	Freeze	6.05	1.12	5.03
		Air	5.82	0.94	4.94
		Solar	5.86	0.84	5.00
6	21	Freeze	6.61	2.34	9.54
		Air	6.36	1.65	9.54
		Solar	6.55	1.99	9.77
	38	Freeze	6.38	1.89	6.81
		Air	6.51	1.44	7.50
		Solar	6.44	1.47	7.70
	60	Freeze	6.06	1.06	4.11
		Air	5.81	0.89	4.27
		Solar	5.83	0.81	4.08

TABLE 2 (continued)

Storage Time (month)	Storage Temperature °C	Drying Method	pH	Anthocyanin Degradation A520/A422	Moisture Content % wet basis
7	21	Freeze	6.56	2.15	11.26
		Air	6.53	1.64	11.07
		Solar	6.60	2.18	11.65
	38	Freeze	6.35	1.89	7.78
		Air	6.34	1.42	7.35
		Solar	6.41	1.59	7.41
	60	Freeze	6.01	1.08	6.32
		Air	6.53	0.89	5.43
		Solar	5.83	0.80	4.70
8	21	Freeze	6.48	2.45	11.40
		Air	6.40	1.74	11.09
		Solar	6.39	2.00	11.58
	38	Freeze	6.22	1.97	7.85
		Air	6.20	1.46	7.35
		Solar	6.76	1.63	7.35
	60	Freeze	5.86	1.10	5.56
		Air	5.63	0.81	5.09
		Solar	5.71	0.78	5.22

237

TABLE 2 (continued)

Storage Time (month)	Storage Temperature °C	Drying Method	pH	Anthocyanin Degradation A520/A422	Moisture Content % wet basis
9	21	Freeze	6.31	2.14	11.73
		Air	6.22	1.54	11.12
		Solar	6.09	1.78	11.12
	38	Freeze	6.08	1.83	7.91
		Air	6.03	1.30	7.38
		Solar	6.08	1.45	7.28
	60	Freeze	5.72	0.96	5.36
		Air	5.50	0.75	4.80
		Solar	5.40	0.91	5.57
10	21	Freeze	6.45	2.10	10.85
		Air	6.40	1.51	10.79
		Solar	6.44	1.79	10.84
	38	Freeze	6.15	1.73	7.05
		Air	6.11	1.29	7.02
		Solar	6.24	1.41	6.95
	60	Freeze	5.79	0.91	5.08
		Air	5.61	0.74	4.50
		Solar	5.80	0.52	5.35

TABLE 2 (continued)

Storage Time (month)	Storage Temperature °C	Drying Method	pH	Anthocyanin Degradation A520/A422	Moisture Content % wet basis
11	21	Freeze	6.65	2.07	10.96
		Air	6.55	1.44	10.56
		Solar	6.60	1.68	10.58
	38	Freeze	6.39	1.68	6.13
		Air	6.33	1.26	6.31
		Solar	6.43	1.34	6.73
	60	Freeze	5.88	0.91	3.57
		Air	5.59	0.72	3.97
		Solar	5.43	0.65	4.27
12	21	Freeze	6.40	1.80	10.35
		Air	6.35	1.29	10.74
		Solar	6.39	1.45	10.56
	38	Freeze	6.20	1.47	6.61
		Air	6.13	1.12	6.52
		Solar	6.20	1.18	7.41
	60	Freeze	5.81	0.63	5.00
		Air	5.58	0.63	4.45
		Solar	5.40	0.45	4.50

[a] Results are average measurements of duplicate samples.

239

TABLE 3
Comparison of Taro Catalase Activity Changes in 6-Month Period[a]

Time (months)	Processing Methods	Storage Temperature (°C)		
		21	38	60
0	Raw	2.600	2.600	2.600
	Freeze drying	0.563	0.563	0.563
	Air drying	0.274	0.274	0.274
	Solar drying	1.313	1.313	1.313
6	Freeze drying	0.162	0.030	0
	Air drying	0.015	0	0
	Solar drying	0.307	0.130	0

[a] The unit is catalase unit/g dried taro flour.

Moisture Change

Tables 1 and 2 show the moisture contents of three types of dried slice and flour samples immediately after drying and storage under three different conditions for 12 months. The higher moisture contents of the air dried and solar dried samples were probably due to some case hardening, thus hindering complete drying. Moisture contents of all of the samples increased with time, except for the air dried and solar dried samples stored at 60° C.

These samples were kept in polyethylene bags without vacuum, partly to test the adequacy of the package and partly to be economical. As in the case of many flours and starches, the equilibrium moisture contents might be reached in the range of 8 to 18 percent, which was not yet the case here.

Eating Quality

Table 4 shows the results of the taste panel tests of dried taro slice samples stored up to 15 months. Significance test by ANOVA was made between samples dried by three different methods but stored at the same temperature. Numbers jointly underlined indicate no significant differences between those samples. Otherwise they are significantly different at $p = 0.05$.

As can be seen from the values of these average scores, none of the dried samples received scores higher than 4.7 before storage and 4.0 after

TABLE 4
Average Scores on Eating Quality of Various Dried Taro Slices
Stored at 3 Temperatures for 0 to 15 Months[a,b]

Storage Time (month)	Storage Temperature °C	Drying Methods			
		Control	Freeze	Air	Solar
		Score on Eating Quality			
0	–	5.87	4.67	3.79	4.17
1	21	5.04	4.02	3.78	3.45
	38	5.15	3.73	3.64	3.89
	60	5.75	4.31	3.92	3.92
2	21	5.59	4.32	3.41	3.80
	38	5.45	3.95	3.91	3.79
	60	5.79	3.42	3.11	2.89
3	21	5.49	4.45	3.99	4.13
	38	5.66	4.61	4.31	4.21
	60	5.05	3.33	3.04	2.84
4	21	5.57	4.32	4.27	3.68
	38	5.72	3.60	3.57	3.41
	60	4.95	3.74	3.10	3.51
5	21	5.39	4.03	3.41	3.64
	38	5.07	3.67	3.38	3.42
	60	5.84	3.86	3.50	3.35
6	21	4.81	4.07	4.20	4.10
	38	4.93	4.25	3.62	3.69
	60	6.33	3.30	3.68	3.39

TABLE 4 (continued)

Storage Time (month)	Storage Temperature °C	Drying Methods			
		Control	Freeze	Air	Solar
		Score on Eating Quality			
7	21	5.76	4.52	3.84	4.14
	38	6.14	4.55	4.02	4.04
	60	5.82	3.96	3.74	3.47
8	21	6.27	4.27	3.96	3.98
	38	6.36	4.40	4.16	3.96
	60	5.97	3.64	3.33	3.53
9	21	5.68	4.42	4.40	3.98
	38	6.07	4.51	3.75	3.79
	60	5.97	3.60	3.00	3.24
10	21	6.22	4.39	3.86	4.11
	38	5.86	4.42	3.64	3.94
	60	6.05	2.77	3.17	2.77
11	21	6.29	4.50	4.12	3.95
	38	5.93	4.32	3.76	3.70
	60	5.80	3.72	3.03	2.97
12	21	6.01	3.86	3.55	3.64
	38	5.98	4.32	3.52	3.40
	60	5.67	3.59	3.43	2.71
13	21	5.90	4.41	3.92	4.00
	38	5.95	4.20	4.02	3.71
	60	-	-	-	-

TABLE 4 (continued)

Storage Time (month)	Storage Temperature °C	Drying Methods			
		Control	Freeze	Air	Solar
		Score on Eating Quality			
	21	5.77	4.23	3.90	3.77
14	38	5.72	4.21	3.72	4.00
	60	-	-	-	-
	21	6.24	3.55	3.76	3.60
15	38	6.14	4.07	3.08	3.59
	60	-	-	-	-

[a] Results are average scores of triplicate runs on a hedonic scale of 7, where 7 = like very much; 4 = neither like nor dislike; 1 = dislike very much.

[b] Numbers joined by underlines indicate samples not significantly different from each other; those not joined by underlines are significantly different at $p = 0.05$, based on analysis of variance.

storage out of a possible maximum of 7.0 The control samples received scores only between 5.0 and 6.2, reflecting the problems of conducting taste panels on this type of product.

Dehydrated taro slices represent an unusual "prototype" of processed carbyhydrate food in that it was never prepared in this manner before. In preparing the samples and conducting the taste panel, several problems were encountered: (1) the relatively bland taste of the rehydrated, cooked slices; (2) the possibility of acridity remaining in the slices if not cooked completely; (c) the change in texture due to drying and storage, resulting also in some loss of flavor and aroma.

Other methods of cooking such as baking, pan frying and adding salt were considered but ruled out, mainly because of the difficulty in maintaining uniformity in samples to be served in terms of degree of "cooking."

The results of taste panel tests of dried taro flour samples stored up to 12 months are shown in Table 5. These samples were served as in paste

TABLE 5
Average Scores on Eating Quality of Various Dried Taro Flour
Stored at 4 Temperatures for 1 to 12 Months[a,b]

Storage Time (month)	Drying Method	Storage Temperature (°C)			
		2	21	38	60
		Score on Eating Quality			
1	Freeze	5.02	5.08	4.19	4.09
	Air	4.72	4.22	4.64	3.95
	Solar	4.09	4.49	4.47	3.56
2	Freeze	4.85	4.95	4.50	4.54
	Air	4.68	4.40	4.60	4.03
	Solar	4.35	4.86	4.92	3.81
3	Freeze	5.11	5.20	5.03	4.48
	Air	4.76	4.95	4.42	3.67
	Solar	4.28	4.67	4.69	3.72
4	Freeze	5.08	5.00	4.98	3.77
	Air	4.03	3.97	4.15	3.59
	Solar	4.45	4.48	4.42	3.19
5	Freeze	5.42	5.26	5.27	4.56
	Air	5.05	4.77	4.53	3.99
	Solar	5.12	5.39	4.88	3.91
6	Freeze	4.97	5.25	5.44	4.48
	Air	4.91	5.43	5.25	3.97
	Solar	5.18	5.25	4.98	4.16
7	Freeze	5.03	4.85	4.91	3.84
	Air	4.74	5.07	4.84	3.50
	Solar	5.52	5.45	5.00	3.50
8	Freeze	5.20	5.23	4.89	3.73
	Air	4.99	4.70	5.05	3.26
	Solar	5.23	5.10	5.19	3.55

TABLE 5 (continued)

Storage Time (month)	Drying Method	Storage Temperature (°C)			
		2	21	38	60
		Score on Eating Quality			
9	Freeze	5.63	5.57	5.36	3.77
	Air	5.15	5.09	4.82	3.21
	Solar	5.18	5.21	5.18	3.24
10	Freeze	5.27	5.24	5.29	3.03
	Air	4.61	4.87	4.70	3.19
	Solar	5.37	5.00	5.00	3.30
11	Freeze	5.31	5.56	5.36	3.05
	Air	4.90	5.08	5.08	2.92
	Solar	4.97	5.41	5.30	3.20
12	Freeze	5.13	5.16	5.20	2.92
	Air	5.28	5.21	5.28	2.80
	Solar	4.95	5.26	5.15	2.59

[a] Results are average scores of triplicate runs on a hedonic scale of 7, where 7 = like very much; 4 = neither like nor dislike; 1 = dislike very much.

[b] Numbers joined by underlines indicate samples not significantly different from each other; those not joined by underlines are significantly different at p = 0.05, based on analysis of variance.

form containing 10 percent flour and 5 percent sucrose (wet weight basis) and cooked to full gelatinization. The control was kept at 2°C. These comparisons were made on the effect of storage temperature on eating quality, rather than on the drying methods. As can be seen from the scoring, the quality deteriorated when samples were kept at 60° C, but those stored at 38°C and below exhibited no changes.

Gelatinization Temperature

The gelatinization temperature ranges of various dried taro samples (fresh versus six months storage) were measured and compared with that of a freshly ground raw taro corm. The range of gelatinization temperature of various samples started at 64° to 67°C, and completed at 71.5° to 73°C. In the ground raw corm, the gelatinization temperature range was 65.5° to 71.5°C. There seemed to be no significant differences among these samples in their gelatinization temperatures.

The gelatinization temperatures of various starches and flours reported in the literature show rice varying from 55° to 75°C, depending on variety (Juliano, 1972), and wheat, varying from 52° to 64°C (D'Appolunia, et al., 1971). Taro starch falls within this broad range (55°-75°C) as measured in this study. Information on the effect of storage on the gelatinization characteristics of various starches and flours was not available.

Energy Value

The energy value of taro was not given in the recent edition of the USDA Handbook (Adams, 1975). In this handbook, dehydrated sweet potato flake was listed as having 3.90 Cal/g, dehydrated potato flake, 3.78 Cal/g, whole wheat flour, 3.79 Cal/g, and cooked rice, 3.97 Cal/g (dry basis). The energy value of six taro samples measured was quite comparable to the energy values of other cereals. On a moisture-free basis, their energy values are:

Sample	Gross energy, Cal/g	Sample	Gross energy, Cal/g
Solar dried, raw	4.03[a]	Freeze dried poi	3.81[a]
Freeze dried, raw	3.94	Freeze dried, cooked	3.77
Air dried, raw	4.00	Air dried, cooked	3.99

[a]Significance tests on these values have not been made due to limited numbers of samples analyzed.

In addition to the various problems in dehydration, processing and quality evaluation, the problem of the acridity factor in taro deserved careful study. This factor must be eliminated by physical, chemical, and/or thermal means before a taro product can be consumed.

SUMMARY AND CONCLUSIONS

Experimental studies of preparing and storing dehydrated taro slices and flour in polyethylene bags at 21°C, 38°C, and 60°C showed that these products underwent small changes in acridity, some degradation of pigment (mainly anthocyanins), some increases in moisture content, and rather large decreases in catalase activity. At the end of 12 months, taste quality of most of the dried samples stored at 21°C and 38°C scored appreciably lower than the control on a 7-point hedonic scale. Those stored at 60°C received a still lower score. Energy values were about the same as those of rice, and slightly higher than wheat. Gelatinization temperature range was in the upper range of rice (66° to 72°C).

The results indicated that these intermediate products were stable at 38°C for 1 year or more. The approach of using these products for remanufacturing into usable forms such as rice or noodles seems feasible.

NOTES

1. Journal Series No. 2280 of the Hawaii Agricultural Experiment Station, University of Hawaii, Honolulu, Hawaii 96822, U.S.A.

REFERENCES

Adams, C.F. 1975. Nutritive Values of American Foods in Common Units. Agr. Handbook No. 456, ARS-USDA, Washington, D.C.

D'appolunia, B.L., Gilles, K.A., Osman, F.M., and Pomeranz, Y. 1971. Carbohydrates. In Wheat: Chemistry and Technology, ed. Y. Pomeranz. St. Paul: Am. Assoc. Cereal Chemists, Inc.

de la Pena, R.S. 1970. The edible aroids in the Asian-Pacific area. Second Intern. Symp. Trop. Root Crops, Honolulu, Hawaii.

Engel, Julien, ed. 1975. Underexploited Tropical Plants with Promising Economic Value. Ad Hoc

Panel Report of the Advisory Committee on Technology Innovation, National Academy of Science, Washington, D.C.

Hans, L. 1963. Catalase. In Methods of Enzymatic Analysis, ed. H.U. Bergmeyer. New York: Academic Press.

Juliano, B.O. 1972. The rice caryopsis and its composition. In Rice: Chemistry and Technology, ed. D.F. Houston. St. Paul: Am. Assoc. Cereal Chemists, Inc.

Plucknett, D.L., de la Pena, R.S., and Obrero, F. 1970. Taro (Colocasia esculenta L. Schott): a review. Field Crops Abstr.

Sastry, L.V.L. and Tischer, R.G. 1952. Behavior of the anthocyanin pigments in Concord grapes during heat processing and storage. Food Technol. 6:82.

Schoch, T.J. and Mayfield, E.C. 1956. Microscopic examination of modified starches. Anal. Chem. 28:382.

Whitney, L.D., Bowers, F.A.I., and Takahashi, M. 1939. Taro Varieties in Hawaii. Hawaii Agr. Expt. Sta. Bull. No. 84, University of Hawaii, Honolulu, Hawaii.

James H. Moy, N. T. S. Wang, and T. O. M. Nakayama: Department of Food Science and Technology, University of Hawaii, Honolulu, Hawaii

4
Starch and Flour from Taro

C. C. Tu
W. K. Nip
T. O. M. Nakayama

ABSTRACT

Taro contains 70 to 80 percent starch on a dry
weight basis. Taro starch granules are unusually
small, with an average diameter of 2 to 5 microns.
A simple method was developed to separate 40 percent of the taro starch. Taro corms of different
varieties used in the preparation of starch were
ground in a meat grinder and further reduced in
size by wet milling using a Fitz Mill equipped with
a 0.069 cm perforated screen. The thin slurry was
filtered under pressure through nylon filter bags.
The filtrate was centrifuged to separate the starch
which was subsequently dried at 45 C. The taro
starch granules were similar to other tuber starches in forming clear, stringy pastes somewhat resistant to enzyme action which indicated that taro
starch can be utilized for both human consumption
and industrial use.
     The residue in the nylon bags after drying was
ground to flour which contained 60 to 70 percent
starch. Products made into noodles, flakes,
cookies, muffins and biscuits were tested by mixing
taro flour with wheat flour. Results indicated
that taro flour can be substituted for part of the
wheat flour required in the pastry preparation.

INTRODUCTION

Taro flour from cooked taro was available about 40 years ago. Since that time taro flour has not been produced in Hawaii.

Taro (<u>Colocasia esculenta</u> (L.) Schott) is one of the important crops in tropical and subtropical areas. Approximately 300 varieties are known (USDA, 1910). Two varieties, Lehua Maoli (wetland red taro) and Bun-long (dryland taro) are commonly cultivated in Hawaii. The wetland taro is used for the production of poi, a native Hawaiian food.

Poi is made by grinding the cooked taro corms, mixing with water and allowing it to ferment (Allen and Allen, 1933). The raw wetland taro is irritating to mucous membranes (Black, 1918; Safford, 1905) and cooking for an hour under pressure is necessary to destroy the acridity. Poi contains about 30 percent solids. It has a shelf life of several days at room temperature. At present, taro is used mainly for poi production. Its use is therefore limited. Taro products which are stable and readily utilized are sought.

Taro contains 70 to 80 percent starch (Payne et al., 1938). The granules of taro starch are very small, 2 to 5 microns in diameter (USDA, 1910), which is approximately one-tenth the size of potato starch. Because of its extremely small size, the starch cannot be easily separated by settling from a slurry.

Taro starch has never been commercially produced. This work was undertaken to develop a simple process for the production of taro starch and flour on a small pilot plant scale.

Because of its low acridity and dryland cultivation, the Bun-long variety was chosen for the production of starch and flour.

TARO ROOT PROCESSING

Taro Flour Production

<u>Peeling and grinding</u>. Peeling of taro corms was done by hand or by an abrasive peeler. The peeler is good for large corms. Loss is larger for the small corms. Peeled corms were cut into small cubes and ground in a coarse meat grinder equipped with a 0.318 cm perforated die. Granules from the grinder were dried by solar drying or oven drying.

The loss of peeling on taro corms is about 25 percent to 30 percent.

Dry milling. 4540 g of the oven-dried, ground corms from each batch were reduced in size by one pass through a Fitz Mill equipped with a 0.160 cm perforated screen and further reduced in size by using a 0.069 cm perforated screen. The fine powder, taro flour A, was collected and weighed. The yield obtained was 4403 g (97 percent).

Taro Starch and Flour B

Wet Milling. 13,620 g of coarsely ground taro corms were suspended in water with a ratio of 1:4 or 1:5 and milled on a Fitz Mill equipped with a 0.160 cm perforated screen.

Filtering. The slurry was filtered by using a Harris bag press (Cascade Metals Corp., San Jose, California). The press consisted of a horizontal frame in which was mounted a series of movable stainless steel plates. When these plates were separated, the spaces between them accommodated nylon bags (38 x 41 cm) containing the slurry from which starch was to be extracted. This was carried out by filling the slurry into nylon bags. The bags were held in position by stainless steel bars and clamps. When the bags were compressed, the extracted starch dropped into a pan located below the bags. The bags were slowly compressed by means of a motor-driven hydraulic cylinder. The compression pressure was about 52.7 kg/cm^2. Nylon press bags for the Harris bag press are available from Cascade Metal Corp., but no specification of bags is given.

Centrifuging. The filtrate (suspension) from the press was centrifuged on a Sharples Super-Centrifuge at 23,000 rpm corresponding to 13,000 xg. Starch granules were collected on the inner surface of the centrifuge cylinder. The effluent containing the water-soluble non-starch saccharides and other water-soluble matter was evaporated to dryness under vacuum. Starch was removed from the centrifuging cylinder using a large spatula and dried at 45°C. The yield of starch obtained was 1907 g (14 percent on ground corms). The residue inside of the bags was removed, dried at 45°C and finely ground in a Fitz Mill with a 0.069 cm perforated screen to fine powder, named taro flour B. The yield obtained was 3180 g (23 percent).

Another batch of 13,620 g of peeled corms was ground, wet-milled and filtered. The filtrate (suspension) containing starch was centrifuged as described above. About 1820 g (13 percent) of starch was obtained. The residue in the bags was mixed with water and ground on a Fitz Mill with a 0.069 cm perforated screen. The thin slurry was transferred into the bags, filtered and centrifuged to separate more of the starch from the residue. The starch was removed from the cylinder with a large spatula, dried at 45°C and weighed. The total amount of starch obtained was 2270 g (17 percent).

The residue in the bags was dried at 45°C and ground to fine powder, named taro flour C.

The processing of taro is summarized in Figures 1 and 2. More than 20 batches were run. The yield obtained from each run was about the same.

New Taro Products from Taro Flour A

Ten grams of taro flour A were mixed with the proper amount of hot water (35 to 45 percent) to make a dough. Synthetic rice and noodles were made from the dough by hand; the products were dried at 45°C. Rice and noodles were also made by incorporating up to 35 percent of soy or mung bean flours. The new products are intended as substitutes for rice and noodles where they are not commonly available.

Use of Taro Flour B in Pastries

Sugar cookies, peanut butter cookies, chocolate chip cookies, brownies, oatmeal cookies, coconut macaroons, sweet muffins, baking powder biscuits, fruit rolls and bread were made with 100 percent taro flour B. A few pastries were made with 50 percent each of wheat flour and taro flour B. The recipes of each product were those in common use (Akahoshi, 1977).

DISCUSSION

In considering potential products, the inherent virtues of this corm crop should be exploited.

Taro and its products are very low in proteins and entirely absent of gluten-like substances. Starch is the main component of taro. Because of

252

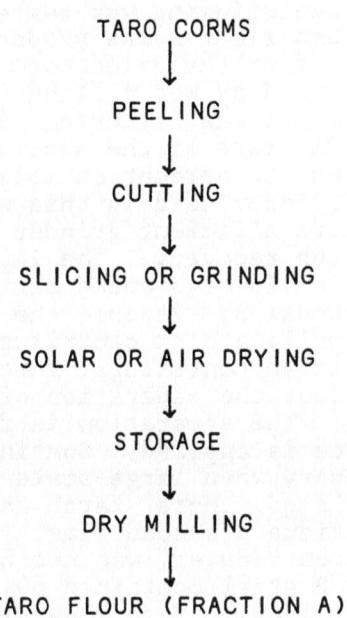

TARO CORMS

↓

PEELING

↓

CUTTING

↓

SLICING OR GRINDING

↓

SOLAR OR AIR DRYING

↓

STORAGE

↓

DRY MILLING

↓

TARO FLOUR (FRACTION A)

Figure 1. Taro flour production.

the low content of proteins in taro, its products
are practically non-sensitive to allergenic patients. Taro products are ideal for hospital diets,
particularly for those who are hypersensitive to
ordinary starch food. This suggests that taro
starch can be a component of food for babies (Glaser
et al., 1964). Its granules are similar to other
tuber starches in forming clear, stringy pastes
somewhat resistant to enzyme action (Goering and
Dettaas, 1972). Minerals and vitamins (Miller,
1931) and an alkaline ash are also present in taro.

Starch and Flour

Taro starch has never been available in Hawaii.
To prepare taro foods other than poi, flour and
starch must be available. A simple process for taro
flour and starch has been developed. The whole
process, including washing, peeling, milling,

filtering, and centrifuging was tested and is shown in Figure 2. Taro flour A was produced by dry milling directly from the dried corm granules. Taro flour B was obtained by wet milling from the residue of a slurry after it was filtered. Starch was isolated from the filtrate of the slurry. The recovery of starch is about 40 percent on solids. This suggests that the grinder used in this work is not efficient. A more efficient grinder could be used to increase starch recovery. The important step in starch preparation is high-speed centrifugation of a thin starch suspension. Because the starch granules are extremely small, a high centrifugal force must be used. A relative centrifugal force of 13,000 xg is needed to effect the separation of starch from a taro suspension. The separation is further improved if a higher force is applied. Continuous centrifugation is necessary when large-scale starch production is contemplated. More starch was obtained by grinding the residue a second time. However, the increase of starch recovery was not high.

Taro flour B still contained 60 percent starch (Tu and Okamoto, 1963). More minerals and fiber are present in taro flour B. Results demonstrated that good pastries were made from taro flour B (Akahoshi, 1977). This suggests that taro flour B is useful for baking purposes. Taro flour B can be considered a main product rather than a by-product in the processing of starch production. Taro flour C should also be a suitable material for baking purposes, as the flour still contains 50 percent or more starch. No pastries were made from taro flour C.

New Taro Food

New taro products--noodles and synthetic rice-- were made. Both products can incorporate up to 35 percent of other plant products such as soy, peanut, or mung bean flour. The texture of the flour with the added products does not show a noticeable difference, but flavor and nutritional value are enhanced. They can be eaten as a daily food. The new products can be readily prepared for eating by immersion in boiling water for a few minutes.

Poi is sticky and viscous. It is easily spoiled. The new products are rigid, non-hydroscopic, and non-sticky. Because of the superior quality of these new taro products, taro in the form of rice or noodles may become a popular food. This will increase the use and production of taro.

254

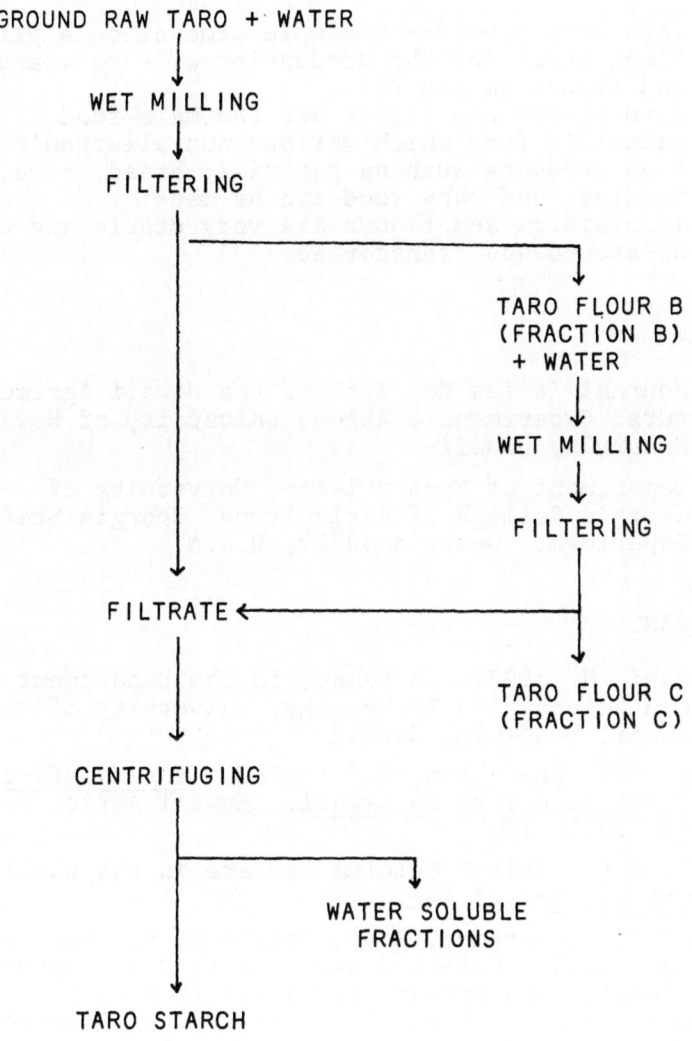

GROUND RAW TARO + WATER

WET MILLING

FILTERING

TARO FLOUR B
(FRACTION B)
+ WATER

WET MILLING

FILTERING

FILTRATE

TARO FLOUR C
(FRACTION C)

CENTRIFUGING

WATER SOLUBLE
FRACTIONS

TARO STARCH

Figure 2.   Taro starch and flour production.

SUMMARY

1. This work provides a simple process on a pilot plant scale for the production of taro starch and flours (A and B).
2. Taro starch and flours are the main food materials from which various non-allergenic food products such as pastries, bread, rice, noodles, and baby food can be made.
3. Taro starch and flours are very stable and can be stored and transported.

NOTES

1. Journal Series No. 2259 of the Hawaii Agricultural Experiment Station, University of Hawaii, Honolulu, Hawaii.

2. Department of Food Science, University of Georgia College of Agriculture, Georgia Station, Experiment, Georgia 30212, U.S.A.

REFERENCES

Akahoshi, N. 1977. A Report to the Department of Food Science and Technology, University of Hawaii, Honolulu, Hawaii.

Allen, O.N. and Allen, E.K. 1933. The Manufacture of Poi from Taro in Hawaii. Hawaii Agric. Expt. Sta. Bull. 70.

Black, O.F. 1918. Calcium oxalate in the dasheen. Amer. J. Bot. 5:447.

Glaser, J., Lawrence, R.A., Harrison, A., and Ball, M.R. 1964. Poi--its use as a food for normal, allergic, and potentially allergic infants. Department of Pediatrics, University of Rochester School of Medicine and Dentistry, Rochester, N.Y.

Goering, K.J. and Dettaas, B. 1972. New Starches. VII. Properties of the small granule starch from Colocasia esculenta. Cereal Chem. 49:712.

Miller, C.D. 1931. Food Values of Poi, Taro and Limu. Bernice P. Bishop Mus. Bull. 37:25.

Payne, J.H., Ley, G.J., and Akau, G. 1938. Processing and Chemical Investigations of Taro. Hawaii Agric. Expt. Sta. Bull. 86.

Safford, W.E. 1905. The useful plants of the island of Guam. U.S. National Mus. Contrib., U.S. National Herbarium 9:416.

Tu, C.C. and Okamoto, R.H. 1963. Determination of starch in sucrose crystal. J. Ag. Food Chem. 11:331.

USDA. 1910. Promising Root Crops for the South: Yautias, Taro and Dasheens. Bur. Plant. Ind. Bull. 164.

C. C. Tu, W. K. Nip, and T. O. M. Nakayama: Department of Food Science and Technology, University of Hawaii, Honolulu, Hawaii 96822, U.S.A.

5

Processing of Root Crop Products, Especially Taro and Cassava, in Western Samoa

Clive Pedrana

ABSTRACT

Taro in Western Samoa is of major importance to the economy both as a stable food and more recently as an export. The development of a stable export market has been difficult due mainly to instability in the domestic market caused by periods of acute shortages when prices rise rapidly and periods of oversupply when prices are depressed below the costs of production.

One method to stabilize taro production is the development of stable processed forms which can be stored for considerable periods of time. The Food Processing Laboratory has been conducting research and development of taro flour-based products as well as frozen taro products.

Production and distribution of a taro flour-based baby weaning food has been underway for some time, and the program is now being extended to the villages. Preliminary baking trials using up to 65 percent taro flour were quite encouraging, but a more practical level of substitution would be at the 10 to 20 percent level.

Combinations of frozen taro and coconut cream have also been developed, but still require market evaluation.

While the technology for development of these products is well established, the costs of taro production in Western Samoa indicate that the utilization of taro flour in bread will require a substantial reduction in the price of taro. Production of baby weaning food and frozen taro products appears to be a more promising avenue for development.

INTRODUCTION

Investigations into the processing of taro in Western Samoa have developed mainly in response to the stimulus of widely fluctuating market prices for taro and, to a lesser extent, cassava.

Research has centered on the development of products based either solely or in part on taro flour or cooked frozen taro and taro slices. In the case of cassava, research has centered on the development of cassava chips as a major component in the livestock feed industry in the country. Small quantities of cassava starch have also been produced and marketed in New Zealand.

The products which have been produced at the Food Processing Laboratory at Alafua are: taro flour, taro-based baby weaning food, taro-based bread, frozen taro slices with coconut cream, frozen baked taro, frozen baked taro slices, deep-fried taro chips, and cassava chips for animal feed. Figure 1 outlines the procedure involved in processing cassava, taro, and breadfruit.

PROCESSING TARO

Taro corms are trimmed, peeled, sliced transversely into slices approximately 0.5 mm thick, then dried in a hot air oven for 12 hours at 145°F (48°C). The chips are then coarse ground through a Fitzmill and then fine ground through a pin mill. The yield of taro flour is approximately 25 percent. The flour produced has been stored for over 12 months in plastic-lined 44-gallon drums, and no problems have been experienced due to either mold deterioration or weevil infestation. The resistance to weevil infestation has been attributed by workers at the University of Hawaii to the presence of calcium oxalate raphides present in the flour. Unfortunately, we have not been able to conduct a full investigation into this particular aspect.

Taro flour produced by this method is the startling product for two further products, baby weaning food and taro-based bread.

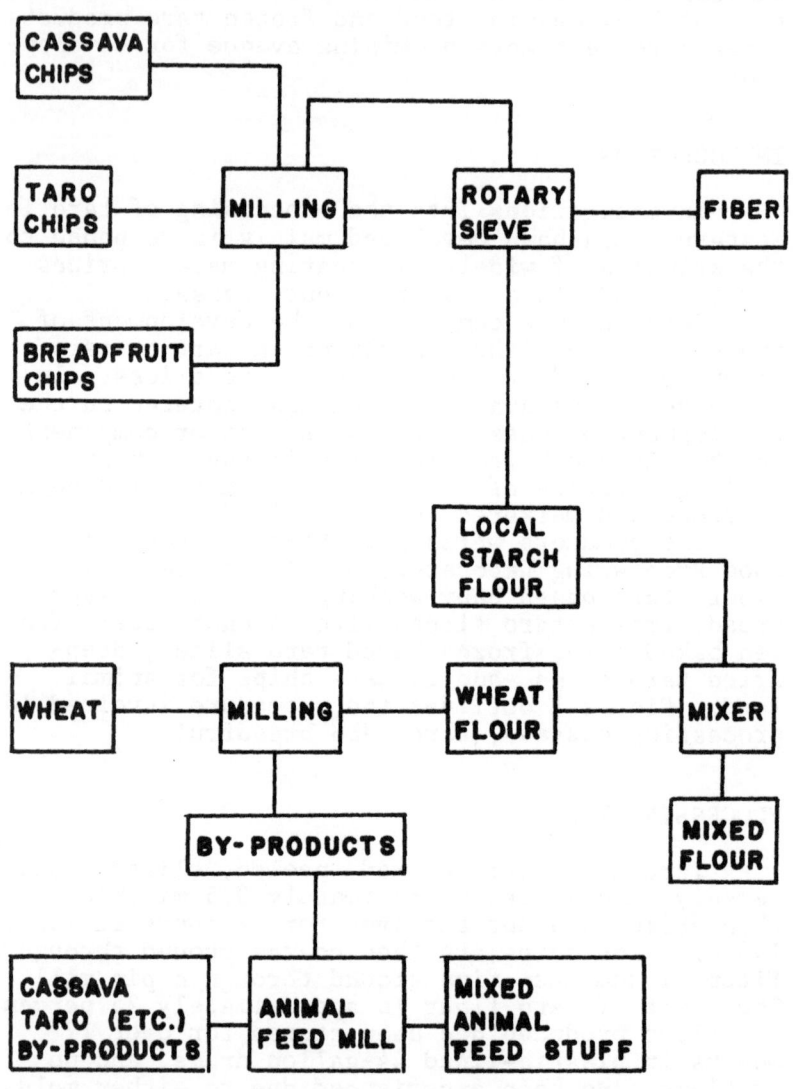

Figure 1. Proposed Process Flow Chart for
Flour Mill in Western Samoa.

A considerable amount of research has been conducted by the University of Hawaii (Payne et al., 1941) on producing taro flour using a process which involved boiling the taro for extended periods of time. This not only made the taro slimy, but also added further water which then had to be removed, thereby adding to the cost of drying.

We have not followed this procedure because the two products in which the taro flour is used are both heated prior to consumption. We have not experienced any irritation in our products due to the raphides of calcium oxalate.

Taro-Based Baby Weaning Food

Development of the baby weaning food was initiated by R. Mawson, previously Food Technologist at the Food Processing Laboratory, and was completed by the author with financial and technical assistance from both Dr. A. Raoult of the South Pacific Commission and Dr. B. Standal from the Department of Food and Nutritional Sciences of the University of Hawaii.

The original formulations of both Mawson and Standal have been modified to suit local availability of raw materials. The two formulations being used in our field trials are set out in Table 1.

TABLE 1
Formulations of Taro-Based Baby Weaning Food

	Formulation I (%)	Formulation II (%)
Taro Flour	17	34
Rice Flour	34	17
Sugar	4	4
Skim Milk Powder	30	30
Coconut Cream (DB)	15	15

The baby weaning food is prepared by mixing the dry ingredients in a Halde mixer and adding in the wet coconut cream. This produces a slightly granular product which is dried to below 7 percent moisture at 140°F (45.5°C) overnight and then placed into laminated aluminum foil packets which are sealed and stored prior to delivery to the Health Department.

The Health Department is currently using this weaning food at the National Hospital as a rehabilitative food. Also, it is being distributed by district nurses to children in villages which have been diagnosed as suffering from sub-clinical malnutrition. Nutritionists from the Health Department have been monitoring the performance of children on the baby weaning food, but these data were not yet available. However, preliminary observations suggest that acceptance by the mothers and growth rates of the children are satisfactory.

One of the major aims in the development of a baby weaning food is that it should, as much as possible, reinforce the development of a child's taste for traditional foods. We consider that the inclusion of both taro flour and coconut cream is consistent with this aim.

In development of the baby weaning food, consideration was also given to how the weaning food was utilized in the village by the mothers. In this regard, one immediate problem was the supply of good quality water for preparing the food. To overcome this we developed a weaning food similar in preparation to porridge which requires heating to a boil before feeding to the infant. In this way the problem of potable water has been overcome, as well as the problem of calcium oxalate irritation.

Future development of the baby weaning food includes expansion of the distribution program when the field trials are completed and development of a commercial pack for sale through normal retail channels.

Taro-Based Bread

The Food Processing Laboratory has investigated the utilization of taro, cassava and breadfruit (Artocarpus utilissima) flour in the local bakery industries. At present we are investigating the establishment of a flour mill which would utilize

262

between 10 and 20 percent of locally produced flours to produce standard bread flour. Three alternative approaches are being investigated:

1. Mixing of imported flour with traditional ingredients and using chemical modifying agents (such as calcium-2 steroyl lactylate) to bind up the starch. This is the approach adopted by the Tropical Products Institute in London.

2. Using imported flour with the addition of local starches and gluten to yield a suitable bread flour mix.

3. Using a blending, sifting plant which could adjust the protein content by physical means.

Preliminary baking trials have been conducted on flours produced by the first two approaches, with local flours being used for up to 63 percent of the total flour. However, with the higher utilization of local ingredients, loaf volume and texture were altered substantially, and this would be difficult to introduce into commercial practice. Inclusion of local flour at 10 to 20 percent does not appear to be a major technical problem, but the high cost of production of taro flour in particular suggests that it will not be introduced commercially for some time. It is interesting to note that, of the three local flours used, the breadfruit-based bread was considerably different in texture from the other two, the breadfruit bread being more of a whole meal bread. Its general acceptance by the public, moreover, was better than the standard wheat breads on the market. There appears to be little research conducted on breadfruit to date. This would appear to be an area where further work is warranted, especially since breadfruit is a tree crop and is adapted to different conditions than the root crops.

Frozen Taro Slices with Coconut Cream

Like many Polynesian islands, Western Samoa has a substantial population which has migrated either temporarily or permanently to New Zealand, Australia, Hawaii, and the mainland United States. Development of taro products which could be exported to these centers offers real market prospects as traditional dishes.

The Food Processing Laboratory has been investigating traditional taro dishes. The process is simply to boil the taro slices until completely cooked, cover with coconut cream, and then freeze to -20°C. Our trials with these products to date have yet to be market tested.

Difficulties have been experienced with "breading up" of taro due to the long cooking times; therefore we have commenced trials on baking taro in the traditional Samoan umu (firepit which uses heated stones).

Frozen Baked Taro and Taro Slices

Development of this product is similar to the frozen taro slices. However, in this case taro is cooked in the traditional Samoan umu and then either sliced and packed into styrofoam trays and frozen to -20°C, or in the case of small taros, packed whole into polyethylene bags and tied with a metal clip.

We are planning to market test these two products in New Zealand through the supermarket chains.

Deep-Fried Taro Slices

The technology of deep-fried taro chips is the same as that for potato chips; however, we have experienced some problems with residual irritation due to calcium oxalate which is apparently not destroyed completely by quick frying.

CASSAVA CHIPS

The processing of cassava chips is well known throughout the world. Our research has centered mainly on the utilization of comparatively high levels (45 percent) of cassava meal in livestock feeds.

At present, research is being conducted into the HCN levels of leaves and tubers of 5 varieties to determine varieties with low HCN.

A number of pig-feeding trials have been conducted to date using the following formulation:

Cassava meal	45%
Cassava leaf meal	17%
Copra meal	30%
Fish meal	7.5%
Salt	0.5%
Vitamin mix	0.2%
DL Methionine	0.2%

In earlier trials the cassava chips were steeped overnight in potassium metabisulphite solution and then drained and dried, but more recently we have switched to straight chipping and drying and then adding DL Methionine to overcome the HCN. However, rations using the first method of production of cassava chips proved superior to chips produced by the methionine method.

REFERENCE

Payne, J.H., Ley, G.L., and Akau, G. 1941. Processing and Chemical Investigations of Taro. Hawaii Agr. Expt. Sta. Bull. No. 86, Honolulu, Hawaii.

Clive Pedrana: Food Processing Laboratory, Department of Agriculture, Apia, Western Samoa

6
Drum-Dried Pineapple-Taro and Mango-Taro Flakes

W. K. Nip

ABSTRACT

Taro is an important food crop in the tropical and subtropical regions. Like the other root crops, taro can be drum-dried into flakes. In the processing of taro, sugar was added in the preparation of the slurry for drum-drying in order to induce thermoplasticity in the finished product. Addition of fruit puree to the cooked taro in the preparation of the slurry for drum-drying has the advantage of adding sugar and at the same time incorporating the fruit flavor into the finished product. This report presents (1) the operation criteria for preparing pineapple-taro and mango-taro flakes; (2) their properties; and (3) the acceptability of the finished products in various forms.

INTRODUCTION

Various root and tuber crops have been successfully drum-dried (Cording et al., 1957, Spadaro et al., 1967; Onayemi and Potter, 1974; Steele and Sammy, 1967; Rodriguez-Sosa and Gonzalez, 1977; Hsu and Chiou, 1971). In the preparation of the slurry for drum-drying, sugar was sometimes added to increase the thermoplasticity in the finished product (Hsu and Chiou, 1971; Spadaro et al., 1967; Hoover, 1973).

Applesauce (Lazar and Morgan, 1966), tomato concentrate (Henig and Mannheim, 1971), and banana (Adeva, Gopez, and Payumo, 1968) have been successfully drum-dried into the form of flakes. In the drum-drying of bananas, sugar and gelatinized corn-starch were added to the puree to adjust the total soluble solid content (Adeva, Gopez, and Payumo, 1968). Recently, addition of modified starch to fruit puree before drum-drying was suggested("Delicate Drying Process Minimizes Fruit and Vegetable Aroma, Color Changes," Food Product Development).

Information on drum-drying properties of combined root crop and fruit mixtures is very limited. Guava-taro and papaya-taro flakes have been developed recently (Nip, 1978). Addition of a fruit puree to taro in the preparation of a slurry for drum-drying has the advantage of incorporating the fruit flavor and color into the final product and at the same time increasing the soluble solids content of the mixture. This paper reports the processing criteria of drum-drying pineapple-taro and mango-taro, their properties, and acceptability of the finished products.

MATERIALS AND METHODS

Material

White taro (Hawaiian cv. "Apii"), pineapple puree and mango puree were used in the preparation of the mixture for drum-drying. White taro was obtained from the Honolulu Poi Company. Pineapple puree was prepared by homogenizing the flesh and core portions of commercial pineapple fruits. Mango puree was obtained by blending the edible portion of mango fruits. The purees were kept in frozen state until used.

Preparation of Mixtures for Drum-Drying

The procedures for preparing the mixture for drum-drying follows the method reported earlier (Nip, 1978). This consists of cooking taro at 121°C for 75 min, and homogenizing the diced taro with the fruit puree at ratios of 2:1 and 2:3 (taro:fruit) by weight in a silent cutter. The total solid content of the mixture was adjusted to about 20 percent with water, and the mixture strained through a 16 mesh

267

screen. The homogenized and strained fruit-taro mixture was then subjected to drum-drying.

Development of the Drum-Drying Procedure

Drum-drying was accomplished with a laboratory double drum-dryer (drum diameter 15 cm and width 19.5 cm) by proper adjustment of the space between the two drums, drum speed, and inlet and outlet steam pressure of the drum.

The Flaking and Packaging Processes

The films obtained by drum-drying the fruit-taro mixtures were collected in plastic bags and flaked by hand. The flakes were sieved through a No. 18 sieve (16 mesh) and packed in airtight bottles.

Bulk Density

Bulk density of the fruit-taro flakes was estimated by weighing 100 g of flakes and measuring the volume they occupied in a graduate cylinder after tapping. The ratio of weight per unit volume was calculated as the bulk density.

Moisture Content

Moisture content of the flakes was determined by drying the flakes overnight in a vacuum oven at 70°C and 23 mm Hg. The difference between the weight before and after drying was calculated as the moisture content.

Equilibrium Moisture Content

Equilibrium moisture content was estimated according to the method suggested by Strolle and Cording (1965). One-gram samples were placed in individual chambers saturated with lithium chloride, potassium acetate, potassium carbonate, ammonium nitrate and sodium chloride solutions with 17, 26, 47, 65, and 75 percent relative humidity at 22°C, respectively. Results were computed as equilibrium moisture content at 22°C, and percent moisture, monolayer moisture free basis (MFB).

Consistency of the Rehydrated Product

Five grams of flakes plus 5 g or 2 g of sugar

were reconstituted with 40 ml or 35 ml of cold water, respectively, by blending the mixture in a Waring blender for 1 min. Consistency of the rehydrated fruit-taro flakes was estimated by means of a Bostwick Consistometer. Results were recorded as the distance in centimeters the slurry will move in 30 seconds.

Taste Studies

Samples prepared for measurement of the consistency of the rehydrated product were also used for tasting studies. Samples were served to 10 taste panelists chosen from the faculty, staff and students of the Department of Food Science and Technology, University of Hawaii, for overall acceptance using a hedonic scale with 9 being the most acceptable, 1 being the least.

RESULTS AND DISCUSSION

Drum-Drying Procedures

By properly adjusting the drum speed, inlet and outlet pressures of the drums, and the space opening between the two drums, continuous films across the drum can be obtained. Operation criteria for drum-drying mango-taro and pineapple-taro flakes are presented in Table 1. From these data, it is obvious that those mixtures with a higher proportion of fruit puree required higher temperatures to drum-dry, as indicated by the inlet and outlet pressures of the drums. These operational figures were lower than those reported for other products (Onayemi and Potter, 1974; Lazar and Morgan, 1966; Adeva et al., 1968) but similar to other fruit-taro flakes (Nip, 1978).

Bulk Density

The bulk densities of the selected fruit-taro flakes are summarized in Table 2. Bulk densities for the 1:2 fruit-taro flakes are the same. However, the mango-taro (3:2) flakes had a higher bulk density than the other fruit-taro flakes (Nip, 1978). This may be due to the higher polysaccharide content of the mango puree as compared to the other fruit purees.

269

TABLE 1
Conditions Used to Drum-Dry Selected Fruit-Taro Mixtures

Composition of Mixture		Inlet Pressure (atm)	Outlet Pressure (atm)	Drum Speed (rpm)	Space Between Drums (mm)
Mango:taro	1:2	1.70	1.22	3	0.8
	3:2	3.20	2.86	3	0.8
Pineapple:taro	1:2	2.04	1.86	3	0.8
	3:2	2.72	2.18	3	0.8

TABLE 2
Bulk Density of Selected Fruit-Taro Flakes

Composition of Mixture		Bulk Density, g/ml Dry Weight Basis
Mango:taro	1:2	0.37
	3:2	0.45
Pineapple:taro	1:2	0.37
	3:2	0.39

Moisture Content

Table 3 presents the equilibrium moisture content (percent MFB) of the mango-taro and pineapple-taro flakes under various percent relative humidity conditions. The equilibrium moisture contents were similar for flakes made from the same fruit-taro ratio. These data also provide information on packaging and storage of these flakes. Based on these results, the percent moisture monolayer (MFB) were calculated. The percent moisture content for calculated monolayer (MFB) values and the values from the wet basis were similar for the fruit-taro (1:2) flakes (Table 4). Large variations were observed for the 3:2 fruit-taro flakes. Based on these observations, it is clear that the percent moisture

TABLE 3
Equilibrium Moisture Content (% MFB) of Selected Fruit-Taro
Flakes at Various Relative Humidities at 22°C

Composition of Mixture		Percent Relative Humidity				
		17	26	47	65	75
Mango:taro	1:2	6.17	7.20	9.68	13.18	17.97
	3:2	5.29	6.73	10.25	15.02	20.58
Pineapple:taro	1:2	5.56	6.87	9.40	12.94	17.96
	3:2	5.45	7.11	10.84	15.93	21.95

TABLE 4
Comparison of Percent Moisture, Monolayer (MFB) and Wet Basis,
for Optimum Stability of Fruit-Taro Flakes

Composition of Mixture		Percent Moisture	
		Monolayer (MFB)	Wet Basis
Mango:taro	1:2	6.67	6.75
	3:2	6.48	5.32
Pineapple:taro	1:2	6.60	6.16
	3:2	7.45	5.10

of the fruit-taro (3:2) flakes required some adjustment, either by storing the flakes in a chamber with
relative humidity at 26 percent or modifying the
operation criteria, to increase the percent relative
moisture content in the finished product. Whether a
range of percent moisture monolayer (MFB) for optimum storage exists would require further storage
studies with flakes with different relative moisture
content.

271

Consistency

Table 5 presents the results on consistency as measured by the Bostwick Consistometer. Reconstituted mango-taro flakes at the ratio of 5 g flakes:5 g sucrose:40 ml water were too watery, as indicated by the distance they moved in the Bostwick Consistometer. The consistency of the reconstituted mango-taro flakes at the ratio of 5 g flakes:2 g sucrose:35 ml water were similar to the other selected fruit-taro flakes (Table 5; Nip, 1978). This indicates that it takes more mango-taro flakes to be reconstituted to the similar consistency of other reconstituted fruit-taro flakes. Consistency of reconstituted pineapple-taro flakes was lower than the reconstituted papaya/guava-taro flakes (Nip, 1977). However, this consistency was rated acceptable as shown in the taste study (Table 6).

Taste Study

Taste study results of the reconstituted mango-taro and pineapple-taro flakes (Table 6) indicated that reconstituted fruit-taro flakes were acceptable to the panelists (between "like slightly" and "like moderately"). The lower ratings for reconstituted mango-taro flakes suggested the improvement on the formulation and/or manufacturing of these flakes may be needed. Results on the reconstituted pineapple-taro flakes were similar to the other selected fruit-taro flakes (Nip, 1978).

TABLE 5
Consistence of Selected Reconstituted Fruit-Taro Flakes

Composition of Mixture		Reconstituted Paste	
		Formula 1[a]	Formula 2[b]
Mango:taro	1:2	20.5	8.0
	3:2	24.0	9.8
Pineapple:taro	1:2	9.7	1.4
	3:2	11.5	2.0

TABLE 6
Average Taste Scores of Selected, Reconstituted
Fruit-Taro Flakes

Composition of Mixture		Taste Scores Reconstituted Paste
Mango:taro	1:2	6.0[a]
	3:2	6.0[a]
Pineapple:taro	1:2	6.4[b]
	3:2	7.0[b]

[a] Formula 1: 5 g flakes + 5 g sucrose + 40 ml water
[b] Formula 2: 5 g flakes + 2 g sucrose + 35 ml water

In summary, mango-taro and pineapple-taro
flakes can be produced by the drum-drying process.
The reconstituted fruit-taro flakes were rated
acceptable as indicated by the taste panelists.
However, research on the storage stability and the
formulation of the reconstituted flakes will be
needed in order to promote the use of these products.

NOTES

1. Journal Series No. 2261 of the Hawaii Agricultural Experiment Station, University of Hawaii,
   Honolulu, Hawaii.

REFERENCES

Adeva, L.V., Gopez, M.D., and Payumo, E.M. 1968.
   Preparation of banana flakes from lakaton (Musa
   sapientum Linn var. Lacatan). Phil. J. Sci.
   97:139.

Cording, J., Jr., Williard, M.T., Jr., Eskew, R.K.,
   and Sullivan, J.F. 1957. Advances in the dehydration of mashed potatoes by the flake process.
   Food Tech. 11:236.

Delicate drying process minimizes fruit and vegetable aroma, color changes. 1976. Food Product Development 10(7):37.

Henig, Y. and Mannheim, C.H. 1971. Drum drying of tomato concentrate. Food Tech. 25:157.

Hoover, M.W. 1973. A process for producing dehydrated pumpkin flakes. J. Food Sci. 38:96.

Hsu, H.Y. and Chiou, K.M. 1971. Instant taro powder, I. Comparison of different dehydrating methods. J. Chinese Agr. Chem. Soc. 9:164.

Lazar, M.E. and Morgan, A.I., Jr. 1966. Instant applesauce. Food Tech. 20:531.

Nip, W.K. In press. Development and storage stability of drum-dried guava- and papaya-taro flakes. J. Food Sci.

Onayemi, O. and Potter, N.N. 1974. Preparation and storage properties of drum dried white yam (Dioscorea rotundata Poir) flakes. J. Food Sci. 39:559.

Rodriguez-Sosa, E.J. and Gonzales, M.A. 1977. Preparation of instant tanier (Xanthosoma sagittifolium) flakes. J. Agr. Univ. Puerto Rico 61:26.

Spadaro, J.J., Wadsworth, J.I., Ziegler, G.M., Gallo, A.S., and Koltum, S.P. 1967. Instant sweetpotato flakes--processing modifications necessitated by varietal differences. Food Tech. 21:326.

W. K. Nip: Department of Food Science and Technology, University of Hawaii, Honolulu, Hawaii 96822, U.S.A.

7
Non-food Applications
of Starch, Especially
Potential Uses of Taro

G. J. L. Griffin

ABSTRACT

Taro literature was examined for information
about taro starch and as a guide to the growth of
interest in the material world-wide. The properties of taro starch are presented, as available
from the literature and supplemented by new observations on thermal stability and microscopy. The
significance of particle size is considered as it
relates to the fields of application of starches
other than as foods, and especially in the plastics
and related industries. The economics of starch
application as a filler-extender is considered and
consideration given to the emergence of special
features of starch fillers not obtained from traditional mineral fillers or wood-flow. This is
exemplified by a detailed examination of the
economics and technology of the starch-filled
biodegradable polyethylene carrier bag plant which
has been operating for 2½ years at Nelson in
England. Process economics between countries are
considered, as given to extensions of the work into
other fields of plastics production now at an
advanced stage of industrial trial.

BACKGROUND

Taro starch has filtered into the technical literature in a very unobtrusive manner. In Walton's (1927) remarkable bibliography of starch covering the period 1811 to 1925, the words taro, dasheen, poi, and Colocasia do not appear in the index, and this author had to skim through the 3500 abstracts to retrieve only two relevant papers, one by Krauss (1921), and another by Ripperton (1923). Ripperton refers to activity in Hawaii, but neither author devotes more than a few words to taro. The excellent section on taro in the National Academy of Sciences (1975) report cites Miller (1927), Whitney et al. (1939), Greenwell (1947), Barrau (1953), and Gooding and Campbell (1961). I am indebted to B. Begley[1] for a pre-publication view of his 1977 draft paper on taro which discloses papers by Rada (1952) and Lennox (1954). Radley (1940) features taro briefly with a photomicrograph (by E. Young) which just resolves the starch as a polyhedral granular material under 3 µm in size. Radley gives the gelation temperature of taro in water, but does not offer any account of the origins of the starch itself. In other words, quite strenuous searching uncovered only ten scientific literature references pre-1961 to taro, and these mostly concerned with agriculture rather than characterizing the starch.

Accepting the proposition that the volume of publications is an expression of willingness to fund research and, therefore, an indirect index of commercial potential, it is interesting to scan the computerized literature systems in search of taro references. The data base with the largest time scale is Dissertation Abstracts (Xerox Corporation), and in the period from 1856 to 1978, "taro" is not listed, but "Colocasia" produces three doctoral theses, De la Pena (1967), Ezumah (1972), and Petterson (1977). For 1971 to 1978, however, Chemical Abstracts yields 39 taro references, and it is interesting to analyze them geographically in terms of originating institutions, thus:

```
USA . . . . .14
   Hawaii  . . 7
   Washington. 3
   Florida . . 2
   California. 2
```

```
Japan . . . . . . . . 9
India . . . . . . . . 4
Egypt . . . . . . . . 2
Portugal . . . . . 2
Cuba, Solomon Isles,
  South Africa,
  Holland, Italy,
  Puerto Rico, Fiji,
  China . . . . . . 1 each
```

The U.S. National Agricultural Library data base
for the period 1970 to March 1978 lists 61 taro
references. Most of these are concerned with
botanical detail, pesticides, plant genetics, or
nutrition. A few have a promotional flavor, e.g.,
"Taro Emerges as a New Commercial Crop" (Murray,
1977), and some get close to actual taro starch
production, e.g., Hsu and Chiou (1971), Standal
(1970), Higashihara, Umeki, and Yamamoto (1975),
Szejtli, Henriquez, and Casteneira (1973), and Moy,
Wang, and Nakayama (1977). This author has not
been able to trace any patents on taro extraction
or application, but his search cannot claim completeness.

Properties of Taro Starch

Apart from occasional descriptions of particle
size, there was little detailed published information on the properties of taro starch itself prior
to the publication of the paper by Higashihara,
Umeki, and Yamamoto (1975). Working with Colocasia
antiquorum Schott var. esculenta Engler (syn. C.
esculenta (L.) Schott var. antiquorum (Schott,
Hubbard and Reider)), they separated their starch
from grated root with 0.2M NaCl solution at pH 8.6,
washed it with water, and finally deproteinized it
by steeping for 24 hours three times in 0.3 percent
NaOH solution. The final product contained 22.4
mg percent of phosphorus and about 14 percent of
amylose. Particle size, by electron microscopy,
was 1 to 2 micrometers (microns), where Radley
(1940) gave 1 to 3 micrometers by optical microscopy. The gelatinization temperature was high
(73 to 74°C) compared with rice (63 to 64°C), and
the solution more opaque. Earlier figures are
quoted by Radley (1940) as part of an argument that
particle size alone does not determine gelation
temperature of starches in water; taro gels at over

70°C by comparison with wheat, potato, and arra-
cacha starch which all gel in the range 55 to 64°C.
Higashihara, Umeki, and Yamamoto (1975) report
that α-amylase could hydrolyze 68 percent of the
taro starch used in their experiments.

Particle Size and Shape

Using material supplied by Dr. J. Moy of the
University of Hawaii, the particle size was rein-
vestigated and found to be 2.5 to 3.0 microns by
analysis of scanning electron microscope pictures
(Figure 1). One source of confusion about the
particle size could reside in the tendency of taro
starch to retain a composite granule configuration
(Figure 2). It is possible to identify these
smooth spheres in the field of the microscope and

Figure 1.   Taro starch (University of Hawaii).
Scanning electron microscope picture at
mag. x2,000 displaying individual
polyhedral particles.

278

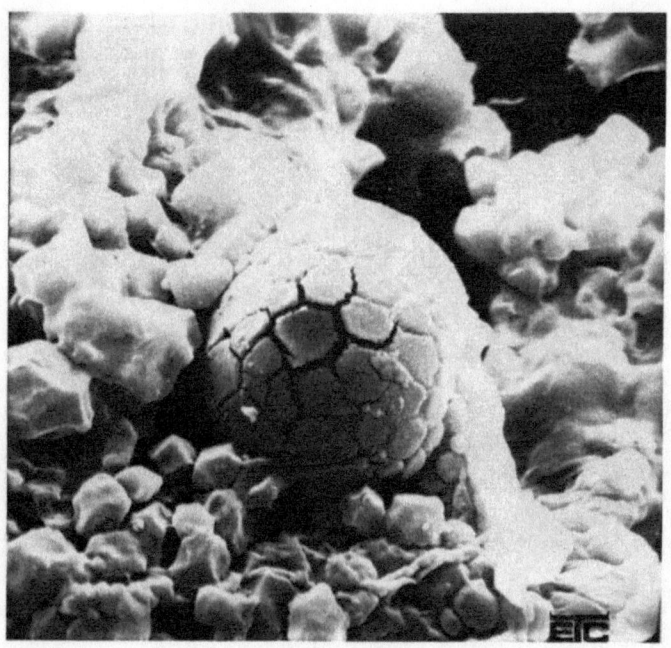

Figure 2. Taro starch with selected field
showing globular structure cracking
into individual grains under heating
action of the beam (mag. x 5,000).

then watch their surfaces fissure into the outlines
of the angular grains under the heating action of
the electron beam. Certainly the polyhedral
granules resemble much more closely the seed
starches such as rice (Figure 3) and maize (Figure
4), rather than the familiar root starches such as
potato (Figure 5) or the stem starch sago (Metroxylon sagus) (Figure 6).

Thermal Stability

    The ability of starches to withstand exposure
to elevated temperatures as a dry powder is of
little interest to the starch industry at present,
but it is crucial to the successful application of
starches as fillers in plastics. This has been
investigated using standard differential thermal
analysis apparatus by Stanton and the results are

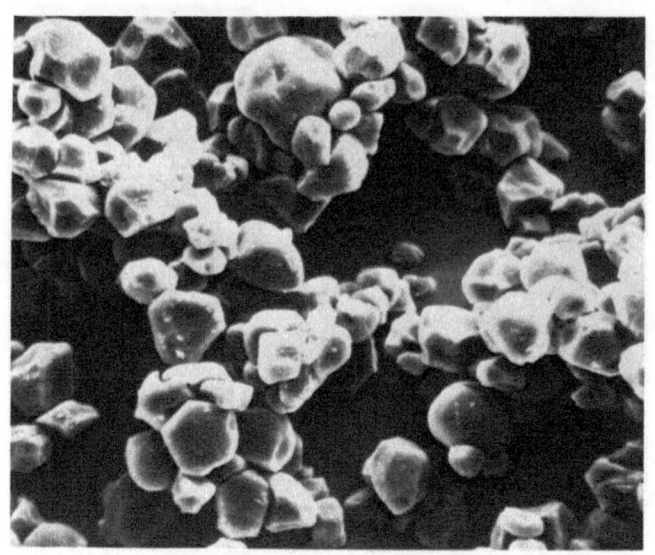

Figure 3.   Rice starch, scanning electron
microscope picture, mag. x 2,000.

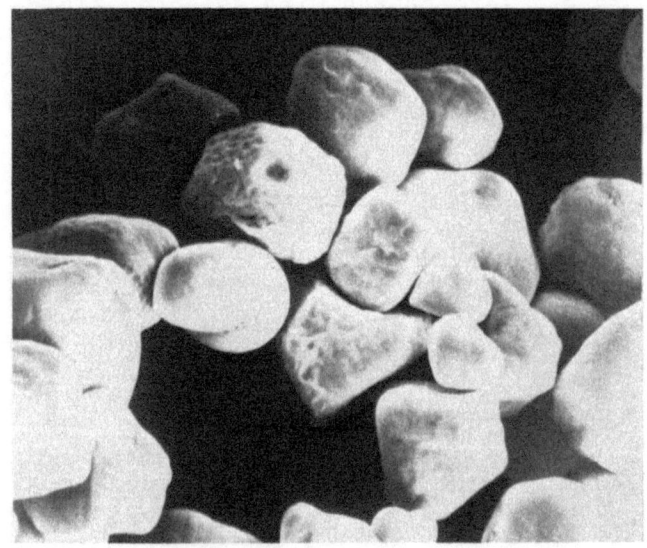

Figure 4.   Maize starch, scanning electron
microscope picture, mag. x 2,000.

Figure 5.   Potato starch, scanning electron
            microscope picture, mag. x 500.

Figure 6.   Sago starch, scanning electron
            microscope picture, mag. x 900.

281

recorded in Figure 7 for a series of common
starches and taro. The odd exothermic behavior of
taro starch when heated in air was immediately
apparent. No other starch in this group of common
commercial materials exhibited the same behavior;
even a sample of 1 micron particle size starch
from Saponaria vaccaria followed the familiar
pattern. A re-run of taro in an atmosphere of
oxygen-free nitrogen showed an inverted curve
(Figure 8), clearly implying that the effect was
one of accelerated oxidation. Further investigation of this aspect is highly desirable, partly in
the expectation of disclosing some special chemical
feature of taro, but equally with a view to check-
ing the level of dust explosion hazard with this
material. As far as plastics applications are
concerned, there is no special difficulty because
the polymers are seldom processed at temperatures
exceeding 230°C, and air is largely excluded.

STARCH AS A COMPONENT OF PLASTICS COMPOSITIONS

Achieving Compatibility

The technical literature contains many hundreds of references to the preparation of chemical
derivatives of starches which owe their origin to
the exploitation of the plentiful reactive groups
on the chemical structure of the starch molecule.
In most cases the procedures described involve
taking the starch into solution followed by
chemical reaction, then precipitation of the deri-
vative, followed in turn by filtration and drying
operations to recover the product. Simple as this
sounds, it proves impossible to create a dry
"ready" product at a cost approaching that of the
common plastics materials because of the expense
involved in chemical reagents, process water,
energy, and handling costs, even though the starting material, starch, may be obtained at one half
of the cost of the polymers. We have to regard the
polymer costs as the datum line in this argument,
and this is difficult with current uncertainties on
petrochemical feedstock prices. It is probably
fair to take a figure of about 55 cents per kilo as
a reasonable value for approximations based on a
low density polyethylene of density 0.918 ton m^{-3}.
Considerations such as these have persuaded
me to concentrate on making starch acceptable to

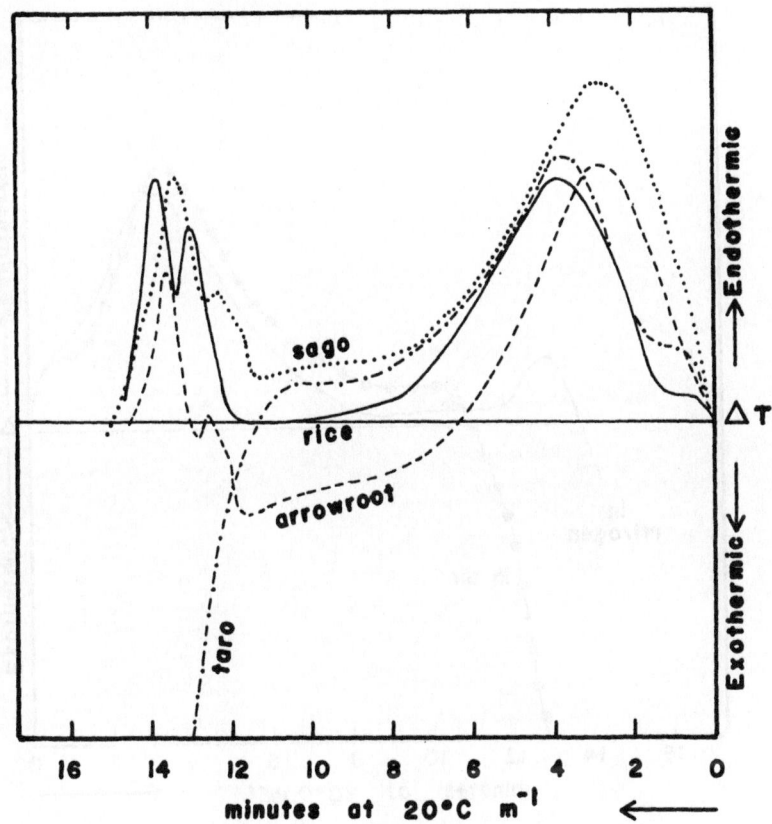

Figure 7. Comparison between the thermal stabilities of three common starches and taro starch, traced from DTA apparatus recordings. The samples were all heated in air.

the plastics industry with the minimum of added cost. Undoubtedly this situation is at its most favorable where whole grain "native" starch is used and only "remedial" action is taken to achieve compatibility. Primarily, the lack of compatibility of normal commercial starch with plastics is determined by the proportion of free water commonly present (9 to 17 percent), and in the physico-chemical antipathy between the starch surface and the polymer. The former is remediable by a straightforward drying operation. The latter

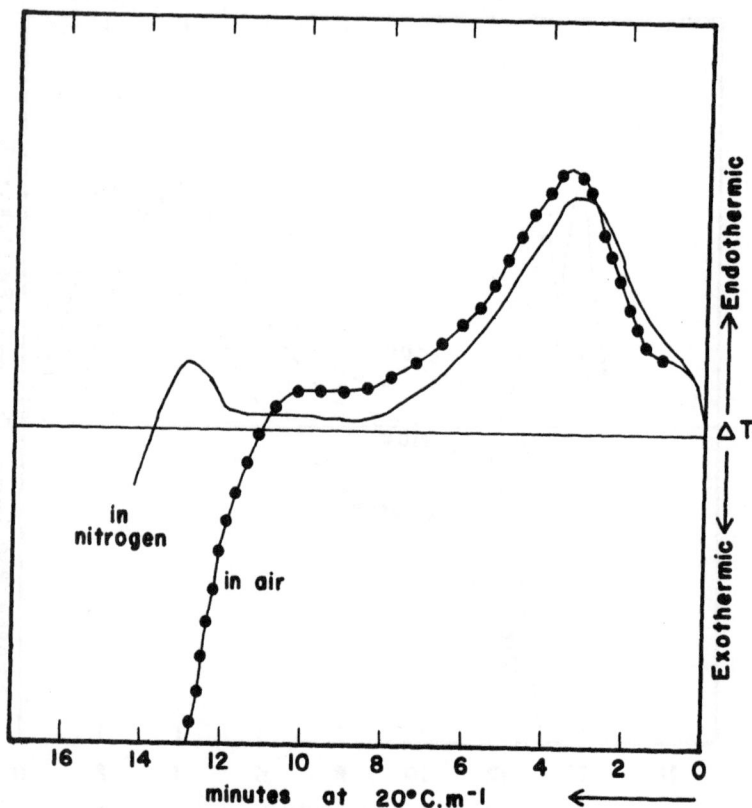

Figure 8. Effect of atmosphere on the thermal
           stability of taro starch, traced from
           DTA apparatus recordings, with the
           samples heated in air and in oxygen-
           free nitrogen, respectively.

offers more difficulty because the polysaccharides
are intrinsically polar and hydrophyllic materials
in contrast to the polymers, which are low in pola-
rity and are essential oleophyllic, or hydrophobic.
This has been overcome by chemically modifying the
starch so as to obscure the influence of the
hydroxyl groups, but the modification is restricted
to the surface layer only, this being the only part
in contact with the polymer matrix. By using this
approach to the interface problem the desired effect
can be achieved with the most minute amounts of
chemical additive. The net effect is to produce a

polymer-compatible starch filler at reasonable cost, and further process economies can be achieved by combining the surface-treating action with the drying process. The use of hydrophobic whole grain starches as plastics fillers is described further in a 1974 patent (Coloroll Ltd., 1974). The physical benefits to be achieved are illustrated by a typical set of test results on injection-molded samples of starch-filled low density polyethylene (Figure 9).

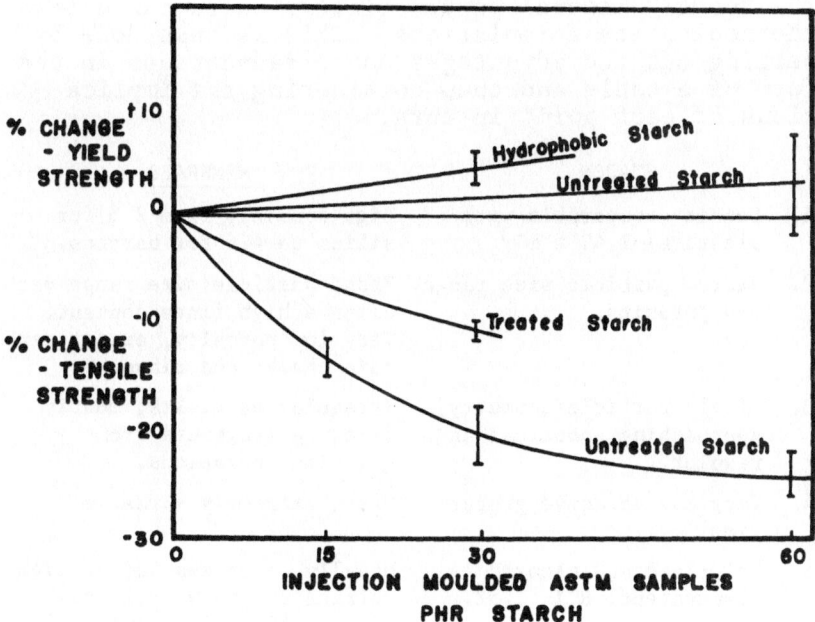

Figure 9. Strength of cornstarch/low density polyethylene composites showing the general increase in yield strength and decrease in tensil strength with increasing starch content. The A samples are made from hydrophobic starch showing the benefit of this pretreatment. B samples use untreated starch. All samples were injection molded to ASTM standard form.

285

It is possible, in a special case, to take the process economics one stage further by using dry milling and air classification to isolate the starch from wheat, thus avoiding the extra energy costs associated with wet milling operations. The product is virtually a "hard" biscuit flour of low protein content, and it has performed very well as a filler for a variety of plastics compounds.

The Case for Using Starch

Having achieved compatibility without an excessive cost burden, it is necessary to justify the use of a filler which is radically different from the "usual" mineral powder fillers encountered in thermoplastics formulations. This is best done by setting out the advantages and disadvantages in the form of a table and then considering the implications of each point in turn.

STARCH	MINERALS
1. Density comparable with plastics--1.49 t m^{-3}	High density, from 2.6 for silica to 4.6 for barytes.
2. Narrow particle size range. Low porosity.	Broad particle size range with often a high fines content. Very low porosity, except certain chalks and clays.
3. Simple particle geometry, approaching spherical and regular.	Irregular particles, mostly fracture fragments from grinding operations.
4. Very low abrasive properties.	Often extremely abrasive.
5. Colorless and almost transparent, R.I. 1.52.	Usually colorless but of high refractive index, e.g. calcite--1.66; wollastonite--1.63; talc--1.59.
6. No significant metallic content.	Transition metals may be released.
7. Thermally stable to 250°C.	Thermally stable to very high temperatures.
8. Not water soluble, but hygroscopic. Also hygroscopic in situ.	Some minerals retain traces of water most tenaciously, but not normally hygroscopic in situ.
9. Low gas permeability.	Very low gas permeability.
10. Biodegradable.	Permanent in a biologically active environment.

286

Considering each of the above points in turn:

Point 1. Because of its low density, starch should be compared with other fillers only on a volume basis which can be seen to almost double its cost effectiveness by comparison with mineral fillers. The common thermoplastics have densities in the 0.9 to 1.2 t m^{-3} region. The attraction of low density fillers in plastics is not always obvious; for example, the benefit of low settling rates in plastisol formulations, or the weight reduction in massive cast resin factory mold units.

Points 2 and 3. The particle morphology of the starches sets them in a special class in the spectrum of available filler materials. No other naturally occurring material can offer such a range of "narrow cut" or monodisperse particles with shapes approximating to spherical. The consequence of this is that the all-important flow properties of filled polymers in their processing stages are disturbed much less by carefully judged starch additions than by any powdered mineral additions. The majority of other organic fillers are fibrous in nature and therefore profoundly influence the rheological properties of liquid systems. Only specialty materials such as graded glass microspheres can equal the performance of starch but the costs are excessive and the products opaque and abrasive. It is also possible to arrange close matches at the formulation stage between product film thickness and starch particle size simply by selecting botanical source. Small additions of such starches can achieve near perfect "antiblocking" action, preventing the mutual adhesion of plastics surfaces.

Point 4. The question of particle hardness is critically important where finished products are going to be worked by cutting tools. This is certainly the case with cast resin systems and filled rigid cast plastics foams. Furthermore, as the speed of modern plastics processing machinery increases, the problem of machine wear becomes more and more important. This is acknowledged by such work as that of Muzzy (1976) on machine maintenance economics. Benefits that could arise from the inclusion of fillers may be largely offset by the increased cost of plant maintenance caused by abrasive mineral fillers.

Point 5. The refractive index of starch is so close to that of certain resins, especially polyvinyl chloride, polyethylene, and cast polyesters, that it is possible to formulate almost transparent

287

compositions. Total clarity is elusive because of the near impossibility of matching optical dispersion coefficients as well as refractive indices, but the products achievable would be perfectly acceptable for agricultural purposes or vehicle roof lights. In close contact situations where narrow-angle light scattering is not obtrusive, the ability of starch-extended polyethylene film to permit the reading of print through film overwrap makes packaging applications quite feasible.

Point 6. The heavy metal content of food grade starches is very low, and no problems of contamination or toxicity should be encountered. Mineral fillers, on the other hand, need to be carefully monitored for toxic contamination, and they may also create another problem in the destabilization of polymers by the introduction of transition metal contaminants into thermally unstable polymers. A well-known instance of this problem has been recognized in the attempts to reinforce polypropylene with asbestos fillers, where the thermal stability of the product can be severely reduced by the accessibility of iron compounds in the filler.

Point 7. The thermal stability of starch has been referred to earlier and the practical working limit of 250°C relates to exposures of the order of a few minutes. Most of the common thermoplastics are processable at temperatures below 230°C so that problems are unusual. There are occasional situations where high internal heat generation occurs on some types of machinery, for example, the high compression extruder screws used on paper coating machines. It is also possible to encounter situations where a mismatch of machine size and molding weight results in material being held at high temperatures for long periods of time in injection molding machines. These situations can be avoided, and it is obviously important that users of starch-extended plastics should be informed of the special nature of the products. Thermal stability problems are not encountered with mineral fillers themselves, but their possible adverse effects on polymers is noted in Point 6.

Point 8. All fine powder fillers can give rise to moisture adsorption problems, and users of pigments, clay fillers, and carbon black will find nothing surprising in handling hygroscopic materials. Starch, however, has a much higher equilibrium moisture content than mineral fillers, and due consideration must be given to this fact. There is an associated small dimensional change which was measured directly in an impressive experimental study by Hellman, Boesch, and Melvin (1951) which showed great differences between starches in this respect, e.g., cornstarch increased in linear dimension by about 2 percent at 10 percent water uptake, as against 10 percent for potato. Under normal climatic conditions, starch embedded in a polymer matrix takes many weeks to approach the equilibrium state and, in the case of low elastic modulus materials such as polyethylene or polyvinyl chloride, the effect is of no consequence. In higher modulus materials, such as polystyrene, there are interesting slow changes in physical properties which have been studied, for example by Sharafi (1977). One positive aspect of the hygroscopic nature of starch is the reduced tendency of the surfaces of starch-extended plastics to accumulate electrostatic charges once they have approached their equilibrium moisture content. This is important in products which have hitherto been troublesome by the attraction of dust.

Point 9. Direct measurement of the absolute gas permeability of starch-filled polyethylene films showed a slight decrease in permeability with increasing starch content. This is as would be expected for a material with such a high degree of crystallinity and is very welcome in the context of packaging films. Mineral fillers, in the main, share this property.

Point 10. Starch is almost indefinitely stable in sterile water, or under exposure to solar UV light. However, it is rapidly hydrolyzed by the amylotic enzymes of microorganisms to yield gums and sugars. Thin layers of polymer are significantly permeable to enzymes, as demonstrated by Griffin and Nathan (in press), and a sufficient proportion of the starch is decomposed under the polymer surface to initiate a progressive breakdown of the material. This process has been described by Griffin (1974) and Whitney and Williams (1976). This work has been largely confined to polyethylene, but some studies have been made on polystyrene, polyvinyl chloride, and polypropylene. Work on these materials by Stanton (1978) made it clear that the initial breakdown of the starch under soil contact conditions makes the plastic film attractive to soil insects, and his work under tropical conditions has been recently repeated in the United Kingdom using captive colonies of woodlice. This gives us the most promising solution so far to the plastics

289

litter problems. Further work by Griffin (1975) demonstrated the importance of garbage-derived fatty materials in the breakdown of buried plastics, and this has led to developments ensuring the slow progressive breakdown of polyethylene itself under soil burial conditions. The details of this work on environmentally acceptable plastics is beyond the scope of this paper, but its significance is clear and the potential applications in agriculture are substantial. A typical recent set of degradation data from a Japanese commercial source is reproduced in Figure 10.

Mineral fillers, obviously, are not biologically active, but they can claim one environmental benefit in that they reduce the calorific value of plastics under incineration by simple dilution. It is worth noting that starch has a similar, if lesser, effect by virtue of being an oxygen-rich polymer itself. Calorific values calculated for a starch/polyethylene mixture are shown in Figure 11.

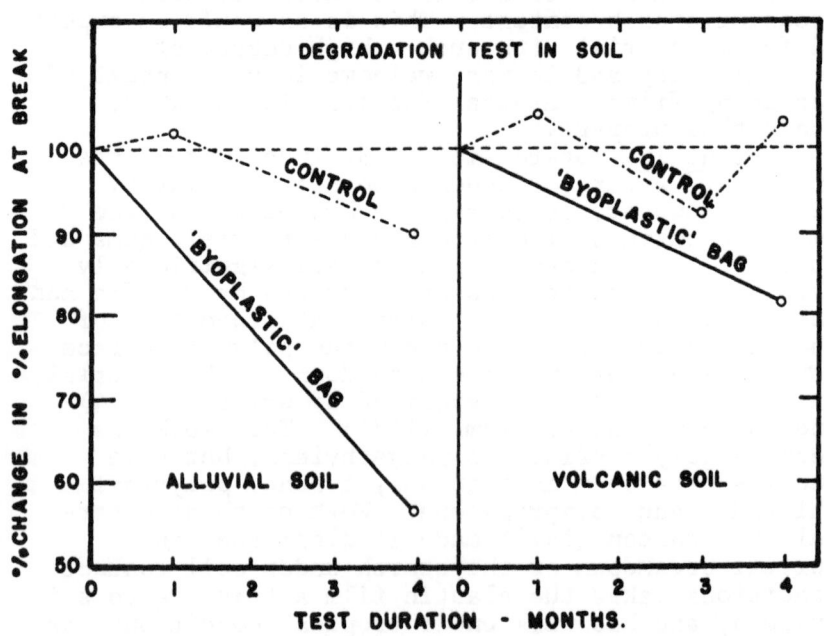

Figure 10.   Biodegradation of starch/LDPE/ unsaturated ester composite film.

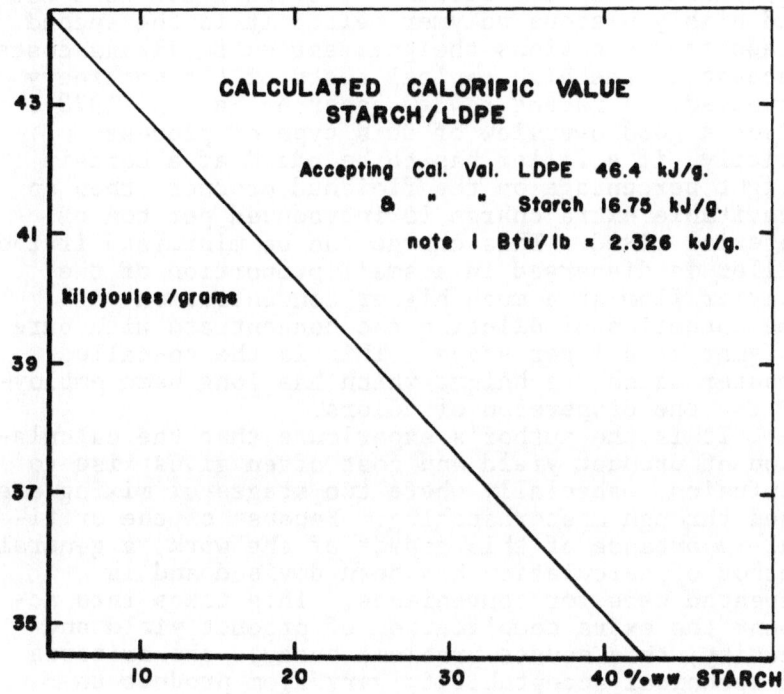

Figure 11. Calculated values of starch/LDPE composites.

Process Economics

When the prices of petrochemical feedstocks jumped sharply in recent years, it was widely believed that the increased cost of plastics articles could be offset by the simple addition of cheap fillers. This belief proved to be largely illusory because cheap minerals seldom become cheap fillers. They have to be crushed and screened, then often surface coated, and finally their addition to plastics systems has to be accomplished by the use of heavy duty melt-compounding equipment. Some benefits can be achieved if the individual situation is carefully appraised beforehand.

We can subdivide the filling activity into two main classes: (1) those formulations where the filler can be added to a system which is still in a relatively low viscosity condition and the processes available relate to those of the paint industry at room temperature and with low energy inputs, and

291

(2) those where the filler has to be added to a hot
and highly viscous polymer melt. It is the second
class that occasions the greatest extra mixing costs
because of the high capital costs of the machinery
involved. A recent review paper by Salden (1978)
gives a good overview of this type of process.
Briefly, if a filler has to be added at a certain
weight percentage on the finished product, then an
inevitable extra charge is introduced per ton of
material mixed. This charge can be minimized if the
filler is dispersed in a small proportion of the
polymer flow at a much higher concentration with
the intention of diluting the concentrate with pure
polymer at a later stage. This is the so-called
"master batch" technique which has long been employed for the dispersion of colors.

It is the author's experience that the calculation of product yield and cost often gives rise to
confusion, especially where two stages of mixing are
used through masterbatching. Because of the critical importance of this aspect of the work, a general
method of calculation has been devised and is
repeated here for convenience. This takes into account the extra complication of product yield and
density; this causes problems because the criteria
of technical acceptability vary from product to
product. Thus a plant line producing flexible
plastic film is judged commercially by its output
in area units per unit of running time at a particular gauge. This will be sensitive to density
variations in the material, while an extrusion line
producing polystyrene sheet for thermoforming into
disposable coffee cups will be judged by the yield
of cups per unit weight of feed material. Thickness of the cups is of less importance than their
stiffness--a property which is actually increased
by the inclusion of starch filler.

Accepting the following symbols,
Q = percent by weight of dry starch in the
finished product
s = cost of the dry starch per ton
P = cost of the polymer per ton
M = cost per ton of the master batch process
D_s = density of starch, usually 1.49 ton/m^3
D_p = density of the polymer in ton/m^3
K = percent by weight of dry starch in the
master batch
the density of any starch/polymer mixture D_m is
given by:

$$D_m = \frac{100\ D_sD_p}{D_pQ + D_s(100 - Q)}$$

Another useful quantity is the ratio by which the density has been changed by the introduction of Q, percent of starch, because this determines the reduction in area yield of sheet material manufactured at a predetermined gauge. This must be set against any cost reduction resulting from the inclusion of starch. Thus,

$$\frac{D_p}{D_m} = \frac{D_pQ + D_s(100 - Q)}{100\ D_s}$$

Now the cost of producing one ton of compound in its finished state, J, excluding molding or extrusion costs, will be

$$J = \frac{(100 - Q) + QS + (100/K)QM}{100}$$

with a starch content of Q percent. If it is necessary to take the density change into account with diminished yield, then the cost must be slightly increased. Thus the cost of producing the same volume of product as would have come from one ton of unfilled polymer, L, would be:

$$L = \frac{((100 - Q)P + QS + (100/K)QM)D_s}{D_pQ + D(100 - Q)}$$

A Working Example of Starch-Filled Plastics Technology

The pioneer production unit for starch-extended polyethylene film was commissioned in September 1975 at the Nelson, Lancashire, factory of Coloroll, Ltd., in England. It comprised a single complete line for the manufacture of shopping bags and is currently due for expansion. The process starts with commercial polyethylene granules and dried hydrophobic starch, and finished printed bags are dispatched from the plant in cartons of 500 units.

As soon as a source of pretreated starch had been organized, the Company decided to place its own masterbatch making line alongside the film extruding unit. The economics of this decision are interesting because in 1975 the subcontract cost of masterbatch manufacture in England would have been about $225 per ton of masterbatch processed using materials supplied. In fact, the small automatic twin-screw continuous compounder installed could be supervised by the same operator as the film extrusion line and its total cost was around $45,000 by comparison with the total plant cost of about $720,000, i.e., only 6 percent of the total. Amortized over 5 years, this represents an equipment cost of $45 per ton of masterbatch produced, leaving a very generous margin for energy, factory overhead, and financing costs when compared with the subcontract figure. This particular plant, being very lightly loaded, has run for 2½ years with no major maintenance costs.

Polymer granules are delivered to the factory by road tanker and transferred pneumatically to a pair of 10 ton outdoor silos. Pretreated starch arrives in 50 kilo multiwall paper sacks which have an inner line of polyethylene film. These sacks are discharged as required into a covered bin equipped with a screw elevator which lifts the starch into a vibratory screw-dosing feeder sited above the hopper of the compounding extruder. Other minor ingredients such as pigment are dealt with in a similar manner. The output from the compounder takes the form of a continuous ribbon which is air-cooled in a forced draught tunnel before being reduced to roughly cubical granules by a continuous cutter. Any increase in the capacity of this plant would require a change in cooling and granulation technique. Dry face-cutting with air cooling of the hot granules wouod suffice up to about 200 kilos per hour, while, above this, face-cutting with a water cooling ring or even submarine cutting would be necessary. Masterbatch granules from the present plant are collected in 250 liter rigid polyethylene drums for transfer across the factory floor to the film extruder. The film extruder is loaded by means of an automatic triple hopper system with dosing valves which can provide a fixed volume ratio between the various granular feed materials. These hoppers can contain virgin polymer, star masterbatch, and recycled material as required. All the hoppers are fitted with level sensors coupled to the storage silos are transfer drums. Careful attention to monitoring the bulk density of the feed materials enables the composition of the finished film to be maintained within working tolerances.

The film itself is produced on a 90 mm single screw machine of advanced design which features internal cooling of the film bubble, as well as modifications to the die and cooling system which make possible the production of a tubular film with longitudinal thick zones which correspond with the handle regions of the finished bags, thus avoiding the need for supplementary reinforcement. The machine has an output of sufficient film to make 1.2 million shopping bags each week, after printing and bagmaking has been done. With individual bags weighing about 20 g, this means a total material throughput of about 1250 tons per annum of which about 6 percent is starch. Figure 12 shows the layout of the masterbatch line, Figure 13 the film blowing unit, and Figure 14 the downstream equipment. Successful production of heavy gauge fertilizer sacks of similar formulation is reported by Coloroll licensees in Brazil, and similar activity is about to commence in Canada.

Figure 12. Continuous compounding line for the manufacture of starch/polyethylene masterbatch at Coloroll Factory, U.K.

Figure 13.　Film extrusion line for starch filled
polyethylene at Coloroll Factory, U.K.

Other Fields of Application Nearing Commercial Production

Successful full scale plant trials have been conducted in England in which pretreated starch has been used as a supplementary extender for the vinyl plastisol used in wallpaper coating. This success arises from the combination of the very favorable rheological properties of starch described earlier with the added attraction of no extra mixing costs.

Figure 14.   Flexographic printing machine on starch-filled
             polyethylene bag line at Coloroll Factory, U.K.

It now seems possible to save 10, and possibly 15
percent of the actual resin used in this process.
The rheological superiority of starch is apparent
from the viscosity plots recorded in Figure 15.
     A similar low viscosity system which is at an
advanced stage of development is the use of starch
as an extender for cast polyester resins, where
useful weight reductions have been achieved in
factory molds containing over 2 kilos of starch.
     Translucent calendered polyvinylchloride sheet
has also been manufactured at a pilot plant scale.
With this system the economic benefits are strong
because this type of polymer is normally batch
compounded in the same production line as the
sheeting unit, obviating the extra costs of masterbatch making.

297

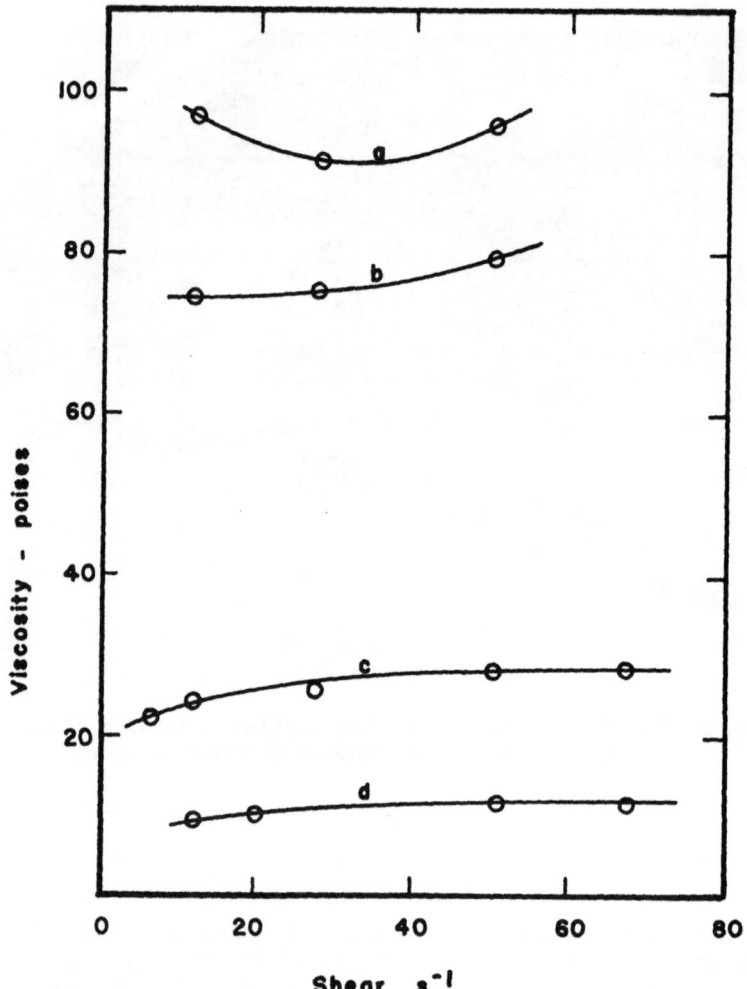

Figure 15.   Effect of added filler on the viscosity of
             vinyl plastisol as used for wallpaper coating.
                  a = precipitated chalk addition
                  b = ground hard limestone addition
                  c = hydrophobic wheat starch addition
                  d = control material as supplied
             Extra filler in a, b, and c was in amounts
             calculated to be of equal volume and based on
             15 percent by weight addition of starch.

ACKNOWLEDGEMENTS

Thanks are due to Mr. C.D. Garg for his assistance with the DTA measurements reported in this paper, and to numerous final year students at Brunel University who have shown sufficient interest in resource conservation and their environment to become involved in this program of work. I am also grateful to Coloroll, Ltd., for permission to describe their manufacturing plant and for general support of the work.

NOTES

1. Bryan Begley, personal communication.

REFERENCES

Barrau, J. 1953. Taro--an Annotated Bibliography. South Pacific Comm. Quart. Bull. 3:31.

Coloroll, Ltd. 1974. Brit. Pat. 1,487,050 and 1,484,833.

De la Pena, R.S. 1967. Effect of different levels of N, P, and K fertilization on yield of taro. Ph.D. dissertation, University of Hawaii, Honolulu, Hawaii.

Ezumah, H.C. 1972. The growth and development of taro (Colocasia esculenta (L.) Schott), in relation to selected cultural management practices. Ph.D. dissertation, University of Hawaii, Honolulu, Hawaii.

Gooding, M.J. and Campbell, J.S. 1961. The improvement in cultivation methods in dasheen and eddoe growing in Trinidad. Proc. of the Am. Hort. Soc., Caribbean Region 5:6.

Greenwell, A.B.H. 1947. Taro--with special reference to its culture and use in Hawaii. Econ. Bot. 1:276.

Griffin, G.J.L. 1974. Biodegradable fillers in thermoplastics. Am. Chem. Soc., Advan. in Chem. 134:159.

_____. 1975. Degradation of polyethylene in compost burial. Preprint volume, 15th Prague Microsympos. Paper G6.

_____ and Nathan, P.S. In press. Enzyme permeability of thin PE membranes. IUPAC Intern. Symp., Stockholm, 1976.

Hellman, N.N., Boesch, T.F., and Melvin, E.H. 1951.
Starch granule swelling in water vapor sorption.
J. Am. Chem. Soc. 74:348.

Higashihara, M., Umeki, K., and Yamamoto, T. 1975.
Isolation and some properties of taro root starch.
Denpun Kagaku (J. Japan. Soc. Starch Sci.) 22:61.

Hsu, H.Y. and Chiou, K.M. 1971. Instant taro powder: a comparison of different dehydrating
methods. J. Chinese Agr. Chem. Soc. 9:164.

Krauss, F.G. 1921. Production of Starch on a Small
Commercial Scale from Root Crops. Hawaii Agr.
Res. Sta. Rept.

Lennox, C.C. 1954. Report to the Trustees of the
Bernice P. Bishop Museum on the Resources of
Waipio Valley, Island of Hawaii.

Miller, C.D. 1927. Food Values of Poi, Taro, and
Limu. Hawaii Agr. Expt. Sta. Bull. 78, Honolulu,
Hawaii.

Moy, J.H., Wang, N.T.S., and Nakayama, T.O.M. 1977.
Dehydration and processing problems of taro.
J. Food Sci. 42:917.

Murray, B.K. 1977. Taro emerges as a new commercial crop. Food Prod. Develop. 11:30.

Muzzy, J.D. 1976. Optimizing screw repair schedules.
Mod. Plastics 53:54.

National Academy of Sciences. 1975. Underexploited
tropical plants with promising economic value.
Washington, D.C.

Petterson, J.H. 1977. Dissemination and use of the
edible aroids with particular reference to Colocasia (Asian taro) and Xanthosoma (American taro).
Ph.D. dissertation, University of Florida.

Rada, E.L. 1952. Mainland Market for Taro Products.
Hawaii Agr. Expt. Sta., Agr. Econ. Rept. No. 13.

Radley, J.A. 1940. Starch and its Derivatives.
Chapman and Hall, London. Photomicrograph No.
61.

Ripperton, J.C. 1921. Hawaiian Starches. Hawaii
Agr. Expt. Sta. Rept.

Salden, D.M. 1978. Melt compounding and compounding machinery. PRI Conference Preprints, Thermoplastics Compounding, London.

Sharafi, M. 1977. Starch/polystyrene composites.
Master's thesis, Brunel University, Uxbridge,
England.

Standal, B.R. 1970. The nature of poi carbohy-
drates. Proc. Intern. Symp. Trop. Root and Tuber
Crops, Coll. Trop. Agr., Univ. of Hawaii,
Vol. I:146.

Stanton, W.R. 1978. The biodegradation of starch-filled plastics in Malaya. "Sago 76" Symp. Vol., Kuching, Malaysia.

Szeijtli, J., Henriquez, R.D., and Casteneira, M. 1973. Content of amylase of different Cuban starch sources. Rev. Cubana Farm., Coden RCUFAC, Series 8:37.

Walton, R.P. 1927. Starches--a List of References. Bull. New York Public Library, January-September.

Whitney, L.D., Bowers, F.A.I., and Takahashi, M. 1939. Taro Varieties in Hawaii. Hawaii Agr. Expt. Sta. Bull. 78.

Whitney, P.J. and Williams, W. In press. Biodegradation of a polyethylene composite. IUPAC Intern. Symp., Stockholm, 1976.

G. J. L. Griffin: Department of Non-Metallic Materials, Brunel University, Uxbridge, Middlesex, UB8 3PH, United Kingdom

8
Processing of
Root Crops in India

N. Hrishi
C. Balagopal

ABSTRACT

Many edible root crops are found in various
parts of India. Those most widely cultivated are potato, cassava, sweet potato, yams, and aroids. Others
are also cultivated, but on a much smaller scale.

Recent efforts toward increased production have
necessitated improvements in post-harvest technology. Methods available in India for the processing
and utilization of cassava, sweet potato, aroids,
yams, and arrowroot are discussed.

Improved processing techniques will make
expanded cultivation of these crops economically
viable.

INTRODUCTION

Many edible root and tuber crops are found in
different parts of India. The cultivated ones are
potato, cassava, sweet potato, yams (Dioscorea
esculenta, Dioscorea alata), aroids (Colocasia spp.,
Xanthosoma spp., Amorphophallus spp.) and Coleus
spp. Further, there are a number of other tuber
crops such as Alocasia spp., Pachyrrhizus spp.,
Helianthus tuberosus, Maranta arundinacea, and Eleo-
charis spp. which are also cultivated, though not in
a systematic way, in many parts of the country.

Though a number of tuber crops have been cultivated, the post-harvest technology on processing

and utilization of these crops (except potato) has
not received much attention. Recent efforts towards
increased production of tuber crops have necessitated opening up new vistas for their processing and
utilization, in addition to programs to improve
existing village level technology. In this communication, various methods and practices available in
India for the processing and utilization of tuber
crops, except potato, are discussed.

CASSAVA

Most cassava is used for human consumption as
fresh tubers by simply boiling, draining off the
water, and consuming the pieces with suitable
sauces. Rapid spoilage due to various causes has
been the impediment to cassava storage. Normally,
cassava cannot be stored fresh for more than 6 to 8
days. Preliminary studies on keeping quality of
tubers, made in the germ plasm collection of the
Central Tuber Crops Research Institute (CTCRI),
revealed that much variability does exist for this
attribute among the different genetic stocks.
Tubers intact after harvest were kept under room
conditions, and the perishability of roots was
assessed after 15 days. Even though a majority of
genotypes showed a high percentage of perishability
within 3 to 4 days, a few genetic stocks were found
to have more than 75 percent of their tubers in good
condition even after 15 days. However, this requires more study.
Both primary and secondary deterioration of
tubers occurs. Primary deterioration is due to the
appearance of blue-black streaks in the root vascular tissue, while secondary deterioration is due to
pathogenic rots, fermentation, and/or softening
of the roots (Booth, 1975). Under normal conditions
of storage, there is an increase in total sugars
followed by a reduction in starch concentration,
HCN content and dry matter. However, the increase
in total sugars would tend to promote microbial
activity. Besides, dehydration and cracking of
tubers during storage permit the easy entry of the
fungus Rhizopus oryzae, which subsequently spoils
the tubers completely (Maini and Balagopal,
1978).
Different methods of preservation of fresh
cassava tubers have been reported such as waxing,
refrigeration, field clamp storage and storing in

303

different materials such as sawdust, coir dust, or peat (Singh and Mathur, 1953; Ingram and Humphries, 1972; Booth and Coursey, 1974). Among the various methods tested, preservation of cassava in moist sawdust seems to be the most promising and economic. Freshly harvested cassava tubers without any damage after retaining the neck portion can be prepared by this method up to 30 days without much biochemical change or deterioration in cooking quality (Balagopal, Maini, and Hrishi, 1978).

The simplest mode of processing, which varies from place to place among the local farmers, is the conversion of tubers into plain or parboiled chips. In the preparation of cassava chips in different localities, no consideration is given to tuber size, shape, or variety. However, the various chips available in the market can be broadly grouped into the following categories:

> "Iritty" - outer skin is removed.
> "Vella" - both rind and skin are removed.
> "Chilta" - both rind and skin are retained.

"Vella" chips usually fetch a higher price than the other chips due to their bright white color. "Chilta" and "Iritty" chips are used for cattle feed. Another type called "Cutter size" is prepared by giving two slanting cuts; these have less moisture and bright color. However, "Vattan" and "Chendu" types of chips are very popular in the market. "Vattan" is circular in shape, whereas "Chendu" is prepared by giving a slanting cut. These chips are used later for human consumption and for manufacture of cassava flour, starch, and other products during the lean months (June through August) when fresh tubers are not available. Parboiled chips are prepared by boiling the slices cut from raw tubers in water for ten minutes and sun-drying. During storage, a number of fungi and bacteria grow on these chips (Balagopal and Nair, 1976). Similarly, the arecanut beetle, Aracerus fasciculatus, causes considerable damage to chips during storage. Adequate care should be taken to see that the entire preparation and storage are done under hygienic conditions. However, at present the chips are mostly dried in the sun and no mechanized process has been devised for curing the chips under hygienic conditions. Therefore, it become important to develop simple and efficient mechanical techniques for peeling, chipping, and drying the slices.

There is urgent need for rapid improvements in processing technology to ensure adequate supplies of

304

quality products. In addition to the manufacture of
food items like sago starch, glucose, and dextrose,
starch is utilized in the textile and laundry industry and other industries, such as manufacture of
adhesives, paper, and hard cardboard. Possibilities
of using cassava to produce industrial alcohol and
beer have been indicated from recent work done at
CTCRI. By mixing grated cassava from parboiled
chips with wheat and groundnut fortified with vitamins, a balanced food item, namely "Kerala Indigenous Food" (KIF), has been produced with the help of
the CARE organization.

Large quantities of cassava tubers are utilized
for manufactured animal (cattle, poultry, pigs)
feed. Spent pulp from the factories is being used
in piggeries, and the dried chips are sometimes
pounded and given to cattle. Attempts at CTCRI to
produce a protein-enriched cattle feed with the help
of a fungus, Rhizopus oryzae, through a fermentation
process appear promising (Balagopal and Maini,
1976). This low-cost fermentation process can
easily be adopted by the small farmers of India.
Large-scale utilization of cassava as animal feed
will greatly help in diverting large quantities of
food grains, now used for feeding cattle, for human
consumption.

SWEET POTATO

Sweet potato is valued as human food for its
high nutritive value, flavor, and digestibility, in
addition to its usefulness as animal feed. Sweet
potato is widely used as food in India after boiling, baking, or frying. Sweet potato flour is often
used in the preparation of biscuits, cakes, puddings, and other preparations. It can be mixed with
wheat flour up to 25% for making "chappaties" and
bread. Low-grade tubers of sweet potato, after
chopping and dehydration, are utilized as feed for
livestock, especially swine.

Sweet potato starch is suitable for sizing
paper and textiles and for use in laundries. In
laundry it is superior to other starches in impart-
ing smoothness and stiffness to fabrics. It forms a
clear, stable gel with high holding capacity, and
forms a useful ingredient of confections and bakery
products. It is employed in the manufacture of
adhesives, dextrins, compositions for insulating
fabrics, coating formulations for dry cells, and in

cosmetics. The spent pulp (pomace) left after the
extraction of starch is pressed, dried, and used as
cattle feed.

Sweet potato flour, prepared by dehydrating and
grinding the tubers, is used as a supplement to
cereal flours in the preparation of bakery products,
puddings, and milk jelly. It can be mixed with
wheat flour up to 25 percent for making chappaties
and bread. It acts as a dough conditioner in bread
manufacture, and functions as a stabilizing agent
in ice cream. It can be used in brewing along with
malt flour, as it contains both starch and amylase.
A mixture of sweet potato flour and groundnut cake
flour in a ratio of 4:1 is superior to rice in
nutritive value. Sweet potato flour can be employed
as a coagulent in slurry thickness in the process of
extracting alumina from bauxite. It is also used as
a molding sand conditioner.

Edible and fermented syrups have been prepared
from sweet potatoes. The tubers are cooked with
water, pulped, and saccharified by treatment with
2 percent malt at 60°C. The solution is clarified
and concentrated to 70 percent solid content. A
glucose syrup is prepared from sweet potato starch
by a process similar to that employed for producing
syrup from tapioca or cornstarch. This is based on
the conversion of sweet potato starch into glucose
hydrolysate by treatment with hydrochloric acid.

Sweet potato forms a valuable raw material for
fermentative production of industrial alcohol, lactic acid, acetone, butanol, vinegar, and yeast. Raw
tubers and pressed juice are employed for the preparation of alcoholic beverages after saccharifica-
tion. Carotene concentrates have been prepared from
varieities rich in carotene. A process for extraction of anthocyanin dyes from peelings has been
patented.

In jelling properties, sweet potato pectin is
comparable to apple pectin. It is obtained as a
by-product from peel and trim wastes of sweet potato
canneries. The starch present is solubilized by
treatment with amylase, and the residue employed for
pectin extraction. The pectin is recovered from the
pulp residue after starch recovery.

OTHER TUBER CROPS

Aroids

Aroids are washed, prepared either whole or in

306

slices, then cooked or baked to eliminate alkaloids or other acrid principals. Even the best variety among the cultivated aroids can cause irritation in the throat or a feeling of discomfort when eaten raw or cooked, due to acridity attributed to calcium oxalate raphides.

Colocasia tubers are peeled, sliced, cooked, and eaten with condiments and adjuncts. Young leaves and stalks are edible and can be cooked and used like spinach or sage. Leaves which are unopened or just about to open are more satisfactory than the older leaves. Tubers of Amorphophallus are used as vegetables or for preparing chips.

Yams

The alkaloid diosgenin and its saponin precursor occur in varying quantities in different species of yams. Diosgenin is abundant in Dioscorea hispida, and its tubers, when consumed in large quantities, cause paralysis of the respiratory organs and even death. The tubers of D. deltoidea are rich in saponins and are used for washing silk, wood, and hair, and as a fish poison. Tubers of D. alata are starchy and can be dried and ground into meal or used as vegetables in the same manner as potato. Their usage is gradually declining owing to the increasing popularity of potatoes. The hill tribes use the tuber as a substitute for rice.

D. bulbifera tubers are used in Kashmir for washing wood and as fish bait. Dried and pounded tubers are used as a medicine for treatment of ulcers, hemorrhoids, dysentery, and syphilis. D. floribunda is used for extraction of diosgenin, which is being used to control ovulation in humans and animals.

Arrowroot

In India, Maranta arundinacea and Curcuma angustifolia are the major arrowroot producing crops. These plants are cultivated to a small extent but not for commercial purposes on a large scale. The starchy flour obtained from the rhizomatous tubers of these plants is valued as food, especially for infants, invalids, and convalescents. It is also employed in the production of biscuits, cakes, puddings, and jellies. Arrowroot starch is also used as a base for face powder and in the preparation of special glues.

Among the tuber crops the processing technology for cassava has been well developed, whereas in the case of yams and aroids not enough work has been done. In view of their importance as food as well as a source of raw material for innumerable industries, research efforts are now being made at CTCRI to develop suitable processing techniques for new value-added products so that cultivation of these crops may become commercially viable.

REFERENCES

Balagopal, C. and Maini, S.B. 1976. Evaluation of certain species of fungi for single cell protein and amylase production in cassava liquid medium. J. Root Crops 2:49.
_____, Maini, S.B., and Hrishi, N. 1977. Storage of cassava roots in moist sawdust. J. Root Crops 4.
_____ and Nair, P.G. 1976. Studies on the microflora of market cassava chips of Kerala. J. Root Crops 2:57.
Booth, R.H. 1975. Cassava Storage. Centro Intern. Ag. Trop. (CIAT) Bull., Cali, Colombia.
_____ and Coursey, D.G. Storage of cassava roots and related post-harvest problems. In Cassava Processing and Storage; Proceedings of an International Workshop, Pattaya, Thailand, 17-19 April 1974. Intern. Devel. Res. Centre IDRC-031e.
Ingram, J.S. and Humphries, J.R.C. 1972. Cassava storage; a review. Trop. Sci. 14:131.
Maini, S.B. and Balagopal, C. 1978. Biochemical changes during the storage of cassava roots. J. Root Crops 4:31.
Singh, K.K. and Mathur, P.B. 1953. Cold storage of tapioca roots. Mysore, CFTRI Bull. 2:181.

N. Hrishi and C. Balagopal: Central Tuber Crops Research Institute, Trivandrum, India

9
Mechanical Devices for Peeling Cassava Roots

E. U. Odigboh

ABSTRACT

Peeling of cassava roots constitutes a major bottleneck in cassava processing. This paper discusses the design and construction of two mechanical cassava peeling devices. The first device is a continuous-process mechanical peeler which consists of a cylindrical knife assembly and a solid cylinder, both mounted parallel and 20 mm apart on an inclined frame. Since an earlier report on this device (Odigboh, 1976a), various improvements have been made. The second device is a batch-process mechanical peeling machine which works on the principle of abrasion. Basically, unpeeled cassava roots mixed with some inert abrasive materials and placed in a vessel are rotated for some time to peel them. Results of performance tests on the prototype device are presented.

INTRODUCTION

The need to mechanize the peeling of cassava roots has been recognized for some time. Manual peeling of the roots is tedious, wasteful, and time-consuming, especially for large-scale factory operations. In fact, cassava peeling constitutes a major bottleneck in cassava processing (Odigboh, 1976b). This paper discusses the design and construction of two types of cassava peeling machines.

THE PROBLEM

Cassava roots vary widely in size and shape
among the numerous cultivars as well as within each
cultivar. While the spindle or carrot-like shape
predominates in most varieties, elongated ovoid and
barrel shapes are common. The roots are irregularly
curved with cross-sections that deviate considerably
from a circle. Variability in cassava peel thickness, texture and strength of adhesion to the root
flesh constitutes another problem. Furthermore,
physical and mechanical properties of the roots
depend on the age of the roots at time of harvest,
which varies from 12 to 48 months.

A mechanical device for effective peeling of
cassava roots must be capable of peeling the roots
of all varieties despite variations in physical and
mechanical properties.

THE CONTINUOUS-PROCESS PEELER

Two views of the continuous-process peeler
prototype are shown in Figure 1. Figure 2 shows the
action zone of the machine.

The machine consists of a cylindrical knife
assembly A, and a roller with roughened surface B,
which are mounted parallel and 20 mm apart on a
frame C, inclined at 15° to the horizontal. The
cassava roots to be peeled and cut into 100 mm
pieces are introduced lengthwise through the feed
channel J at the elevated end of the space between
A and B. The machine is powered by a 1 hp single
phase electric motor.

When the machine is started, A and B both rotate clockwise at 200 rpm and 88 rpm, respectively,
while the cassava root pieces rotate counterclockwise and move down the incline as the knives of A
progressively peel them. There is a continuous flow
of the cassava root pieces which enter one end of
the machine unpeeled and come out the other peeled.

The pusher system F, shown in Figures 1 and 2,
was originally intended to help push the cassava
pieces along as reported earlier (Odigboh, 1976a).
This has been discarded since the cassava pieces can
move along properly without it. Also, a new
arrangement, shown in Figure 3, is used to provide
necessary pressure on the roots for effective peeling. It comprises a chain of 12.7 mm diameter mild
steel rollers, secured such that their axes of
rotation are perpendicular to the roots' path of

310

(a) Knife side view

(b) Roller side view

Figure 1. The continuous-process cassava peeler. A - knife
assembly; B - roller; C - frame; D - bevel gear speed redu-
cer; F - pusher system; G - worm gear; H - meshing gear;
K - electric motor; L - side guards; J - feeding channel;
P - freely rotating pipe

Figure 2. Knife assembly A and roller B
as mounted on Frame C.

travel through the machine and so offers no restric-
tion to the forward motion and revolution of the
roots. Hinged only on one side, individual rollers
can move up and down with enough independence of the
others to allow for the different root sizes, while
at the same time providing the necessary peeling
pressure.

It is easy to visualize that the system described here can handle all cassava root sizes. Since
the cassava roots are conical in shape, they are
easily cut into straight pieces which then become
practically truncated cones. A truncated cone of
cassava laid on the space between two parallel
rollers nearly in contact is thus contacted all
along its length as shown in Figure 4. The diameter

312

of 92 mm for the knife assembly A and roller B and
their separation of 20 mm were chosen on the basis
of the usual root sizes of the major cassava cultivars, to ensure that the tail ends of the roots are
not crushed instead of peeled. As A rotates, it
brings the knives into position to perform the peeling operation, while B serves to position the roots
for peeling. Because both A and B rotate clockwise
but at different speeds, they cause a fast counterclockwise revolution of the truncated cassava cones
about their long axis, thereby facilitating peeling.

Figure 3. View of peeler showing peeling
           pressure arrangement.
           P - Peeling pressure chain of rollers

313

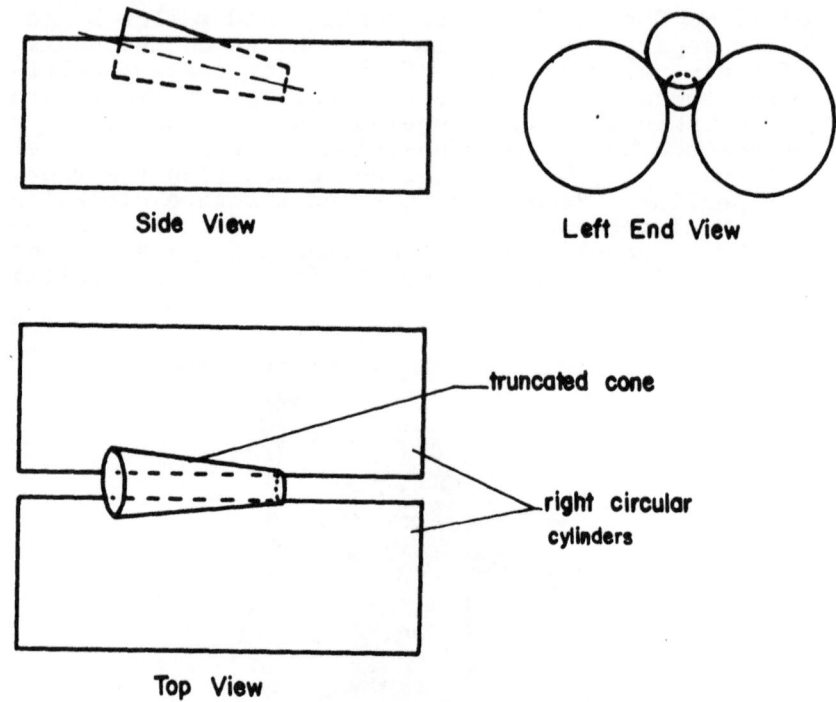

Side View

Left End View

truncated cone

right circular cylinders

Top View

Figure 4. Locating a truncated cone between two parallel cylinders nearly in contact.

Performance of the Continuous-Process Peeler

As detailed earlier (Odibgoh, 1976a), the machine can peel mixed sizes of cassava with an efficiency of about 75 percent, but it performs better with sized lots of cassava roots. When set up for a specific range of root sizes, the peeling efficiency is over 95 percent. On the average, the machine has a through-put rate of 185 kg/h. However, hand trimming is not eliminated, especially for unsized lots of cassava roots and for cassava root sizes below 40 mm diameter. Furthermore, the requirement that the roots be cut into approximately 100 mm straight lengths and then sized is time-consuming and therefore a serious disadvantage.

THE BATCH ABRASION PEELER

While watching performance tests on the continuous-process peeler, a colleague suggested an
investigation into the possibility of peeling cassava by abrasion. Consequently, the batch abrasion
cassava peeling machine was designed and constructed.

Figure 5 gives orthographic views of the prototype batch peeler. It consists simply of a 210L
drum mounted eccentrically on a shaft on a frame.
The drum is rotated at 40 rpm, through V-belt drives
from a 0.56 KW, 1440 rpm electric motor. A 200 x
150 mm opening cut near one end of the surface of
the drum provides for the introduction and removal
of cassava roots from the drum. The opening provided with a gasketed cover is watertight when
closed. The beauty of the design is its simplicty
of construction; an assembly of one drum, one shaft,
one electric motor, and a speed reduction V-belt
drive system mounted on a framework is all that is
required to give the complex side-to-side and up-
and-down motions of the drum to do the peeling.

Operation Principle

To operate, unpeeled roots are loaded into the
drum. A quantity of some inert abrasive material is
mixed with the roots. The drum, with its charge of
root and abrasive mixture, with or without added
water, is rotated until the cortex is completely
abraded.

The speed of rotation is critical for proper
operation of the machine. Shown in Figure 6 are
some of the forces acting on a cassava root within
the drum. Due to the eccentricity of mounting,
different points of the drum surface are at different radial accelerations depending on their radial
distances from the axis of rotation. The centrifugal force resulting from the speed of rotation must
not exceed the weight of a root at any point of the
drum. Otherwise, the root would be pressed against
the drum's inner surface and carried round and round
without a relative motion. For peeling to occur,
the rpm must be such as to allow the roots to move
relative to the drum or, at the worst, to cascade
within the drum. The angle θ was determined
experimentally as 20.5°. The actual design angle
of inclination used is 30°.

315

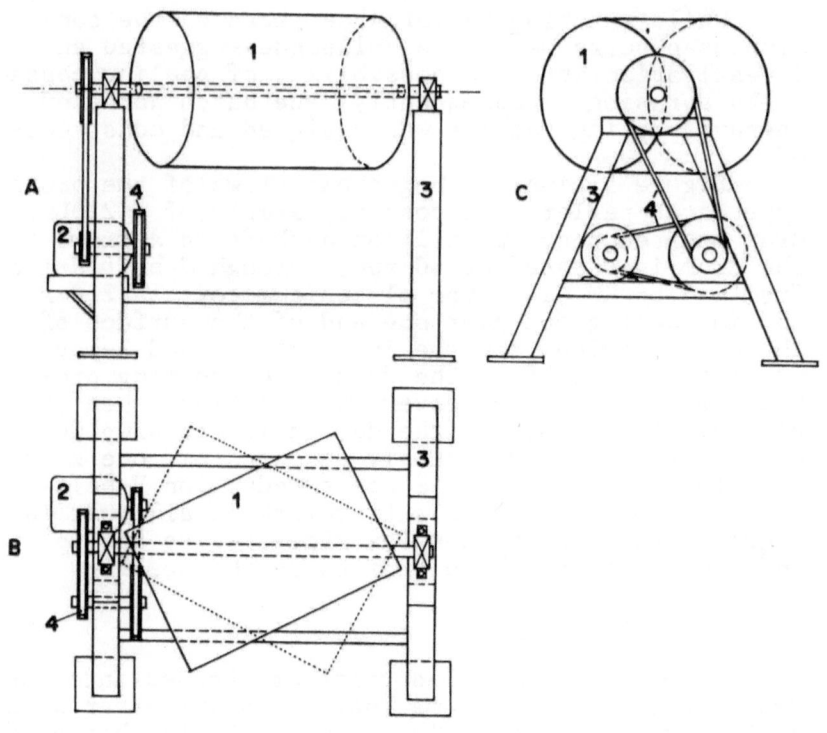

Figure 5. Orthographic views of prototype
abrasive cassava peeler.

A - front view; B - top view;
C - side view; 1 - drum;
2 - electric motor; 3 - frame;
4 - belt drives

Preliminary Performance Tests

Peeling effectiveness of the machine may be
assessed by the thoroughness of peeling in minimum
possible time. Peeling effectiveness depends on the
abrasive used, the quantity of cassava in the batch,
the ratio of abrasive to roots, and the quantity of
water used. Performance tests were designed to
determine the optimum combinations of these factors.
Table 1 presents some results of the tests.

316

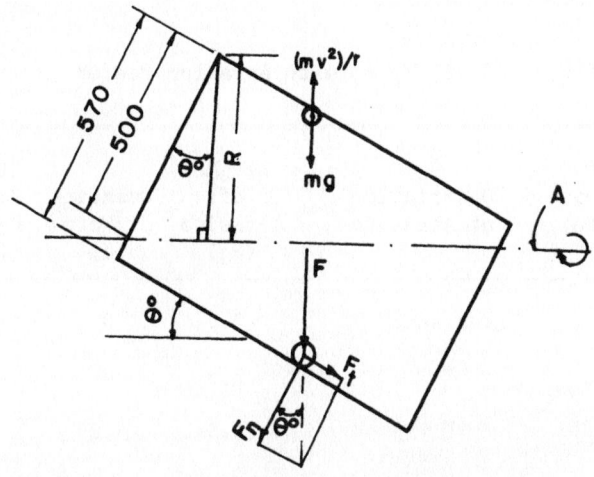

Figure 6. Some of the forces on a particle within the rotating drum.

A $\quad$ = axis of rotation
$\theta°$ $\quad$ = drum slope (=3°)
R $\quad$ = 500 cos $\theta°$ (=433 mm)
mv^2/r = centrifugal force
mg $\quad$ = weight of a root
F $\quad$ = $(mv^2/r + mg)$
F_n $\quad$ = F cos $\theta°$

DISCUSSION OF RESULTS

In general, thorough peeling was achieved. The abrasive got to all parts of the root and abraded the cortex without noticeable loss of the root flesh, irrespective of the variability in the root shape and size. Figure 7 shows a photograph of the peeled roots, illustrating the thoroughness and uniformity of peeling over the root contours.

However, the peeling rate was not very high. As evident from Table 1, peeling with added water was faster than without water. Quarry sand, being sharper and rougher, proved a better abrasive than river sand, and gave the fastest rate of peeling. Apparently, the best weight ratios of roots to abrasives were between 4:1 and 5:1. The critical speed which allows this to happen occurs when the centrifugal force is less than the weight of the

317

TABLE 1
Summary of Preliminary Tests with Abrasion Peeler

Load of Cassava Roots Per Batch (kg)	Description of Abrasive	Weight of Abrasive (kg)	Quantity of Water (1)	Time to Complete Peeling (min)
10	River sand	2.0	0	40
10	"	2.5	0	35
10	"	2.5	5.0	30
20	"	2.5	5.0	35
20	"	5.0	5.0	30
20	50% river sand +50% quarry sand	5.0	5.0	25
20	Quarry sand	5.0	5.0	20
20	Quartziferous rough gravel	2.5	5.0	20
20	Quarry stone passing through 4.8 mm (3/16") and retained on 1.6 mm (1/16"0 sieves	5.0	5.0	15
30	"	5.0	7.5	20
40	"	5.0	7.5	25
40	"	7.5	7.5	22
40	Quarry stone + 20% quarry sand	7.5	7.5	18

Figure 7. Cassava roots peeled with
batch precision peeler.

materials within the drum. Thus, as shown in Figure
6,

$$(mv^2)/r < mg$$

or $(2\pi rn)^2 < gr$

and $\quad n \quad < (1/2\pi) \; \sqrt{} \; (g/r)$

For the specific geometry and dimensions of the
prototype machine, the highest value of rpm satis-
fying this inequality at all points of the drum
surface is obtained at the largest radial distance
of \$ = 433 mm as shown in Figure 6. The critical
rpm may therefore be computed as

$$n = (1/2\pi) \; \sqrt{} \; (g/43.3) = 45.5 \text{ rpm}$$

The actual design speed is 40 rpm, as already
stated.

In addition, the inclination of the drum
resulting from the eccentric mounting must be enough
to allow the roots to roll down the slope as the
drum moves up and down and from side to side. Thus,
the frictional resistance must be less than the

319

component of weight and centrifugal force down the incline. That is, referring again to Figure 6,

$$\mu \ (W+(mv^2)/r)\cos \ \theta < (W+(mv^2)/r) \ \sin \ \theta$$

or $\mu < tg \ \theta$

where μ = coefficient of friction.

It was observed that the abraded peels were finely divided, and tended to foul and thereby dull the abrasives as peeling progressed. This fouling or dulling effect was most pronounced for the river sand and least for the coarser quarry stone. It was also observed that the fouling of the abrasive was more pronounced when the quantity of water was so small as to cause a pasty mixture of the abrasives and the peels. On the other hand, an excessive quantity of water tended to float the cassava roots, with a resultant reduction in the effectiveness of abrasion. Another observation was that the finer the abrasive, the smoother the surface of the peeled roots. But the faster rate of peeling possible with the coarser abrasives is considered an advantage greater than the disadvantage of relatively rougher surface of the peeled roots.
These observations indicated possible refinements on the machine to achieve faster peeling rates.

REFINEMENTS ON THE BATCH PEELER

Following the preliminary performance tests, certain refinements were made on the drum unit of the peeling machine to improve peeling effectiveness. The inner surface of the drum was covered with a sheet of 5 x 5 mm rhombic expanded metal so that the cassava roots would be in contact with the abrasive surface of the expanded metal instead of the relatively non-abrasive drum surface. Since, as explained previously, the abraded peel of the roots was finely divided, it would readily fall through the expanded metal onto the drum beneath, thereby minimizing the fouling of the abrasives. Two arrangements of the expanded metal shown in Figure 8 were experimented with. In the arrangement shown in Figure 8a, the expanded metal was fastened on the inside of a framework of 6 mm diameter rods. Thus, the resultant cylinder of expanded metal was concentric with, and at a separation of, about 6 mm

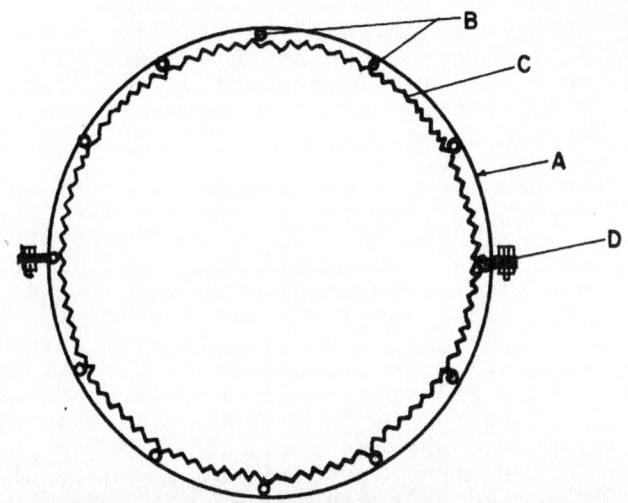

Figure 8a. Cross-section through modified drum unit showing
details of modifications. A - Drum half; B -
framework of rods; C - expanded metal; D - flanges.

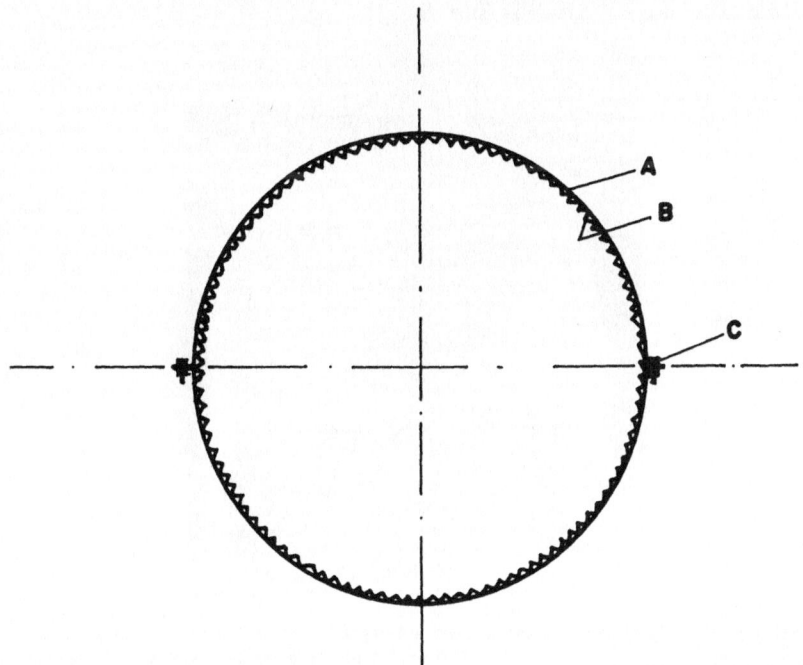

Figure 8b. Cross-section through drum unit showing details of
modifications. A - Drum in two halves; B - expanded metal; C - bolted flanges.

321

from the drum inner surface. In the arrangement shown in Figure 8b, the expanded metal was attached directly on the inside surface of the drum. Use was made of adequately coarse abrasives which therefore projected to about the same surface, as formed by the expanded metal.

The drum itself was perforated with apertures small enough to prevent loss of the abrasives through them. The portion of the shaft within the drum, which was a 48 mm diameter galvanized iron pipe, was also perforated. One end of the shaft was sealed. The other end was fitted with a small ballbearing holding a small polyvinyl chloride water hose, through which water was continuously sprayed onto the mixture of cassava roots and abrasives. The water would continuously wash off and carry away the abraded peels through the perforations in the drum. A shield at one side of the drum prevented splashing of water. The drum was cut in two halves and held together only by bolting through flanges that were provided, as shown in Figure 8, to facilitate disassembly for thorough cleaning of the unit. The modified prototype batch peeler is shown in Figure 9.

Figure 9. Prototype abrasion cassava peeler with loading/unloading gate F open to show expanded metal covering; A, perforated drum; B, frame/stand; C, belt-drive cover; D, water hose.

322

Performance Tests on Peeler with Modified Drum Unit

To assess the effectiveness of the refinements on the drum unit, tests similar to those discussed previously were performed. Table 2 gives the summary of the tests.

Discussion of Test Results

The objectives which prompted the modifications on the drum unit were mostly realized. Specifically, the expanded metal mesh effectively contributed to the peeling of the cassava roots, as shown by the result of tests conducted without abrasives. Water sprayed onto the cassava roots and abrasives continuously washed off the abraded peels and carried them away through the performations in the drum. Fouling of the abrasives was prevented, and the

TABLE 2
Summary of Tests on Modified Prototype Peeler

Load of Cassava Roots Per Batch (kg)	Description of Abrasive[a]	Weight of Abrasive (kg)	Water Flowrate (1/min)	Time to Complete Peeling (min)
20	–	–	1.0	25
20	1	2.5	1.0	15
20	1	5.0	1.0	10
20	2	5.0	1.5	7
30	2	5.0	2.0	12
40	2	5.0	2.0	16
40	2	7.5	2.0	13
50	2	7.5	2.0	18
50	50% each 1&2	7.5	2.0	17

[a] 1 - Quartziferous rough gravel
   2 - Quarry stone passing through 4 mm and retained on 1.6 mm sieves

323

abrasives were easily recovered for re-use. It was easy to observe when peeling was completed, since the peels did not obstruct the view. As a result of these improvements, peeling rate was substantially increased (contrast results in Table 2 with Table 1). However, the expanded metal arrangement of Figure 8a was not as effective as that of Figure 8b. This was because the abrasives that could be utilized effectively were too large for the purpose.

The higher the water flow rate, the more effective the peeling process. But the necessity of large quantities of water may prove a disadvantage. It would, of course, be possible to filter and recycle the water. In any case, the peeled roots were clean and required no further washing, as was the case when water did not run continuously.

Use of a mixture of various sizes of abrasives was found more effective than abrasives of one size range. It was determined that the maximum load of cassava roots per batch was 50 kg, at which the capacity of the peeling machine was about 180 kg/h, discounting loading and unloading times. It might be possible to increase the capacity by using a bigger electric motor and a bigger drum unit.

SUMMARY AND CONCLUSION

As an overall evaluation, it may be stated that the batch abrasion peeler seems to be a better answer for elimination of the drudgery involved in peeling cassava roots. It does a better job than peeling with hot lye, which poses the problem of partial cooking and gelatinization of outer layers of the useful root flesh. The abrasion peeling machine eliminates the need to cut the roots into straight lengths or to size the roots prior to peeling as for the continuous-process peeler, since it can handle cassava roots of all sizes and shapes. The peeling achieved is so thorough that hand trimming is virtually eliminated. If the duration is closely watched, peeling is achieved with negligible loss of useful root flesh. This is an important advantage when it is considered that manual peeling as done in gari factories leads to losses of 10 to 15 percent of the useful root flesh.[1]
The abrasion machine is simple to construct and can, with or without modifications, find application in peeling other root and tuber vegetables for institutional feeding or for public eating places.

NOTES

1. D.D. Ibe, Technologist in Charge, University
   of Nigeria, Nsukka, personal communication.

REFERENCES

Odigboh, E.U. 1976a. A cassava peeling machine:
development, design, and construction. J. Agric.
Eng. Res. 3:361.

_____, 1976b. Mechanization of Nigerian
cassava production and processing: research needs
and interests. J. Inter. Agric. Eng. 31:20.

*E. U. Odigboh: Agricultural Engineering Department, University
of Nigeria, Nsukka, Nigeria*

10
Cassava Chips Processing and Drying: A Cassava Chipping Machine

E. U. Odigboh

ABSTRACT

No effective or generally applicable storage technology exists to prevent the fast post-harvest deterioration of cassava roots. So they must be processed soon after harvest. When intended for feedstuffs, non-traditional food industries, or export, the roots need to be processed into chips to facilitate conversion into a dried product with a better keeping quality. This paper describes the design of a chipping machine and some drying experiments on the chips. The results showed that although all drying occurred in the falling-rate period, sun-drying of machine-produced chips was fast and at a rate that remained nearly constant until a moisture content of about 20 percent (wet basis). For some hand-prepared 2.5 x 5 x 50 mm chips tested, drying was much slower with more pronounced variable-rate drying characteristics.

INTRODUCTION

Mechanization of Nigerian cassava production and processing--in particular research needs--were discussed in an earlier paper (Odigboh, 1976). However, that paper did not stress sufficiently the problem posed by the highly perishable nature of fresh cassava roots. Several authors (Booth, 1975; Booth and Coursey, 1974; Ingram and

327

Humphries, 1972) have reported on various attempts to solve the storage problems. Still, no effective or generally applicable storage technology exists to combat the fast post-harvest deterioration of cassava roots. They must therefore be processed soon after harvest into some form of dried products, especially when intended for feedstuffs, non-traditional food uses, or export.

To facilitate drying, the roots need to be made into small chips. In his listing of commercially available machinery for cassava processing, Ingram (1972) admitted that no standard machinery of European manufacture for chipping of cassava is readily available. Manurung (1974) reported on Malaysian and Thai cassava chipping machines which were described as satisfactory. This paper describes a cassava chipping machine, designed and built at the request of the National Root Crop Research Institute at Umuahia, Nigeria. Some sun-drying experiments on the chips are also discussed.

THE CHIPPING MACHINE

The cassava chipping machine is shown in Figure 1. The machine consists of an assembly of three knives, A, similar in construction to the assembly of a wood planer (Figure 2). The 450 mm long knives are held in slots in a 93.5 mm diameter cylinder, such that the projection of the knives above the cylinder.surface is adjustable. The knife assembly, A, is mounted horizontally on a frame. Parallel to A, a second cylinder, B, 56 mm in diameter, is mounted higher and at a 2.5 mm gap from A (Figure 2). The hopper, C, holds the cassava roots, R, which are fed lengthwise into the chipping zone between A and B. The roots are cut into 150 to 200 mm lengths to reduce their natural curvature, and so facilitate the horizontal positioning of the roots when they meet the knives. The hopper, C, is partitioned along its length to match the cut root lengths and along its width to accommodate one or two root diameters (Figure 1). Thus the hopper consists of compartments in which 150 to 200 mm lengths of cassava roots are stacked one on top of the other to facilitate their flow down to the knives for chipping.

The knives, A, rotate clockwise at 375 rpm, and are belt driven from a 1-hp electric motor. Cylinder B is rotated counterclockwise at 95 rpm

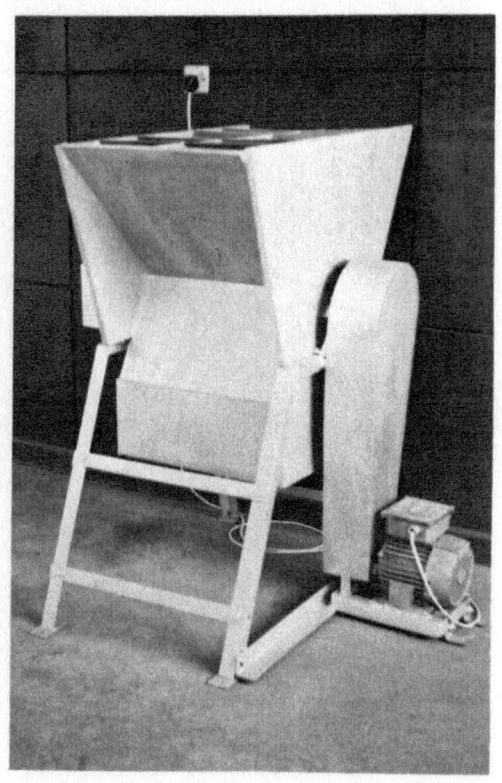

Figure 1. The cassava chipping machine.

by direct gearing to A. The cassava root in the
chipping zone, in contact with A and B, also rotates counterclockwise, as the knives of A progressively remove chips from it. The weight of the
roots stacked on top, aided by the position and
counterclockwise rotation of cylinder B, provide the
pressure to push the root onto the knives for chipping. The bottom part of the hopper, D, stops the
chips as they are flung from the knives and directs
them for collection below.

329

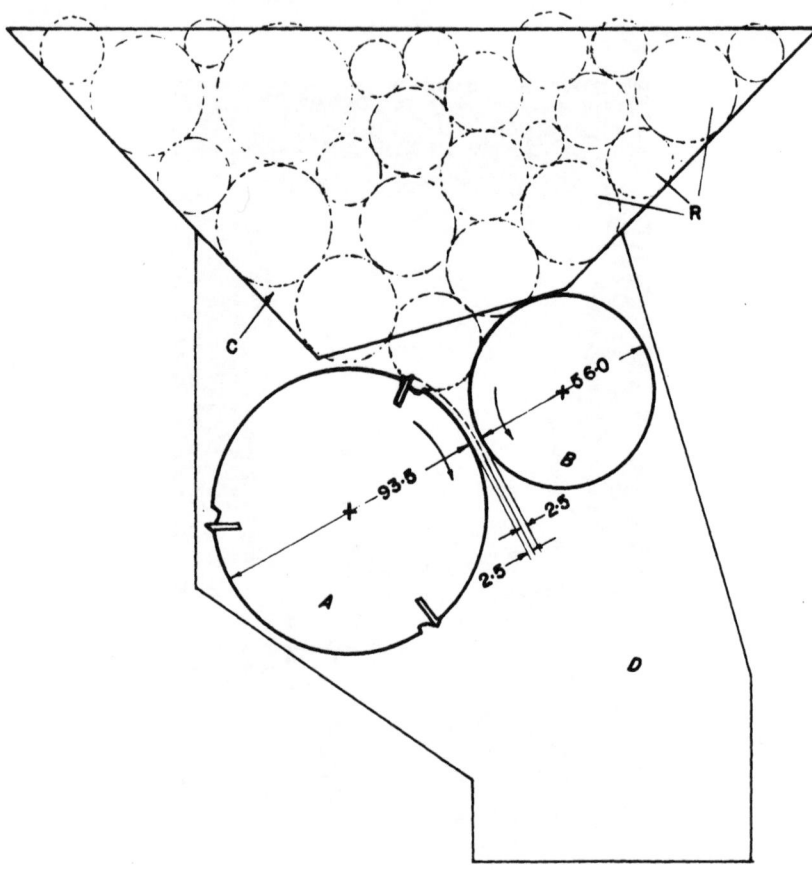

Figure 2. Sketch of a cross-sectional view of the machine showing knife assembly A, roller B, hopper C and chip collector D and cassava roots R.

Machine Performance

Dimensions of chips from the machine depend on:

1. The rpm of the knives;
2. Projection of knives above the cylinder surface;
3. Pressure of cassava roots against the knives during chipping;

4. Orientation of roots in the chipping zone relative to the knives; and
5. Varietal and other properties of the cassava roots.

Set up as described--with the knives projecting 2.5 mm above the cylinder surface, rotating at 375 rpm, and maintaining a minimum of three cassava root pieces stacked in each compartment of the hopper-- the machine produced chips 3 to 5 mm wide and 2 to 3 mm thick. Chips which were 150 to 200 mm long when cut broke up into 30 to 75 mm lengths when flung from the knives. The percentage of finer particles ranged between 1 and 3 percent. Figure 3 shows the machine-produced chips in a collection tray. The machine chipping process imparted a spongy texture to the chips (Figure 3). This probably had some advantageous effects on the drying of the chips. The machine had an average through-put of 225 kg of chips per hour under the conditions stated. The machine can be modified easily for manual operation to make it more suitable for use by individual

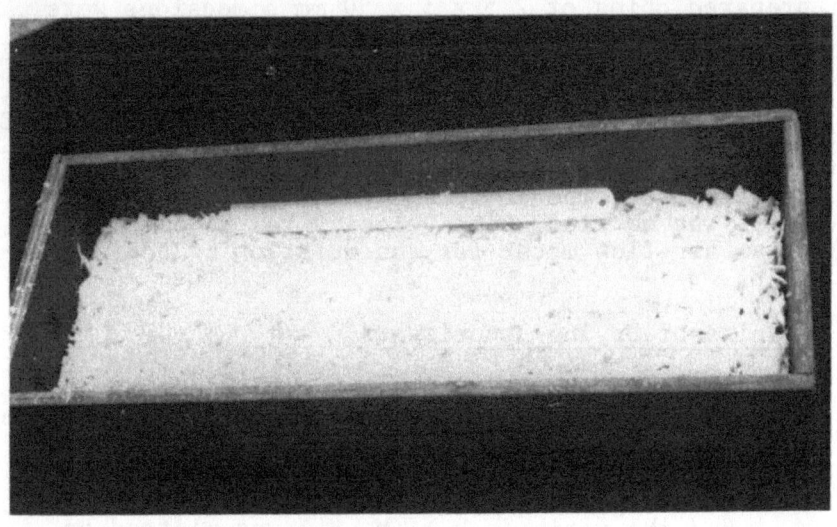

Figure 3. Machine-produced chips in a collection tray.

farmers. However, it was primarily intended to aid local small- to medium-scale industries involved in cassava-based feedstuff production, non-traditional food uses, or export.

DRYING EXPERIMENTS

Eighteen-month-old cassava roots of the 60444 variety from the University of Nigeria Faculty of Agriculture farms were used for the experiments. The peeled roots had the following approximate analysis:

Component	Percentage
Moisture content (wet basis)	62.40
Protein	0.16
Fat	0.09
Ash	0.14
Carbohydrate (starch)	37.21 (by subtraction)

Chips produced by the machine as well as hand-prepared chips of 2.5 x 5 x 50 mm dimensions were tested in drying experiments. Chip samples were dried in an oven at 28°C to compare with sun-drying results. Chips were spread in a single layer on drying trays for both the sun-drying and oven-drying tests (Figure 4). Results of the experiments are presented in Tables 1 and 2. Table 1 also gives the ambient temperature and relative humidity. There was no wind; wind velocity was not measurable on an air-flow meter for the duration of the experiments.

Results of Drying Experiments

The behavior of the cassava chips during the drying experiments are presented in Figure 5. Given that the moisture content dry-basis at a given time was denoted as M, the ultimate constant moisture content as M_e, and the initial moisture content as M_o; when the ratio $(M-M_e)/(M_o-M_e)$ was plotted on semilogarithmic paper against time for the machine-produced sun-dried chips, the points fell on two separate straight lines MM and NN (Figure 6). This seems to indicate the existence of two falling-rate periods in the moisture content range 1.5 to 0.15

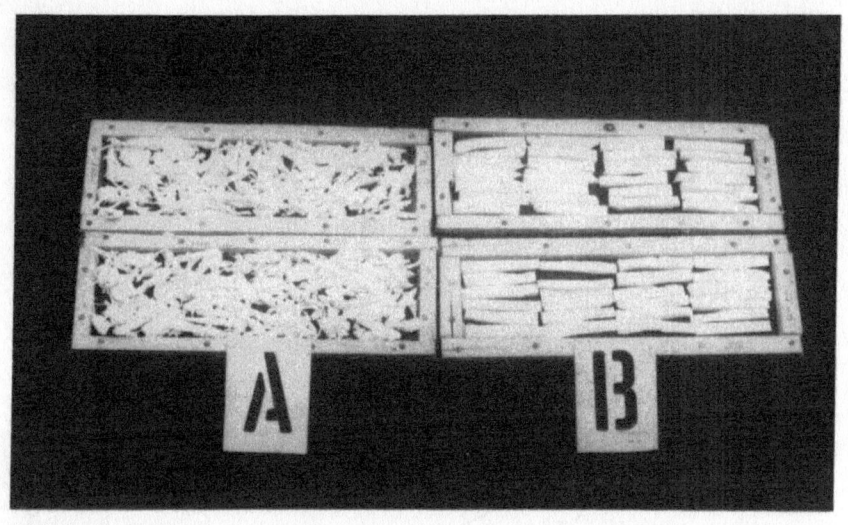

Figure 4.   Chips in test-drying trays.
A - machine-produced
B - hand-prepared

g/g dry matter. The validity of this treatment may
be questionable since the drying conditions of temperature, relative humidity and even air velocity
were not constant. The ambient relative humicity
was lower and the temperatures higher during the
later stage of sun-drying. This may explain the
slope of the NN portion of the plot in Figure 6.
Still, the chips apparently dried entirely in the
falling rate period. Also, the machine-produced
chips dried appreciably faster than the hand-prepared chips. Its rate of drying in the sun remained
nearly constant until a moisture content of about 20
percent wet basis was realized, while the hand-prepared chips exhibited more pronounced variable-rate
drying characteristics. Machine-produced chips
dried down ultimately to only 50 percent under the
same conditions. The faster sun-drying rate of the
machine-produced chips was probably due to the
spongy texture imparted to the chips by the chipping
process. Chips dried at a slower rate in the oven
(Figure 5) even though the oven temperature was set
at about the same value as the average outdoor

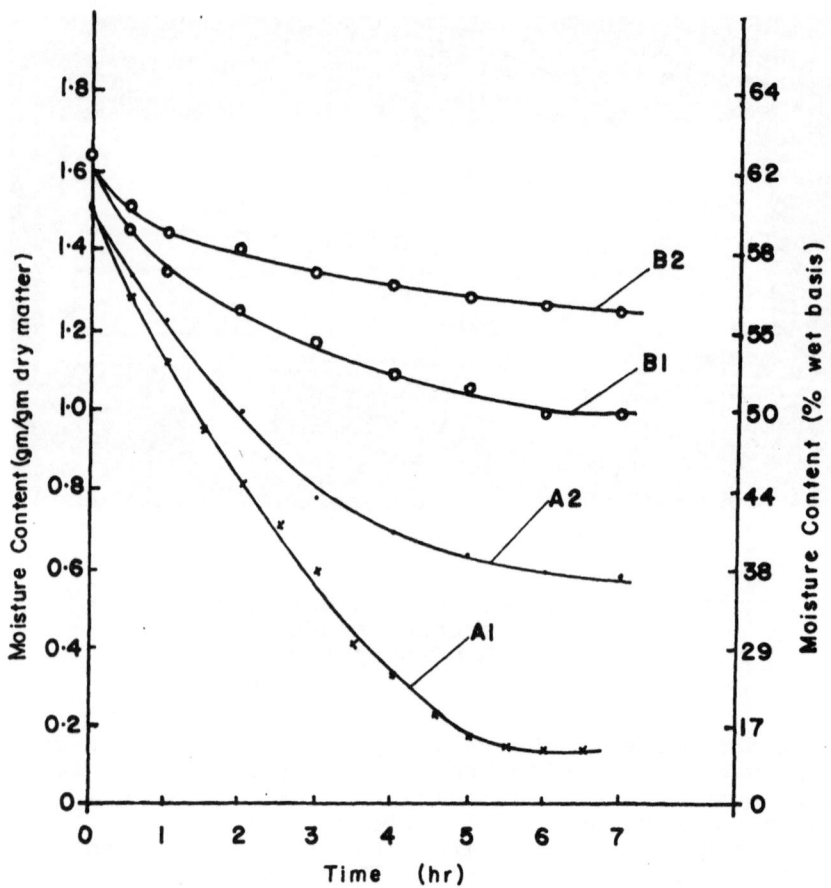

Figure 5. Drying curves for cassava chips.

    A - Machine-produced chips
    B - Hand-prepared chips
    1 - Sun-dried
    2 - Oven-dried

TABLE 1
Sun-Drying of Cassava Chips[a]

Time	Ambient Temperature (°C)	Ambient Relative Humidity (%)	Machine-Produced Chips: Sample A Weights (g)	Hand-Prepared 2.5 x 5 x 50 mm Chips: Sample B Weights (g)	Sample A Average Moisture Content (g/g dry matter)	Sample B Average Moisture Content (g/g dry matter)
9.10 am	25	72	34.2	40.4	1.51	1.64
9.40 "	25		31.0	37.3	1.28	1.44
10.10 "	27	71	29.3	35.8	1.15	1.34
10.40 "	27		26.5		0.95	
11.13 "	28	63	24.8	34.4	0.82	1.25
11.40 "	27		23.3		0.71	
12.12 pm	28	58	21.6	33.2	0.59	1.17
12.40 "	29		19.2		0.41	
1.10 "	30	55	18.3	32.0	0.34	1.09
1.42 "	29	16	16.5		0.21	
2.10 "	29	52	15.9	31.4	0.17	1.05
2.42 "	28		15.6		0.15	
3.10 "	28	52	15.5	30.5	0.14	0.99
3.40 "	27		15.5		0.14	
4.10 "	28	56		30.5	-	0.99

	Average Temperature 27.7°C	Average Relative Humidity 59.9%	Bone Dry Weight 13.6 g	Bone Dry Weight 15.3 g		

[a] Experiment conducted on August 10, 1976.

temperature. Also, oven-dried chips were no longer white, having acquired a brown tinge. In contrast, sun-dried chips remained white even after several hours in the oven at 100°C for moisture content determination. To study this further, samples of wet chips were put in the oven at various temperatures ranging from 60°C to 150°C, and the time taken for chips to acquire pronounced browning at each temperature was recorded. Observation indicated that chips became brown probably because of scorching. Hand-prepared chips were found to be more sensitive to browning than the machine-produced chips. The observed sensitivity of chips to browning is presented in Table 3.

TABLE 2
Oven-Drying of Cassava Chips

Time (hr)	Oven Temperature (°C)	Machine-Produced Chips: Sample A Weight (g)	Hand-prepared 2.5 x 5 x 50 mm Chips: Sample B Weight (g)	Sample A Moisture Content (g/g dry matter)	Sample B Moisture Content (g/g dry matter)
0	28.0	40.8	38.9	1.52	1.63
0.5		37.9	37.0	1.34	1.50
1	27.8	36.0	36.1	1.22	1.44
2		32.4	35.5	1.00	1.40
3	28.1	28.9	34.6	0.78	1.34
4		27.4	34.1	0.69	1.30
5	28.0	26.6	33.8	0.64	1.28
6		25.8	33.6	0.59	1.27
7	28.0	25.6	33.3	0.58	1.25

Average Oven Temperature 27.98°C	Bone Dry Weight 16.2 g	Bone Dry Weight 14.8 g

TABLE 3
Sensitivity of Chips to Browning

	Time Required to Brown (min)							
	Oven Temperature (°C)							
Sample	60	70	80	90	100	110	120	150
Machine-produced chips at 61% moisture content, wet basis	74	61	39	31	19	16	11	6
2.5 x 5 x 50 mm hand-prepared chips at 62.4% moisture content, wet basis	56	35	21	14	10	9	7	4

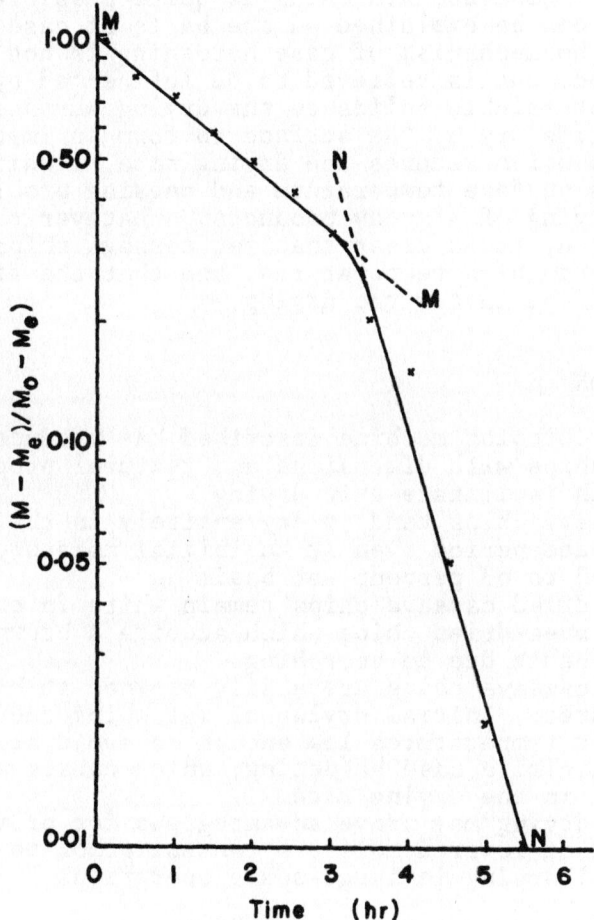

Figure 6. Falling-rate periods of machine-produced cassava chips during sun-drying conditions of Table 1.

Browning could be due also to some chemical reaction. However, scorching is quite possible since it can be explained on the basis of case hardening. The mechanism of case hardening is not fully understood, but is believed to be influenced by migration of soluble solids to the drying surface. These solids dry at the surface to form an impermeable skin which reduces the drying rate, leading to a rise in surface temperature and causing problems in the drying of starchy products. Whatever the explanation, it is clear that wet cassava chips are sensitive to high temperatures, and that the finer the chips the faster the drying.

CONCLUSION

The chipping machine described has produced cassava chips with dimensions and textural properties which facilitate chip drying.

Cassava chips tend to dry entirely in the falling-rate period even at an initial moisture content of 60 to 63 percent wet basis.

Sun-dried cassava chips remain white in contrast to oven-dried chips which acquire a brownish tinge probably due to scorching.

Wet cassava chips are easily browned at high temperatures. Initial drying of wet chips should be done at temperatures low enough to avoid browning and minimize case hardening, which causes a reduction in the drying rate.

Sun-drying may prove advantageous for providing chips of lowered moisture content prior to commercial drying in large-scale operations.

REFERENCES

Booth, R.H. 1975. Cassava Storage. CIAT Publication, Series EE16. Centre Internacional de Agricultura Tropical, Cali, Colombia.

_____ and Coursey, D.G. 1974. Storage of cassava roots and related post-harvest problems. In: Cassava Processing and Storage; Proceedings of an Interdisciplinary Workshop, Pattaya, Thailand, 12-19 April 1974. Int. Dev. Res. Centre, IDRC-031e.

Ingram, J.S. 1972. <u>Cassava Processing: Commercially Available Machinery</u>. Tropical Product Institute, London.

_____ and Humphries, J.R.O. 1972. Cassava storage: a review. <u>Trop. Sci</u>. 14:131.

Manurung, F. 1974. Technology of cassava chips and pellets processing in Indonesia, Malaysia and Thailand. In: Cassava Processing and Storage; Proceedings of an Interdisciplinary Workshop, Pattaya, Thailand, 12-19 April 1974. Int. Dev. Res. Centre, IDRC-031e.

Odigboh, E.U. 1976. Mechanization of Nigerian cassava production and processing: research needs and interests. <u>Agr. Engr</u>. 31:20.

ACKNOWLEDGEMENT

The work described in this paper is part of a project funded by a grant from the University of Nigeria Senate Research Grant Committee.

E. U. Odigboh: Agricultural Engineering Department, University of Nigeria, Nsukka, Nigeria

11
The Role of Women in Cassava Processing in Nigeria

C. Ebun Williams

ABSTRACT

In the transformation of agriculture from a subsistence to a market economy, it is necessary to pay attention to the impact of socioeconomic changes on farm women. It is important to identify tasks which are heavy for women and apply technology to alleviate the strain.

Since women process most of the food crops produced in the rural areas, it is desirable for policy makers to plan capital-saving machines to ease the drudgery in their work in order to create and encourage more women in rural areas to participate in food processing ventures and thus help in increasing food available in the market.

The objectives of this paper are:

1. To investigate the processing of cassava into gari (farina) in some selected villages in Oyo State of Nigeria. In investigating this aspect of gari production, I shall endeavor to look into the problem of introducing a capital-intensive labor-saving machine with a view to comparing it with the traditional way of processing cassava into gari.

2. To make recommendations on increasing the profit in gari and cassava production.

CASSAVA PROCESSING INTO GARI IN SELECTED VILLAGES OF
OYO STATE

Nigerian women have always played an important
role in the economic activities of their country.
Apart from their traditional roles of child rearing
and housekeeping, they have played a vital role in
farming and agricultural processing, trading, and
handicraft industries. Development of the modern
economy has led to the opening up of new economic
opportunities for women. One important area in
which women have taken an active part in national
development is in the growing and processing of cassava into gari, a staple food for many Nigerians.
The process is long and highly labor intensive,
but its advantage is that it can be carried out in
the home along with household chores.

Traditional Method of Processing Cassava into Gari

There are seven stages in cassava processing.
These include: uprooting cassava plants, transporting tubers from farm to home; peeling and washing
the tubers; grating, pressing and fermenting the
pulp; and, finally, roasting.

Uprooting Cassava Plants

In the selected villages where the study was
conducted (Francoise, 1977), women were not seen
uprooting cassava plants because the research was
conducted during the dry season; however, some women
who answered the questionnaire claimed to have up-
rooted cassava during the rainy season. During the
period of study, women hired laborers to do this
work. It took about three hours to uproot two
hundred plants of cassava. The work was done early
in the morning between 6 am and 9 am to avoid the
heat of the sun. Cassava was processed as soon as
it was harvested to avoid green discoloration, as
this affects the whiteness of the gari produced.
Storage of cassava was done by not harvesting it.
The uprooting was alternated with the transportation
of cassava tubers home.

Transporting of Cassava Tubers

Women and children transport cassava tubers
from farm to village on their heads. Collecting
cassava from the farms sometimes alternated with

the peeling of tubers because of the long hours involved in walking to and from the farms. The produce was conveyed to the village in small calabashes (gourds) which held about twenty medium-sized tubers at a time. For this reason, the time spent on evacuation of produce from the farm was rather prolonged.

Peeling of Tubers

Peeling of tubers was done with knives. The task was not difficult, but tedious, and was done with the cooperative efforts of the cassava owner, her friends and relatives. Because cassava is a highly perishable crop, mutual assistance and work exchange are usually involved in the peeling process. Peeling was the most time-consuming task in gari processing, because it was a social activity; the women did not hurry to finish because the longer they stayed, the more they conversed, chatted and laughed. It was observed that the task took from dawn to dusk.

Washing of Peeled Tubers

Some women washed the peeled tubers if they were a bit dirty after the peeling, some others did not wash them at all. Water was a constraint in the area of study, as there were no wells or pipe-borne water. Water was obtained from streams which dried up during the "harmattan," a dry season during which winds blow for days from the desert regions to the north.

Grating of the Peeled Roots

Grating of clean tubers was done mechanically. The machines were owned by enterprising male farmers who supplemented their earnings by operating mechanical graters in the villages. It was observed that the machines were either in a state of disrepair, or were too old and needed replacements, as they broke down very frequently and sometimes skipped grating chunks of cassava. These chunks of cassava were put back in the machine gradually during the grating, thus keeping the women busy during this operation. Moreover, the frequent breakdowns of the machine accounted for the long hours spent in grating, even though it was mechanically done.

Fermentation and Pressing of Cassava Pulp

This process consisted of putting the grated
cassava roots into a bag and pressing it with a
weight (e.g., stones).[1] Three things happened in
this process. The pulp was sufficiently drained of
water, thus making it dry enough for roasting.
Secondly, the water so drained reduced the content
of hydrocyanic acid (HCN) and other acids derived
from hydrolysis. Thirdly, fermentation occurred
in the cassava pulp.[2]

Roasting of Dried Pulp

Roasting was the most difficult part of gari
processing for many reasons. Women alternately
sieved and roasted moist pulp.   Sometimes, the
sieving was done by a friend or a relative, but the
roasting was always left to the owner, as the quality of gari depends on the skill of the person who
roasts it. Some women who processed more than one
bag of gari at a time with a view to selling it at
Ibadan (Oyo State capital) hired women laborers who
helped them during the roasting process. These
laborers processed one bag of gari in three days
at N 1.20, or N 1.00 per day when fed by the owner
during the operation.[3] In two of the cases, the
laborers and their children were allowed to eat
some of the gari during the period of processing.
It was estimated that one laborer and her children
could eat one "Kongo," which costs N 0.30 to N 0.40
during the period of survey.[4] This roughly brought
the labor cost of roasting one bag of gari to
N 4.00.
Roasting of gari was the most disliked aspect
of gari processing. This was because the traditional equipment used made it uncomfortable. Clay pots
are buried in the ground longitudinally, supported
by clay masonry. Since the clay pot has a small
orifice, the worker comes in contact with the steam
all day. If the inlet of the pot is bigger (like
using half of the pot), the steam becomes less
intensive with an increased outlet. During roasting, moisture is removed and the starch jells,
giving a dry, free-flowing granular product called
gari. Heat treatment breaks down residual unreacted
cyanide and drives off lingering hydrogen (Government of Nigeria, 1975). Efficient heating is essential for cyanide removal. Gari production involves
complicated stages which gradually reduce the total

amount of cyanide in cassava roots, thus making consumption of gari safe.

From the above it can be seen that traditional gari production is laborious and time consuming. It took an average of about 90 hours to process a bag of gari weighing about 103 kilograms. The most time consuming tasks in the processing were:

1. Peeling of tubers: 65% of the total hours spent on the whole operation; and

2. Roasting of pulp: 24%. Though the women spent more time in peeling, they complained more about the continuous heat generated by steaming wet pulp while roasting.

In four of the villages, an in-depth study was done to determine the total number of hours spent on processing one bag of gari (Table 1). Time spent varied mainly with the nearness of village to farms,

TABLE 1
Total Number of Hours Spent to Process One Bag of Gari

Tasks	Villages					
	Kokogi (hr)	Igbonla (hr)	Akolo (hr)	Okosse (hr)	Average (hr)	% of Total
Transport	3.0	2.75	6.0	5.5	4.3	4.64
Peeling	60.0	54.5	71.0	55.0	60.11	64.70
Washing	2.0	2.0	3.5	1.5	2.25	2.42
Grating	7.75	7.5	7.0	7.6	7.46	1.59
Pressing	2.6	2.5	3.75	2.16	2.76	2.98
Toasting	22.5	22.5	20.0	23.0	22.0	23.68
Total	97.92	85.75	105.25	88.78	92.92	100.0

Source: Francoise, 1977.

number in the groups peeling cassava, and the skill
and help involved in roasting of pulp.

COSTS AND RETURNS IN SMALL-SCALE GARI PRODUCTION

    Costs of production (Table 2) should include
not only how much was paid out for raw materials and
hired labor by the owner, but also her labor input.
The owner's labor cost is often omitted in estimates
of costs and returns. It was observed in the study
that women did not take their labor costs into
account, hence they continued to process gari even
when they were doing so at a loss. In computing the
following tables the owner's labor cost was computed
at rates paid to househelp per day in urban centers
in 1976. The amount was meager, but it represented
what would be paid in cities for uneducated persons.
    Table 3 shows average net revenue per bag of
gari processed. Two villages processed at a loss
in the year 1976. Comparisons in average production
and average net revenue between the years 1974-1976
showed that while the cost of cassava had doubled
between 1974 and 1975, the average net revenue did
not follow the same trend (Table 4). The women
found it difficult to double the price of gari at
the early stages of inflation, for fear that the
lower income wage earners who were their major pur-
chasers would abstain from buying and this would
mean a total loss of revenue. Gradually by 1976
there had been an adjustment in gari price.

Production of Gari in 1977

    Prior to the foregoing study, an experiment on
gari made with cassava planted by the women was also
in progress in the same location. The experiment
was initiated by the author, and by pleas from the
women for help in maintaining steady employment
throughout the year. The women complained of the
high cost of living and low-unprofitable seasonal
gari production, both of which were forcing them to
migrate to urban areas.
    Four villages--Apata Oloro, Kokogi, Alojo, and
Iporin-Sangodeyi--participated in groups of 13, 12,
and 8 women, respectively, each group owning com-
munal cassava farms. Table 5 shows area planted,
weight of tubers harvested, weight of bags of gari
processed from the tubers, and efficiency of cassava conversion into gari.

345

TABLE 2
Average Total Cost to Produce One Bag of Gari in 1976

| Villages | Fixed Cost Depreciation in One Year[a] (₦) | Labor Cost (₦) | Variable Cost | | Total Cost (₦) | Ratio of Cassava Cost to Total Cost % |
			Other Cost (₦)	Cassava Cost (₦)		
Igbonla	1.69	4.50	3.41	14.72	24.32	60.5
Akolo	1.69	4.50	3.34	14.49	24.02	60.3
Olorunda	0.80	4.50	2.75	12.48	20.53	60.7
Agbaaakin	2.36	4.50	3.14	14.63	24.63	59.4
Kokogi	1.02	4.50	2.70	12.53	20.75	60.3
Akosse	1.43	4.50	3.14	12.83	21.90	58.5
Poube	2.30	4.50	2.64	16.89	26.33	64.1
Alojo	0.82	4.50	3.38	15.86	24.56	64.5
Iporin Sangodeyi	4.75	4.50	3.33	10.00	22.58	44.2
Onissa	1.85	4.50	3.00	12.50	21.85	57.2
Average Total Cost of All Villages	1.87	4.50	3.08	13.69	23.15	59.1

[a] Fixed costs represent the depreciation cost within one year of the materials used by women to process gari, namely utensils such as wheat flour bags, baskets, locally woven sacks for pressing cassava pulp, jute sacks for marketing gari, knives for peeling cassava tubers, calabashes, toasting pots, sieving materials and brooms for toasting. Fixed cost was calculated using the straight line method:

$$\text{Depreciation} = \frac{\text{Cost-Salvage Value}}{\text{Life Span}}$$

with the salvage value being zero.

TABLE 3
Average Net Revenue Per Bag Per Village in 1976[a]

Villages	Gross Revenue (₦)	Total Cost (₦)	Net Farm Revenue (₦)	Net Farm Income Plus Opportunity Cost (₦)
Igbonla	29.52	24.32	5.20	9.70
Akolo	23.86	24.02	-0.16[b]	4.34
Olorunda	27.38	20.53	6.85	10.35
Agbaakin	30.19	24.63	5.56	10.06
Kokogi	26.50	20.75	5.75	9.75
Abosse	24.65	21.90	2.75	6.75
Poube	29.44	26.33	8.11	7.11
Alojo	25.07	24.56	-0.51[b]	5.01
Iporin	40.00	22.58	17.42	21.52
Onissa	30.00	21.85	8.15	12.65
Average	28.66	23.15	5.51	9.82

[a]Apata Oloro women did not process gari in 1976.
[b]Processed gari at a loss.

TABLE 4
Average Net Revenue Per Bag from 1974 to 1976

Production	Average Gross Revenue (₦)	Average Cassava Cost (₦)	Average Total Cost (₦)	Average Net Revenue Plus Labor Cost (₦ 4.50) (₦)	Average Net Revenue (₦)
1974	13.83	7.11	14.18	4.22	-0.28
1975	25.60	15.32	24.07	5.03	1.58
1976	28.68	13.69	23.15	9.82	5.51

Source: Francoise, 1977.

TABLE 5
Conversion Efficiency of Cassava into Gari in the Four
Experiment Villages, 1977

Village	Member-ship	Land Area Planted (ha)	Average Weight of Tubers Harvested (kg)	Average Total Wt. of Gari Sold (kg)[a]	Efficiency of Cassava Conversion into Gari Ratio	%
Apata Oloro	13	1.50	15,778	1,362	11.58	8.6
Kokogi	12	1.00	11,809	1,095	10.79	9.3
Alojo	8	0.75	5,904	560	10.54	9.5
Iporin-Sangodeyi	8	0.50	2,976	273	10.90	9.2

[a] Weights of the gari bags varied from 91 to 112 kg, averaging
103 kg, the figure used for the above computation.

Cost of Producing One Bag of Gari in 1977

Inflationary trends increased the cost of producing gari in 1977. Table 6 shows that cassava for
making one bag had risen from an average price of
₦ 13.69 to ₦ 20.00 in 1977, an increase of about
70 percent. The mean total cost of processing a
bag of gari at ₦ 23.15 in 1976 became ₦ 50.25 in
1977.

This experiment was unique around this area
because women do not normally farm; they only help
as farm hands during the harvesting of crops like
maize. For this reason, it became necessary for
them to employ labor to do the strenuous work.
Labor had become scarce, and consequently expensive,
even in villages and hamlets. Because of this and
other reasons (like social journeys undertaken by
the women in connection with their extended families) the cassava plots were often neglected.

Group labor was used by all the women at
various stages, especially in the processing of the
gari. In villages where the group contained more

elderly women, they employed surplus laborers, or their husbands, to work on their plots for a fee.

The opportunity costs in the study were solely responsible for any profits that might have accrued from the sale of the product. These included group labor engaged in farm operations such as some planting, weeding, harvesting cassava tubers, and making gari. Table 7 shows the opportunity costs contributed by each group.

In spite of the deficit in the average gross revenue, two of the groups which were moderately conscientious about the experiment offset their total outlay by 75 percent with the contribution of their group labor. All four groups had again planted cassava for the 1978 harvest. It is hoped that the success of the first two villages (together with another successful experiment on maize production) will spur the groups into making more effort toward improving their production system.

TABLE 6
Average Total Cost of Producing One Bag of Gari in 1977 (Around Fashola Area)

Villages	Fixed Cost Depreciation In Year (N)	Labor Cost (N)	Other Variable Cost		Total Cost (N)	Percentage of Cassava Cost to Total Cost	Percentage of Labor Cost to Total Cost
			Other Cost[a] (N)	Cassava Cost (N)			
Apata Oloro	1.69	17.01	2.60	20.00	41.39	48	41
Kokogi	1.05	18.07	2.60	20.00	42.35	47	44
Alojo	0.82	26.05	2.60	20.00	49.92	40	52
Iporin-Sangodeyi	4.75	39.99	2.60	20.00	67.34	30	59
Average Total Cost for All Villages	2.10	25.60	2.60	20.00	50.25	41	49

[a] Other costs include firewood, and oil used in processing.

TABLE 7
Net Revenue Per Bag Per Village in 1977

| Village | Gross Revenue (₦) | Total Cost (₦) | Net Farm Revenue (₦) | Opportunity Cost | | | Net Farm Income (₦) |
				Labor (₦)	Cassava (₦)	Total (₦)	
Apata Oloro	49.28	41.39	7.89	8.79	20.00	28.79	36.68
Kokogi	49.00	42.35	6.65	7.01	20.00	27.10	33.66
Alojo	49.20	49.92	0.72	4.50	20.00	24.50	23.78
Iporin Sangodeyi[a]	41.70	69.34	27.64	9.50	20.00	29.20	1.56
Average for All Villages	47.29	50.78	14.54	6.43	20.00	27.40	23.92

[a] Iporin Sangodeyi planted cassava in September 1976 (late in the season just before the end of the rains) and harvested in August 1977, before the maturity of tubers. The village was at great disadvantage in many ways. First, there was a long spell of drought soon after the cassava stems sprouted, followed by two occasions when cattle grazed on their farm. Because of neglect of the farm, they spent much money paying for weeding (five times).

PROCESSING BY MECHANIZED METHOD

As stated earlier, the process of gari making is very labor intensive, productivity is low, and it is sometimes unprofitable. Its advantage is that it affords women a chance to carry out a number of other activities in the home. If, however, it is to make a more satisfactory contribution to rural income and to provide food for the rapidly growing urban population, productivity will have to be improved. It is in realization of this that efforts have been directed toward mechanizing gari production.

Initiatives to mechanize the process have been based on the erroneous premise that in order to improve the productivity of gari processing, it is necessary to establish a highly capital intensive unit employing expensive foreign machinery. This paper does not intend to go into the mechanics of how such processing machinery works, but suffice it to say that there are about three of these machines

in the country. The Newell Duaford processing
plant, which costs about ₦ 45,000, was established
by the Federal Institute of Industrial Research at
Oshodi. A similar machine was established by the
Ido Cooperative Farming and Produce Marketing
Society at Elere near Ibadan costing over
₦ 1,000,000. It is obvious that this kind of large
capital investment cannot be undertaken by any
except the wealthiest farmers. Introduction of such
large, capital-intensive machines has raised a number of important issues. At Oshodi, it was found
that large numbers of women are needed to peel the
cassava. These women are therefore performing the
same process as in domestic gari production.

Also, the slower method of fermentation used
in villages completely hydrolizes the toxic HCN, a
process which is not complete with machine processing. Thus, traces of poisonous HCN in mechanically
produced gari are left. Thirdly, what happens to
the large numbers of rural women who depend on the
production of gari for their livelihood? It is not
yet clear whether the gari produced by these
machines will be cheaper; but undoubtedly the quality will be more uniform and hygienic.

The fourth issue is that there is no place for
the small domestic producers in the industrial
process. It is clear that such a large and expensive machine will require a large area from which
its raw materials and labor will be obtained. These
are getting difficult to come by in the face of
higher wages in secondary and tertiary industries.
In order to feed the machine at high efficiency
level, it would require a guaranteed and steady
supply of good quality cassava transported over a
large area, which will be difficult to obtain under
the small-scale farming system and the difficult or
inadequate infrastructure existing at present in the
country.

Finally, introduction of large factories is
likely to make redundant large numbers of women who
derive their income from traditional production
processes without providing them with any alternative sources of income. They would either become
purely cassava producers with a lower rate of
return or join the army of the unemployed, resulting
in social disruption.

The contention should therefore be that this
type of capital intensive machinery should not be
introduced without a careful study of the consequences.

RECOMMENDATION FOR INCREASING THE PROFITABILITY OF CASSAVA PROCESSING

It is not yet clear that capital intensive mechanization is the ideal, most economic way to develop cassava processing. The author contends that simple intermediate technology might provide an alternative. Such intermediate technology involves using low-cost machines which can meet the needs of peasant farmers, e.g., introducing more effective graters and improved toasters.

In order to make efficient use of such machines, farmers (the gari producers) should be encouraged to pool their resources through formation of cooperative organizations (such as is being encouraged by the Department of Agricultural Extension Services). Also, farmers should be backed by supply of improved and high yielding cassava cuttings; fertilizers, pesticides, and fungicides; credit; a guaranteed market; and an efficient extension service.

These experiments with cooperative organizations that grow their own raw materials and use small-scale gari production equipment have shown that an increase in the profitability of processing cassava into gari can be achieved under good management.

REFERENCES

Agboola, S.A. 1968. The introduction of and spread of cassava in Western Nigeria. Niger. J. Econ. Soc. Stud. 3:369.

Akinrele, I.A., Ero, M.I.O., and Olatunji, F.O. 1971. Industrial Specifications for Mechanized Processing of Cassava into Gari. Federal Institute of Industrial Research Technical Memorandum, Lagos, Nigeria. Federal Ministry of Industry Tech. Pub. No. 26.

Federal Republic of Nigeria. 1975. Third National Development Plan 1975-80. Volume II. Project Summary, Lagos, Nigeria. Central Planning Office, Federal Ministry of Economic Development.

Francoise, Favi. 1977. Women's role in economic development: a case study of villages in Oyo State. Master's thesis, University of Ibadan, Nigeria.

Jones, W.O. and Akinrele, I.A. 1976. Improvement of Cassava Processing and Marketing. National Accelerated Food Production Programme, Nigeria.

Williams, C.E. 1974. Rural women and national development in developing countries. Paper read at the Seminar on Prospects for Growth in Rural Societies with or Without Active Participation of Women, New Jersey, December.

Williams, C.E. 1976. Women's group in the social and economic development of rural areas: case studies of selected pilot villages. Paper presented at the Seminar on Self Reliant Development, Freetown, Sierra Leone, September.

C. Ebun Williams: Department of Agricultural Extension Services, Faculty of Agriculture, University of Ibadan, Ibadan, Nigeria

Section V
Economic Analysis
of Tropical Root Crops

1
Taro Processing in Hawaii: An Economic and Historical Perspective

Bryan W. Begley

ABSTRACT

Taro is processed and sold in several forms in Hawaii today. These include poi, chips, and kulolo. Poi is the ground form of taro with some water added, chips are similar in shape though not in taste to potato chips, and kulolo is a semi-hard product made from taro, coconut, and sugar. In the past the list also included flour and bread. Taro is also sold in Hawaii in supermarkets in its fresh form, sometimes described as table taro.

This paper focuses mostly on the processing of taro into poi. The reason for this is fairly simple. The bulk of taro grown in Hawaii is processed into poi and only a small proportion of the crop is sold in the fresh form or ends up as chips or kulolo.

Whatever the role and importance of poi in the past, the figures in this study suggest that, except where processors have sources of high quality taro, poi making is not a particularly profitable business today.

INTRODUCTION

Taro in Hawaii is mostly eaten in the form of poi. Poi is made today by boiling or baking the roots or tubers, removing the skin, trimming, and

357

then grinding and straining by machine, adding in the process some water. The end product is a smooth paste.

The processing of taro into poi is not unique to Hawaii, but it is uncommon in other areas of the world. In the South Pacific, where taro is a very important staple, the tubers are most often baked or steamed. In Asia taro is often an ingredient in pastries or puddings.

Baking taro or eating poi with meat, fish, or seaweed are both practiced in Hawaii but by different groups of Polynesians. The Samoans or Tongans who are largely concentrated in one or two areas on the island of Oahu buy fresh taro from supermarkets and bake it before eating. On the other hand, Hawaiians or those who are part-Hawaiian buy poi in its ready-made processed form. Though both methods of using taro are found in Hawaii, the relative importance of each is not the same. Over 95 percent of taro grown in Hawaii is made into poi.[1]

When Captain Cook landed in Hawaii 200 years ago, poi was probably the major staple of the Hawaiian people. Today it is considerably less important than rice or potatoes; nevertheless, it is still used regularly by many people with some Hawaiian blood. In the course of a series of interviews with people who use poi (see Begley, Spielmann, and Vieth, in this volume), many respondents expressed concern at the price of poi today. Clearly such comments merit some attention. It was therefore decided to allocate some time and resources to looking at the economics of taro processing in the hope of providing information on the resources or factors that are needed, the costs involved, and the major problems of processing or "milling" taro. In summary, it was anticipated that this study would throw light on what role processing played in the part of the total system which begins with the raising of the crop in Hanalei or Waipio on the outer islands and ends up with poi on the food tables of Hawaiians in Waianae or Kalihi.

It should be pointed out at this juncture that this study cannot be represented as the final word on poi processing in Hawaii. The basis for such a statement would be a complete survey of poi processing throughout the State, an exercise which was not carried out for a variety of reasons. The geographically dispersed location of poi mills, the lack of detailed records in most instances, and the reluctance or refusal to disclose details of operations by a few key millers would have made such an

exercise not only very costly and difficult, but
ultimately, perhaps impossible to execute. Nevertheless, through the assistance given over a long
period of time by the majority of those involved in
processing--even when in certain instances they were
unable to supply detailed information--it was possible to establish or describe the processing of poi
in Hawaii, some of the key problems of the processing industry, and the costs involved in creating
what is undoubtedly a unique product.

In view of the importance of poi as a traditional staple for the Hawaiians and other Pacific
island people, it would seem essential that at
least some attention be given to the history of poi
and poi-making in the islands. Changes that have
occurred in the structure of the poi industry over
the last 25 years are then singled out for particular discussion before some comments are made about
the changes in acreage and total production of taro
over that period. Then the location of the mills
still producing poi on each island are pinpointed
and some summary comments made on the flow of supplies of taro to the processors.

In quick succession the process and the plant
and manpower likely to be used in the production of
poi in a "model" poi factory are then covered. A
representative model of a poi mill operation is set
out, and the cost per pound of producing poi estimated. To do this, a certain scale of operation
was chosen and the cost of plant, equipment, raw
material, and labor was then calculated for a certain production level. The various costs were
classified according to whether they were fixed or
variable. Finally, various unit costs of production
figures were calculated and explained in some
length, with particular attention being paid to
recovery rates of taro after cooking and trimming.
Recovery rates have a considerable influence on the
profitability of processing.

TRADITIONAL POI-MAKING IN HAWAII: THE PROCESS AND
THE PEOPLE

Modern practices contrast markedly with the
traditional methods used and the time involved in
the making of poi before Captain Cook discovered
Hawaii in the 18th century. Handy (1940), in describing the process used by the Hawaiians, noted
that they first cooked the taro corm in an "imu," or

outdoor cooking oven. This was an open pit or hole
in the ground in which logs were burnt and on which
carefully selected stones were laid and heated. On
top of these stones, the washed taro corms, together
with sweet potatoes, yams, and other items, were
placed, covered with mats and leaves, and cooked
for 2 to 6 hours. With the arrival of Europeans,
the Hawaiians greatly speeded up this process by
switching to boiling taro in kerosene tins.

After baking or boiling, the outer skin was
removed and then a thorough job carried out of
removing "every vestige of peeling or flaws," as
Mary Kawena Pukui (1967) describes it. Using a
stone pounder, the corms were then mashed and moistened with water. When necessary, different varieties of taro could be combined according to strict,
time-sanctioned recipes. In order to get the poi to
the right consistency, the right proportion of water
had to be added and the poi kneaded or squeezed until all lumps were removed. Poi when made by
Hawaiians comes in many colors, reflecting the
varieties used.

With the arrival of Europeans, many changes
occurred. As noted earlier, there was the switch
from cooking in imus to boiling in 5-gallon cans,
but the greatest change was the commercialization
of poi-making. Poi-making passed out of the hands
of the Hawaiian family and into the hands of commercial operators. This did not happen immediately
after the arrival of Captain Cook, nor did traditional poi-making methods ceased with the coming
into being of poi "mills" or processing plants.
Mary Kawena Pukui's description of poi-making,
complete with imu and poi-pounder, refers to events
occurring in the 20th century. But inexorably,
poi-making passed out of the hands of Hawaiians as
the islands were settled by immigrants from Asia and
the mainland. As Crawford (1937) pointed out,
by the end of the 19th century the Chinese were
milling 80 percent of the poi in small backyard
factories dotted around the islands. Today, poi-
making is manufactured in a few small factories
mostly owned by descendants of the Japanese who came
to Hawaii to serve in the plantations.

MILLING TODAY

Twenty-five years ago, there were 32 processors
or mills making poi in the islands. Today there are

10 mills still in operation, of which at least 9 are able to obtain sufficient supplies to manufacture poi throughout the year.

Accurate figures on production of poi are not available from processors and could only be obtained at considerable expense by carrying out surveys of all stores selling poi throughout the islands. Nevertheless, it is possible to make a rough estimate of how much poi is being produced today. These figures indicate that the total consumption, as well as per capita consumption of poi in the islands, has probably halved at least over the last 25 years. This only parallels what the figures on production reveal. The average of land under taro between 1950 and 1965 dropped by 50 percent, and by another 10 percent between 1965 and 1975. The most significant change occurred on Oahu where the equivalent of today's total production acreage was virtually eliminated between 1948 and 1975. We have only 460 acres (186 ha) under wetland varieties of taro in the State today. Even when the acreage under dryland methods is added in, the total area is less than 500 acres (just over 200 ha). In 1948 there were 1000 acres of taro in the State.

Location of Mills

The number and location of poi mills on each island was set out in an earlier position paper and is as follows:[2]

TABLE 1
Distribution of Poi Factories in Hawaii

Island	No. of Factories	Site
Hawaii	1	Kukuihaele--Hamakua Coast
	1	Hilo
	2	Kona
Maui	1	Kahului
Oahu	2	Honolulu
	1	Haleiwa
Kauai	1	Kapaa
	1	Waimea

Supply Lines

In general two flow lines of raw product can be traced to the poi factories. The first is from farms to mills on the same island, the second to mills on another island. While the majority of poi mills draw on local supplies from the same island, over 60 percent of taro grown on Kauai, Maui, and Hawaii moves to Oahu. With over 80 percent of the State's population living in the City and County of Honolulu, clearly Oahu provides a large market for poi; two processing plants are located in Honolulu and one in Haleiwa to supply it. The seven local factories on Kauai, Maui, and Hawaii process approximately 40 percent of the taro grown in the State.

Thirty years ago Oahu had by far the largest acreage under taro. Today only a few acres remain, and so over 95 percent of the raw material required by the three Oahu poi mills must be shipped in by barge.

As noted earlier, the poi millers on Kauai, Maui, and Hawaii depend basically for supplies on local farmers. In some cases this involves simply buying from farmers a mile or so down the road, but in other cases millers must procure taro from farmers located over 30 miles (50 km) from the mill. Hilo and Kona millers, for example, buy a proportion--and in some cases the bulk--of their supplies from Waipio Valley, 40 or 50 miles from the plants. Much of the Maui taro, whether destined for Oahu or for local use, is trucked 40 miles.

Modern Poi-Making

Though modern poi-making involves using machinery, it is not a completely mechanized operation. The process is as follows. Taro corms are first cooked in a pressure cooker for several hours, then allowed to cool for periods of between 6 and 10 hours. This cooking facilitates peeling and results in the elimination of a chemical agent which causes severe irritation to the skins of most people who handle taro.

After removal from the pressure cooker, taro undergoes two operations which are carried out manually. The first is removal of the outer skin and the second is trimming, which is carried out by using a small knife. The clean, thoroughly washed taro is then hoisted up by conveyor belt or by hand

to where it is ground. The actual grinding is
done by machine, but the taro is pushed manually
into the grinder using a wooden pounder. Grinding
may be a single or double process. In some mills
the taro undergoes a rough grind first, then a
smooth grind. In other mills only a single grind is
carried out. Finally, to ensure the right consistency and remove extraneous solid matter, the poi is
centrifuged and water added.

The smooth, viscous poi is then ready for
packaging, often carried out with the aid of a
machine. Poi is packaged in various sizes, 12, 16,
28, and 32 ounces (392, 448, 784, and 896 g, respectively) being the most common, and then delivered
in the early morning to retail stores.

PLANT AND MANPOWER NEEDS

Forty years ago Payne, Ley, and Akau (1941)
carried out a considerable amount of research into
the processing of taro. In papers and reports they
described the equipment they purchased for their
work. Now many years later, in visiting poi factories throughout the State, one notes how much of the
same equipment is still in use. In fact, some of it
dates from at least that period, the late 1930s. A
bagger, conveyor belt, or centrifuge may have been
added, but from cooking to grinding the equipment
in use appears to have changed little.

The equipment is housed in a building of 2000
square feet (185 m^2), and often much less. Used in
the first phase, that of preparing the taro, are a
boiler, one or more cooking retorts, one or two long
stainless steel trays for fine-trimming, and a moving belt to lift up the peeled and fully trimmed
material prior to grinding.

For the second phase, that of grinding and
refining taro into poi, the equipment used includes
one or two grinders, a centrifuge, and a vat where
the poi is collected immediately after being manufactured. From this vat the viscous or semi-liquid
poi is pumped through a stainless steel line into
a semi-automatic bagger which ejects pre-set quantities of poi into plastic bags.

Not all plants have two grinders, not all
plants have a centrifuge, and not all plants have a
conveyor belt, but all use boilers, cooking retorts
(or ovens), trimming trays, a grinder, a bagger, a

pump, and plenty of buckets. Also, all use considerable volumes of water, particularly during trimming and grinding.

In the mills that were visted on Oahu and the
outer islands, between 5 and 7 people were involved
in making poi. A couple of people--one of whom
might be the owner or manager--loaded the sacks into
the retorts and boiled the taro well ahead of time.
In the evening or early morning, the peeling, trimming, grinding, and bagging began. In the mills
visited, an average of 34 man-hours were expended
in processing a little over a ton of taro (2000
pounds = 909 kg). Usually this meant between 5 and
7 people working 4 to 5 hours. Though individuals
in the industry hold professional qualifications
and the operation requires careful supervision, it
does not appear that operators need particular
engineering or processing expertise in order to
supervise normal operations at a poi plant.

In terms of overall size, it would seem that
compactness is the best term to describe the
various poi mills. All appeared to fall within the
1200 to 2400 square foot range. All could neatly
be fitted into a corner of any of the taro patches
from which the raw product comes.

Estimating Costs of Processing

Determining the actual cost of producing a
package of poi is difficult for many reasons. The
first, and perhaps the major one, arises because of
the difficulty of deriving an actual overhead cost
figure. Equipment in use and even the buildings, in
some cases, are anywhere between a year and 70 years
old. But the cost at which a boiler or grinder were
installed in 1905 or 1935 bears little relationship
to what the same piece of equipment might cost today. Yet not to include some cost for a piece of
equipment which has already had several lifetimes
and by now has been depreciated many years ago, and
the depreciation allowance possibly long spent,
would be misleading.

Another problem arises because, in general,
records are not kept of the amount of taro bought
and the poi produced week by week, month by month,
over an entire year. This is not to suggest that
millers are not keenly sensitive to differences
between taro processed in summer and winter, or
aware of degree of variation in the quality of taro

supplied, or to the relationship between quantity processed on any given night and the manpower needed. But millers do not generally keep detailed figures of processing on a cumulative weekly basis.

Basic Assumptions and Procedure

In order to cope with the problem of working with equipment of variable age, estimates of the replacement costs of the standard items likely to be purchased if one were setting up a poi factory today were used. For the purpose of this exercise, equipment and plant were assumed to be installed in a 2000 square foot building. A standard rental fee covering land and buildings was used. Given this building and plant, costs were then estimated for processing 3000 pounds (or 1½ tons = 1818 kg) of fresh taro four times a week. This could be viewed as output of a medium-sized poi plant in the State.

COSTS OF MAKING POI IN A MODEL PLANT

Explanation of Fixed Costs

Amortizing plant. The figures used in this study, shown in Table 2, apply to equipment considered basic and essential for processing, and were obtained from the processors or millers or from staff associated with the Food Processing Department of the University of Hawaii. The individual items were amortized over 5 or 10 years depending on the likely life span. Whether or not the cooker or bagger or any other piece of plant and equipment had 5 or 10 years of life depended on whether moving parts or motors were involved. Equipment with moving parts or motors was considered more likely to need replacement and was therefore given the shorter life span of 5 years.

In calculating the annual cost of a piece of equipment and ultimately the amount to be charged for a week's operation, the salvage value was usually charged at 5 percent of the purchase price. But with items such as the vans, forklift, and trucks, the salvage value was increased to 10 percent of the original cost. This was because the resale value was considered to be higher in the case of these particular items.

TABLE 2
Fixed Costs of Making Poi

Items				Weekly Cost	%
Rental (2,000 sq ft of land plus building @ $0.30/ft/month				$125.00	2.23
Equipment:	To Purchase New ($)	Amortized Over	Salvage Value		
1 boiler	10,000	10 years	5%	$ 20.00	.36
2 cookers	5,000 ea	10 years	5%	20.00	.36
2 moving belts	3,000	5 years	5%	11.50	.20
1 rough grinder	1,500	5 years	5%	6.00	.10
1 smooth grinder	2,000	5 years	5%	7.50	.13
1 centrifuge unit	4,000	5 years	5%	15.00	.27
1 vat	500	10 years	5%	1.00	.02
1 pump	700	5 years	5%	3.00	.05
1 bagger	10,000	5 years	10%	36.00	.64
2 trimming trays	600 ea	10 years	5%	2.25	.04
1 fork lift	8,000	5 years	10%	30.00	.54
2 vans	7,500 ea	5 years	10%	30.00	.54
1 flatbed truck	7,500 ea	5 years	10%	30.00	.54
Misc. stainless steel baskets, scales, pallets, replacement parts	5,000	5 years	5%	25.00	.45
				$362.25	6.47
Insurance (2% of total fixed costs per annum)				15.35	.27
Working capital reserve				400.00	7.15
TOTAL				$902.60	16.12

Working Capital Reserve

Some explanation of the provision of $400 for a working capital reserve is necessary. Normally the period for which costs are estimated is a year. In such a case adding a provision for 3 months for working capital appears reasonable. Proportionately, the magnitude of the figure seems appropriate when the other items are calculated on the basis of a year. But with poi a period of a week seems to fit in more naturally with how millers actually operate. Though millers may have at the back of their minds the need to carry 3 months working capital on hand, provision of a reserve for a week seems more in line with other figures used in this case. An actual 3 months working capital reserve was considered to be in the region of $5,000.

Notes on Variable Costs

Raw materials. The cost of purchasing fresh taro is the largest single variable cost, accounting for over 25 percent of variable costs (Table 3). Millers do not pay a standard price throughout the State. In some areas, they pay considerably more than 13 cents a pound (28.6 cents/kg). But 13 cents a pound, which is the farm gate price paid by the largest processor, can be viewed as the minimum price paid by millers. Though paid by the pound, farmers actually stack their taro in 80- or 100-pound (36 to 45 kg) sacks which are then collected at the farm gate. In some cases farmers actually deliver directly to an agent or even to the miller.

Trucking and Shipping Charges

Where the final destination is Honolulu or Haleiwa or even Kahului, trucking charges are incurred by the miller. On top of this, freight charges must be paid to ship taro from Kauai, Maui, or Hawaii to Honolulu, Oahu.

Because the distances involved are so variable, a flat figure of .002 cents per pound has been charged to cover trucking costs. This again must be seen as a minimum charge--in many cases millers must pay considerably more in order to move taro into their factories.

Wages and Salaries

A team of 5 to 7 people consisting of trimmers, a grinder, and a packer, working anywhere between 5 and 7 hours in the evening or early morning can handle quite variable quantities of taro. If the taro is free from disease, of good size, partially trimmed when it arrives at the plant, and relatively clean, 5 people working 5 hours (5 x 5 = 25 hours) can handle a ton (2,000 pounds). But if the taro is diseased and very small in size, 7 people working 5 to 6 hours (7 x 6 = 42 hours) will be needed to handle a ton.

In the face of such variation, which is a fact of life for most millers, it has been decided to choose a set number of hours during which 6 tons of taro could be processed. Given that processing will take place 4 times a week, then the provision is

TABLE 3
Variable Costs of Making Poi

Items	Weekly Costs	%
Raw materials--6 tons of fresh taro @ $.13/lb	$1,560.00	27.89
Wages plus 39% fringe benefits: 160 man-hours (4 x 40) @ $5/hr plus 39% fringe benefits	1,112.00	19.88
Salary--annual 1 @ $20,000 plus 39% fringe benefits	556.00	9.94
Utilities--electricity, water, steam fuel oil for boiler, telephone, gasoline	175.00	3.13
Freight on 6 tons shipped from Outer Islands @ $.078/lb	940.00	16.80
Trucking charge (for moving 6 tons 30 miles) @ $.002/lb	24.00	.43
Packaging supplies (plastic bags, boxes, etc.)	250.00	4.47
Miscellaneous (overalls, knives, etc.)	200.00	3.57
Total variable costs	$4,817.00	86.11
Total fixed costs (Table 1)	777.60	13.89
TOTAL COSTS	$5,594.60	100.00

for a total of 160 (4 x 40) hours being expended in a week. As mentioned above, the quantity of taro processed in that time should be 6 tons or 4 x 1,500 pounds each night.

The salary figure included in variable costs is conservative. However, both the wage and salary figures are increased by 39 percent to allow for the payment of social security, insurance, and other fringe benefits.

ANALYSIS OF COSTS AND RETURNS

General

The variable component of total costs is of
much greater significance than the fixed costs component. Variable costs account for over 80 percent
of total costs. In fact, the variable component is
likely to account for an even higher proportion of
total costs. Much of the equipment in use as well
as the buildings have been written off many years
ago and are no longer a charge on the operation.
This means, in effect, that while poi millers must
take into account some fixed costs charges they are
primarily concerned about items that make up variable cost.

In this situation attention can really be
focused on cost of raw material, wages, salaries,
freight charges, and utility costs. As noted
earlier, the cost of raw material is a major expense. So also are wages and transportation
charges--especially for poi mills located on Oahu
which incur trucking and shipping costs. The ultimate profitability of the poi, therefore, is dependent on the prices paid for taro supplies, labor,
and transportation. When the cost of raw material
is a significant expense item, then the amount of
quality product bought becomes quite important. If
a processor has access to good quality taro then
the cost of raw material, labor costs and transportation charges will all be significantly lower than
when the miller buys poor quality taro. A measure
of the quality of taro being processed is the recovery rate. This can be simply defined as:

$$\frac{\text{Total quantity of poi produced}}{\text{Total quantity of taro purchased}} \times 100 \text{ percent}$$

If the quality of taro is high, a recovery rate of
between 60 and 70 percent can be expected. If the
taro is average, the recovery rate will probably
range between 50 and 60 percent. But if the taro is
low, with much evidence of disease and many small
corms, the recovery rate drops well below 50 and
perhaps on some occasions below 40 percent. One
way of explaining the significance of this ratio is
to contrast two recovery rates.

If the recovery rate for two production runs,
for example, is 70 percent and 35 percent, respectively, and the same quantity of poi is needed in
each case, then in the first run 2,000 pounds of

taro will be used and in the second 4,000 pounds.
In the second run not only will raw material costs
be doubled but labor and utility charges will also
be markedly increased. Having introduced the concept of a recovery rate, this might be an appropriate point to look at the unit costs of producing
poi shown in Table 4.

Unit Cost of Production

Clearly the recovery rate is important. It
influences not only the total raw material used and
the labor needed but also shipping and trucking
costs. When more raw material has to be processed,
then transportation costs rise as well as raw
material and labor costs.
In Table 4 the recovery rate is shown in column
1 on the left, and set out across the table are the
unit costs of producing 1 pound of poi using various
combinations of costs.

Recovery Rate

In general, it might be expected that the
closer a poi factory is to the source of raw material the higher the recovery rate, and conversely,
the further the factory is from the farmers the
lower the recovery rate. The situation is in fact
more complex.
Trucking, shipping, and storing taro appear to
lower recovery rates so that millers on Oahu are
likely to find it difficult to achieve a recovery
rate of 50 percent. But even on the outer islands
a 60 or 65 percent recovery rate may be difficult
to achieve. The basic reason seems to be that
good quality large taro is not consistently available. The quality of the taro raised by the
farmers is critical in determining recovery rates
in the mill, and disease and small taro corm
problems are facts of life that millers have lived
with for many years. In some cases, and at particular times during the year, problems are worse than
at other times. But they are always present.
Waipio Valley farmers, for example, for the last
four or five years have had serious disease problems and, as a consequence, millers who depend for
supplies on this valley can anticipate at best
average recovery rates of 50 to 60 percent. But
the miller located at Kukuihaele--overlooking Waipio
Valley--is likely to obtain a better recovery rate
than the miller in Honolulu, using raw material from

TABLE 4
Unit Cost of Producing Poi

Columns: (1)	(2) Total Costs		(3) Variable Costs		(4) Total Costs Less Manager's Salary		(5) Variable Costs Less Manager's Salary	
Recovery Rates in Percentage (based on quantity of 6 tons)	Oahu	Outer Islands	Oahu	Outer Islands	Oahu	Outer Islands	Oahu	Outer Islands
40	1.16	1.03	1.00	.81	1.04	.85	.89	.69
50[a]	.93	.78	.80	.65	.84	.68	.71	.55
60	.78	.65	.67	.54	.70	.57	.59	.46
70	.67	.55	.57	.46	.60	.49	.50	.40
80	.58	.48	.50	.40	.52	.43	.44	.35

[a] As a reference point it is assumed that a 14-oz packet of poi is sold by the processor to the retailer for approximately 48 cents (equivalent to a wholesale price of 55 cents for 1 pound or 16 oz of poi).

Formula: $\dfrac{\text{Costs (\$)}}{\text{Quantity Recovered (Pounds)}}$ = unit cost per pound the same source, simply because the former processes the taro closer to actual harvesting. Taro processed on Oahu is probably between 2 and 5 days old when processed--simply because of transportation--and disease has made greater inroads into the corms in that time.

Total Costs

The basic difference between Oahu and outer island costs arises because of shipping charges. Oahu processors have to transport their taro to Honolulu by barge. This charge has been deducted from the total cost figure applied to outer island processors and other costs used.

Variable Costs

In view of some earlier comments about the age of the buildings, plant, and equipment in use, it would seem appropriate to calculate a unit cost using only variable costs. This has been done, but to keep matters in perspective the relatively small contribution of fixed costs to total costs is shown by comparing the unit cost figures in column 2 with those of column 3 in Table 4.

Once again the freight charge on taro shipped to Oahu is deducted from the variable cost on the outer islands.

Deducting Salaries

Wages must be paid to hourly workers--though not, of course, to family members--but salaries need not be paid. Therefore, the salary of the owner has been deducted from the total as well as the variable costs in columns 4 and 5 of Table 4, and unit costs then derived.

GENERAL DISCUSSION AND CONCLUSIONS

If we assume that the miller is paid approximately 55 cents for 16 ounces (448 g) of poi, then we have a reference point for studying the unit cost of production figures. An explanation of how these unit cost figures are derived might be in order at this point.

Recovery rates on 6 tons of raw material, which is assumed to be a week's throughput, are first calculated at 40, 50, 60, 70, and 80 percent

recovery levels. Then the cost figures are taken
from Tables 2 and 3. Finally the actual qualities
of poi produced at various recovery rates are set
against various cost combinations and unit cost
figures derived. An actual calculation is shown
below:

Step I: 70 percent recovery rate on 6 x
2,000 lb = 8,400 lb of ready-to-
sell poi.

Step II: Total costs on Oahu for processing
6 tons = $5,594.60.

Step III: Total Costs ÷ Total Output at 70%
Recovery Rate = $5,594.60 ÷ 8,400 =
.67 cents = unit cost of producing
1 lb of ready-to-sell poi at 70%
recovery rate on Oahu.

Having demonstrated how unit costs are calculated, the next stage is to look at the unit costs
associated with various recovery rates. The unit
costs for recovery rates of 40 and 50 percent are
set out in Table 5.

TABLE 5
Various Cost Figures with Differential Recovery Rates

	At 40% Recovery Rate ($)	At 50% Recovery Rate ($)
Total Costs		
Oahu	1.16	.93
Neighbor Islands	1.03	.78
Variable Costs Only		
Oahu	1.00	.80
Neighbor Islands	.81	.65
Total Costs Less Manager's Salary		
Oahu	1.04	.84
Neighbor Islands	.85	.68
Variable Costs Less Manager's Salary		
Oahu	.89	.71
Neighbor Islands	.69	.55

If it is assumed at this point that plant and equipment in the model are used only for the purposes of making poi, then at recovery rates of 40 and 50 percent as shown above, the full cost of producing a pound of poi on the neighbor islands or on Oahu is not covered when the processor receives the equivalent of 55 cents. In fact, the figures suggest that poi millers on the neighbor islands—whose costs are lower than those on Oahu because of lower transportation costs as noted earlier—just manage to recover variable costs, less the manager's salary, when the recovery rate is 50 percent.

To place these figures in perspective it should be mentioned that most of the millers interviewed considered themselves fortunate to consistently obtain 60 percent recovery rates. A 60 percent recovery rate would probably be above the average attained throughout the State. Seventy percent recovery rates are possible but infrequent, and 80 percent is attainable only if a processor mills or processes taro grown in his own fields and enforces rigorous quality control during harvesting.

Yet the figures in Table 6 suggest that a 70 percent recovery rate is necessary in order for millers on the neighbor islands—those with lowest costs—to recover all of their costs. Of some importance is the fact that this group of millers does cover its variable costs with a 60 percent recovery rate.

For processors on Oahu the figures imply a much gloomier picture, given present costs and present prices, than what applies to neighbor island processors who are closer to the sources of raw material and who, in addition, do not have to barge their taro. Even if Oahu processors achieve recovery rates of 80 percent, it appears that they still fail to recover all their costs. What is perhaps of more significance is the fact that Oahu millers require a greater than 70 percent recovery rate—which is probably still close to impossible, given the circumstances under which they operate—in order to recover variable costs.

In summary, if the objective is to recover the total costs of the operation, then these figures suggest that poi processing is not a profitable operation unless recovery rates are high (greater than 70 percent). Given the poor quality of taro generally available and the deterioration that takes place as a result of shipping taro from the

374

neighbor islands to Oahu and then storing it for several days, it appears quite difficult to achieve high recovery rates, particularly on Oahu.

If the objective is to cover variable costs, this is possible with a 60 percent recovery rate on the neighbor islands. On Oahu the recovery rate has to exceed 70 percent to achieve this.

But while in general these figures suggest that poi-making is not very profitable, certain facts should not be lost sight of, which probably modify somewhat the actual situations under which poi processors operate.

In the first place most of the plants making poi are using equipment that is 10 to 70 years old. Therefore the costs charged in the fixed costs section are overstated. These costs really apply to a company setting up to produce poi in a new factory. Because fixed costs comprise less than 15 percent of total costs, the profitability picture is

TABLE 6
Various Unit Cost Figures with Differential Recovery Rates

	At 60% Recovery Rate ($)	At 70% Recovery Rate ($)	At 80% Recovery Rate ($)
Total Costs			
Oahu	.78	.67	.58
Neighbor Islands	.65	.55	.48
Variable Costs Only			
Oahu	.67	.57	.50
Neighbor Islands	.54	.46	.40
Total Costs Less Manager's Salary			
Oahu	.70	.60	.52
Neighbor Islands	.57	.49	.43
Variable Costs Less Manager's Salary			
Oahu	.59	.50	.44
Neighbor Islands	.46	.40	.35

375

affected, but less seriously than would be the case if the main component of costs, variable costs, were overstated.

Secondly, it has been assumed in the model that only one product, poi, is made and therefore that all costs, both of a fixed and variable nature, must be fully charged against this product. Such, in fact, is not always the case. In certain instances other products are made, and therefore at least some of the costs for equipment, power, and other items can be partially set off against these other production activities. Though it would be difficult to calculate how much fixed or variable costs are reduced by these activities, some adjustment could and should be made when the plant is used for making several products.

Finally, certain processors do grow their own taro and use it in processing. It is reasonable to charge a cost for the taro used, but not necessarily the same price that other processors pay for their raw material. An added benefit in this situation is that the grower-processor may insist on far tougher quality control standards than other processors might like to enforce.

In conclusion, these facts suggest that the costs presented in this report possibly overstate the actual costs. But, given the gloomy picture presented in Tables 4, 5, and 6, some explanations seemed called for simply to explain how those still making poi mange to remain in business.[3] But given the average actual recovery rates that processors get--which is probably closer on average to 50 than 60 percent--and the high proportion of costs accounted for by variable cost items such as raw materials, wages and transportation, it appears reasonable to suggest that overall, modern poi processing, while probably profitable in some instances, is not a highly profitable commercial activity.

NOTES

1. An officer of the Department of Agriculture estimated that 3 percent of taro production was sold as fresh or table taro. A 2 percent allowance is used to cover taro used for taro chips and other non-poi products.
2. Bryan W. Begley, "Taro in Hawaii: Study of a Root Crop Delivery System," unpublished position paper, 1977.

3. Comments from several of these still remaining
   in business suggested that poi-making is certainly not profitable when the quality of taro
   is poor. Further, they suggest that relocating
   or setting up a new poi plant would not appear
   to be justified given the present prices and
   costs.

REFERENCES

Crawford, D.L. 1937. Hawaii's Crop Parade.
    Honolulu: Advertiser Publishing Co.
Handy, E.S. Craighill. 1940. The Hawaiian Planter:
    His Plants, Methods, and Areas of Cultivation.
    Bulletin 161, Bernice P. Bishop Museum, Honolulu.
Payne, J.H., Ley, Gaston J., and Akau, George.
    1941. Processing and Chemical Investigations of
    Taro. Hawaii Agricultural Experiment Station
    Bulletin No. 86.
Pukui, M.K. 1967. Poi making. In Polynesian
    Cultural History. Honolulu: Bishop Museum
    Press.

*Bryan W. Begley: Department of Agricultural and Resource
Economics, College of Tropical Agriculture, University of
Hawaii, Honolulu, Hawaii*

2
The Implications of Cassava Processing and Marketing for Other Root Crops

Truman P. Phillips

ABSTRACT

All tropical root crops have experienced increases in production during the last 15 years, but only cassava has experienced substantial development of new markets. In cassava the main developments are export markets as an animal feed ingredient in the European Economic Community (accouting for approximately 10 percent of production, or as an industrial starch in Canada, Japan and the United States (accounting for less than one percent of production.) The diverse markets for cassava are examined in an attempt to identify factors which would be relevant to the market diversification for other tropical root crops. Additionally, new market possibilities for cassava (i.e., single cell protein and ethyl alcohol production) are examined and discussed in terms of their inference for other tropical root crops.

Discussion of the size of cassava production and processing activities illustrates that small scale production and processing can successfully supply even export markets (viz. Thailand's cassava starch export industry). The discussion also illustrates the possibility of small and large scale processing activities working together (viz. the chipping (small scale) and pelleting (large scale) plants in Thailand).

Conclusions reached are that there are unexploited markets available to other tropical root

crops, and that these markets need not depend on
large scale processing. However, realization of
these markets may require processors, entrepreneurs
and governments to appreciate that tropical root
crops can be more than subsistence crops.

INTRODUCTION

World tropical root crop production is estimated to be approximately 174 million metric tons
planted on approximately 20 million hectares (Table
1). If it is assumed that typical plants are of
the order of 1 to 4 hectares by families of 4 to 7
people, it may be estimated that 20 to 175 million
people are directly involved with the production of
tropical root crops.

In terms of area and production cassava is
clearly the dominant tropical root crop, accounting
for 50 percent of the area and 60 percent of production. The dominance of cassava is no doubt
explained by its adaptability to a wide range of
conditions, its ease of production, and its risk-
minimizing attributes. The markets for cassava are:
as a human food, 88 percent; an animal feed ingredient, 11 percent; and an industrial starch, 1 percent (Phillips, 1974). The latter two markets are
located primarily in developed countries and are
supplied by Thailand.

In developing countries, tropical root crops
provide about 6.8 percent of the daily per capita
consumption of calories (Table 2). In Africa the
figures rises to 21.6 percent, and in the South
Pacific the figures exceed 27 percent. Table 2,
however, understates the importance of tropical
root crops to the extent that consumption level of
foodgrains in developing countries is maintained by
importing 15 percent of consumed cereal and cereal
products. Excluding rice, food cereal imports are
more than 20 percent of consumption (calculation
based on Halder, 1976). If a distinction is made
between imported and domestically produced calories,
tropical root crops provide approximately 8 percent
of domestically produced calories in all developing
countries. By geographic regions the figures on
the percent of domestically produced calories
provided by tropical root crops are: Latin America
7.8 percent; Near East, 2.6 percent; Far East 4.1
percent; and South Pacific 34 percent.

TABLE 1
1975 Area and Production of Tropical Root Crops

a. Area (1000 ha):

	Africa	Latin America	Near East	Far East	South Pacific	Total	%
Potatoes	343	1,034	339	875	-	2,591	12.7
Sweet Potatoes	939	367	4	1,059	106	2,475	12.2
Cassava	5,823	2,663	234	2,677	20	11,417	56.1
Taro	726	-	2	17	33	778	3.8
Yams	1,961	32	88	7	15	2,103	10.3
Roots & Tubers NES	518	129	-	317	23	987	4.9
Total	10,310	4,225	667	4,952	197	20,351	100.0

b. Production (1000 mt):

	Africa	Latin America	Near East	Far East	South Pacific	Total	%
Potatoes	2,039	8,951	4,206	8,445	-	23,641	13.6
Sweet Potatoes	5,539	3,379	94	8,764	560	18,336	10.5
Cassava	42,844	32,201	1,128	27,643	221	104,037	59.7
Taro	3,569	-	59	90	262	3,980	2.2
Yams	19,279	291	260	30	200	20,060	11.5
Roots & Tubers NES	1,446	811	-	1,674	390	4,321	2.5
Total	74,716	45,633	5,747	46,646	1,633	174,375	100.0

Source: FAO, 1976.

TABLE 2
Total Per Capita/Day Calorie Supply and Amount Supplied
by Tropical Root Crops

	1961-65	1972	1973	1974
Developing Countries	2106	2165	2121	2170
Roots and Tubers	156	148	146	147
%	7.4	6.8	6.9	6.8
Africa	2049			
Roots and Tubers	464	445	445	448
%	22.6	21.3	21.6	21.6
Latin America	2349	2470	2470	2491
Roots and Tubers	177	179	163	159
%	7.5	7.2	6.6	6.4
Near East	2299	2424	2430	2457
Roots and Tubers	56	52	53	52
%	2.4	2.1	2.2	2.1
Far East	2019	2054	1984	2053
Roots and Tubers	72	64	66	67
%	3.6	3.1	3.1	3.3
South Pacific[a]	2208	2388	2390	2375
Roots and Tubers	707	671	666	655
%	32.0	28.1	27.9	27.6

[a]Author's calculations

Source: FAO, 1977a

    As indicated, tropical root crops have made a
substantial, but often overlooked, contribution to
the diets of people in tropical developing countries.[1] Until the early Sixties the export market
for tropical root crops was limited to food and
industrial grade starches. However, the creation
of the European Economic Community (EEC), and the
subsequent market for cassava as an animal feed
ingredient, revealed the commercial potential of

tropical root crops. The development of the cassava animal feed industry, based on small scale production and processing, is interesting because it demonstrates that cassava can be a "cash crop." Prior to the development of the EEC animal feed market, cassava was thought only to be a subsistence crop and probably an inferior good.[2] Since other tropical root crops are also characterized as being subsistence crops, the point of interest then becomes whether the transformation of cassava from a subsistence crop to a cash crop can be applied to other tropical root crops. Thus an attempt will be made in this paper to describe key economic and marketing factors which have contributed to the transformation of the cassava industry, and to relate these factors to other tropical root crops.[3]

CASSAVA POST-PRODUCTION SYSTEM

The cassava post-harvesting system may be characterized by a series of concentric circle segments radiating from the central circle, production (Figure 1). The figure illustrates the number of different products which can be produced from cassava and the alternative paths which can be used to produce a final product. For example, there are four routes which can be used in the production of flour:

harvesting → rasping → soaking → flour
harvesting → chipping → drying → flour
harvesting → storage → rasping → soaking
→ flour
harvesting → storage → chipping → drying
→ flour

Schematically the cassava system is representative of the general tropical root crop post-production system; however, the other tropical root crops do not now have the diversity of processes and outlets. In fact, the main processing route for many of the other tropical root crops is to be consumed as a human food directly after harvesting.

It is the multitude of processing channels and markets which distinguish cassava from other tropical root crops. Therefore, the following section contains an examination of the existing and potential cassava markets and an evaluation of some of the unique features of each market and their importance to other tropical root crops.

382

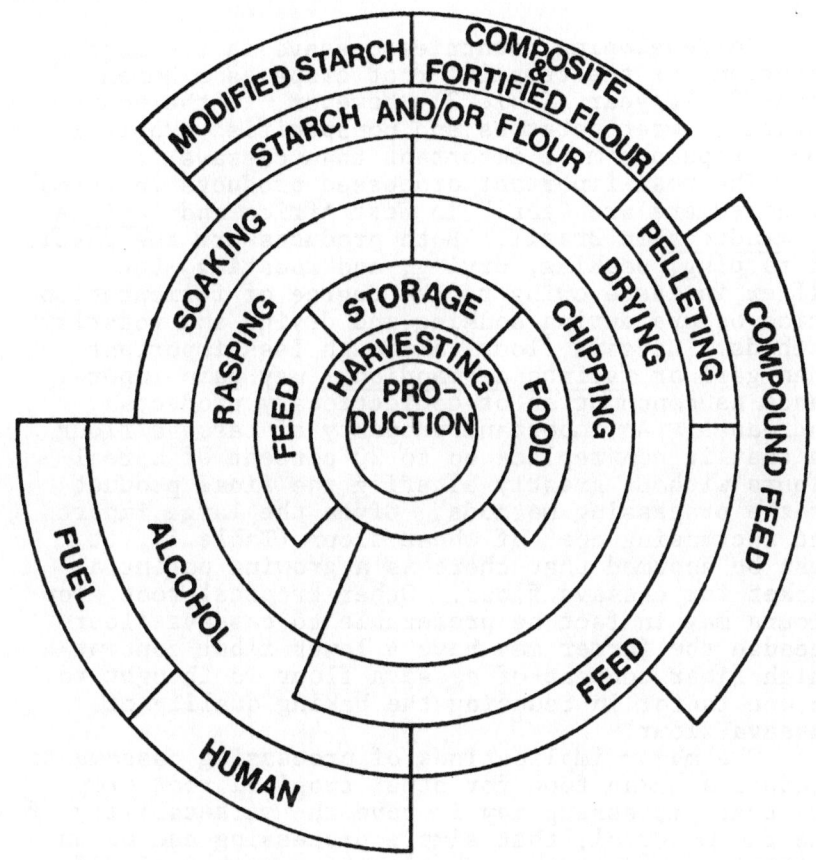

Figure 1. Cassava post-production system.

CASSAVA MARKETS: HUMAN CONSUMPTION

Because of the HCN content of many cassava
varieties, cassava is usually consumed in a processed rather than fresh form. The essential aspect of
most food processing techniques is the rasping,
soaking, and drying of cassava, which permits much
of the HCN and water to be removed from the fresh
cassava. An interesting, but perhaps originally
unintentional, aspect of processing is that the
final product is denser and less perishable than
fresh cassava. Furthermore, the processing activity
often enables the producers to realize economic
benefits from both the production and processing of
cassava.

In developing countries cassava is the most
important of the tropical root crops as a human
food--31 kg/year (Table 3)--however, in the South
Pacific, sweet potatoes and nonspecified tropical
root crops are more important than cassava.

The most important processed products in terms
of consumers are "gari" in West Africa and farinha
de mandioca in Brazil. Both products are the result
of rasping, soaking, drying, and roasting, but
differ in taste owing to the degree of fermentation
which occurs during soaking, and drying and roasting
methods. Cassava flour, although less important
than gari or farinha de mandioca, may gain importance as consumption of confectionary products
increases. An important property of cassava flour
is that it can replace up to 20 percent of cereal
flours without greatly altering the final product
or the processing methods. Given the large import
and increasing cost of wheat flour (Table 4), it
must be assumed that there is a growing potential
market for cassava flour. Other tropical root crop
flours may in fact be preferable to cassava flour
because the former may have a lower fiber content
(high fiber content of cassava flour is thought to
be one factor in reducing the baking quality of
cassava flour).

The major implications of processing cassava to
produce a human food for other tropical root crops
are that processing may improve the marketability of
the raw material; that simple processing can be an
initial step in the production of a number of different final products; and that processing on a
small scale provides an activity which enables the
producer to add value to the raw agricultural product. Thus the small-scale processing of any
tropical root crop should have similar benefits.
The implication to be drawn from the cassava experience is not that other tropical root crops should
be used to produce gari or farinha de mandioca,
although this is technically possible in many
instances. The commodities produced must meet
local needs.

Given the size of developing country wheat
flour imports, and assuming that 20 percent can be
replaced by tropical root crop flours, it may be
calculated there is the following potential for
tropical root crop flour.[4]

Africa	139,000 tons
Latin America	137,000 tons
Near East	282,000 tons

TABLE 3
1975 Per Capita Demand for Tropical Root Crops (kg/yr)

	Roots and Tubers	Potatoes	Sweet Potatoes	Cassava	Yams	Unspecified Tropical Root Crops
Africa	188	5	11	102	33	13
Latin America	83	22	8	36	1	2
Near East	25	18	--	5	1	--
Far East	27	6	7	13	--	1
South Pacific	311	1	127	26	44	122
Total Developing:	64	9	7	31	6	4

Source: FAO, personal communication

TABLE 4
Import of Wheat Flour by Developing Countries and Selected Regions

A. Quantity (100 metric tons)

	1969	1970	1971	1972	1973	1974	1975	1976
Developing Countries	3261	3388	3678	2878	3102	2986	3505	3340
Africa	467	548	575	552	503	548	799	693
Latin America	668	694	752	684	737	692	689	697
Near East	808	1014	1174	910	1059	1006	1358	1412
Far East	1222	1031	1074	623	702	642	588	464
South Pacific	90	95	97	102	95	91	64	66

B. Value ($10^6)

	1969	1970	1971	1972	1973	1974	1975	1976
Developing Countries	338	334	379	312	458	743	969	816
Africa	58	66	72	74	88	143	235	176
Latin America	77	75	84	78	106	147	184	173
Near East	63	81	105	84	134	251	351	338
Far East	126	99	105	61	111	178	177	108
South Pacific	11	10	10	12	16	20	18	18

Source: FAO, 1975, 1977c.

Far East   93,000 tons
South Pacific   13,000 tons

CASSAVA MARKETS: STARCH

Cassava starch has dominated the trade of
tropical root crop starch (Table 5), but the heading, starch, conceals the difference in quality and
type of starch. For example, arrowroot is specifically used for production of certain types of biscuits, and no other starch seems to have the desired
properties. While tropical root crop starches have
general similarities, they have specific properties
which are not shared by all tropical root crops, as
noted below.[5]

Cassava starch: The unswollen grains are
roughly circular with concentric rings and
usually a hilum. The size is approximately
15 to 25μ in diameter. Gelatinized cassava
starch, commercially traded, is three times
larger than unswollen starch, and has
saucer-like shapes with no regular markings.
The center is usually dark.

Sago starch: Similar to cassava starch, ranging from 20 to 60μ.

Potato starch: Composed of large oval or
conchoidal grains with oyster-shell markings
of less than 100μ, and smaller rounded or
flattened grains approximately 15μ in size.
A visible hilum is located near the end of
the grain. The cross seen under polarized
light is centered at the hilum.

Arrowroot starch: Constitutes both the largest
(135μ) and smallest (7 to 12μ) starches;
similar to potato starch.

In Canada, Japan, and the United States, tropical root crop starches have through time captured
a relatively smaller proportion of the total market.
Tropical root crop starches have not maintained
their market share because developing countries have
not been able to provide a steady supply of quality
starch; tropical root crop starches are not always
competitively priced, owing in part to transportation costs; and developing countries do not normally
produce the specialized starches which have become
more important. The last point is an extremely
important limitation on the export of tropical root
crop starches, because developing countries do not
generally have the capability to produce highly

TABLE 5
United States, Canadian, and Japanese Imports of Specific Tropical Root Crop Starches (1000 metric tons)

	1960	1961	1962	1963	1964	1965	1966	1967	1968	1969	1970	1971	1972	1973	1974	1975
United States[a]																
Cassava	127.0	139.1	74.0	110.9	133.5	162.4	154.5	137.9	87.9	88.4	93.8	82.6	64.3	49.0	74.4	38.4
Arrowroot	2.8	2.1	2.7	2.6	1.9	2.2	1.4	1.6	1.6	1.4	1.6	1.5	0.6	0.5	2.1	1.1
Potato	3.2	2.5	1.1	12.4	3.5	12.9	0.7	0.7	0.5	0.4	1.4	2.3	6.0	16.8	12.1	6.5
Canada[b]																
Cassava	2.0	1.8	1.6	1.6	3.0	4.4	5.8	9.1	7.2	6.6	9.1	4.2	2.6	4.1	0.8	2.0
Tapioca	0.7	0.8	0.7	1.2	0.9	0.7	0.6	0.7	1.0	0.9	0.6	0.7	0.7	0.5	0.6	0.6
Arrowroot	0.3	0.3	0.3	0.4	0.4	0.6	0.4	0.5	0.6	0.8	0.5	0.2	0.7	0.8	1.1	1.0
Potato	2.9	1.3	1.6	2.1	3.8	6.7	4.3	3.1	3.5	6.2	9.0	1.3	1.2	2.2	2.2	1.7
Japan[c]																
Cassava											50.3	47.0	50.6	71.8	139.7	71.1
Sago											21.0	17.5	12.5	14.6	15.9	16.9

a U.S. Department of Commerce
b Statistics Canada
c GATT, 1977

specialized modified starches which are being
increasingly used by the food processing industry.
The specialized modified starch trade is constantly
changing, thereby suggesting that substantial
research and development are required to maintain a
share of the market. Given the relative unimpor-
tance of the world market for tropical root crop
starch- it is doubtful that developing countries
will (or should) place much emphasis on the research
and development of modified tropical root crop
starches. Therefore, it is anticipated that tropical root crop starches will maintain those markets
which utilize their specific physical and chemical
properties, but it is unlikely that there will be
an increasing demand for such starches.

A potentially more fruitful market is the
domestic market for starch. Cassava has secured a
place in at least two domestic markets, Brazil and
Malaysia. However, in the former case there are
still substantial imports of maize starch (1495
metric tons valued at $1.2 million in 1974),[6] which
in some instances could be replaced by domestically
produced starches. Domestic demand for starches
should grow with increased industrialization (particularly textile, paper, adhesive, and some forms
of mining), but it may require government promotion
of the use of domestically produced starches.
Failing this type of promotion, industrialists may
be tempted to use readily available starches from
developed countries.

CASSAVA MARKETS: ANIMAL FEED

Cassava became an animal feed ingredient
basically because the Common Agricultural Policy of
the European Economic Community (EEC) raised the
price of feedgrains, making it economical for EEC
animal feed manufacturers to use energy rich feeds
such as cassava and protein rich feeds such as soybean meal to produce a superior quality cereal-type
product.[7] Imports of cassava began in the early
sixties and have expanded to 3.7 million metric
tons (worth about $400 million c.i.f. Rotterdam)
(Table 6). Dutch imports account for approximately
53 percent of total imports, while West German and
Belgium/Luxembourg imports account for an additional
40 percent of total cassava imports. In fact,
Thailand provides more than 90 percent of the Dutch,
French, and West German requirements (GATT, 1977).
Currently, theonly other supplier of note is

TABLE 6
European Economic Community Imports of Cassava,[a] 1962 to 1975 (1000 metric tons)

	1962	1963	1964	1965	1966	1967	1968	1969	1970	1971	1972	1973	1974[b]	1975[b]	1976[c]	1977[c]
West Germany	366	387	462	520	702	533	481	548	591	479	387	420	431	484		
France	23	20	18	17	16	n.a.	n.a.	n.a.	n.a.	38[b]	139[b]	159[b]	164	146		
Netherlands	1	5	17	76	96	159	237	444	502	599	650	700	1088	1233		
Belgium—Luxembourg	23	72	105	100	70	113	127	212	268	278	293[b]	204[b]	394	449		—
Total European Economic Community:	413	484	602	714	884	805	845	1204	1410	1750	1850	1900	2077	2312	3441	3700

[a] Phillips, 1974.
[b] GATT, 1977.
[c] Estimated from Bangkok Bank Monthly Review, 1977.

Indonesia (in 1975, supplying slightly more than 300,000 tons). Originally, Thai cassava faced strong competition from cassava supplied by Angola, Brazil, Indonesia, People's Republic of China, and Tanzania. However, only Thailand, and to a lesser degree Indonesia, were able to provide continuously the EEC requirements. In 1962 the product was either cassava meal or chips, but in 1969, when Thailand acquired pelleting equipment, the product changed to pellets.

The growth of the EEC cassava market is in reality a testament to the ability of Thai farmers and processors to respond to a hitherto unthought-of market. The Thai cassava export industry was encouraged and supported by the West Germans, and benefited from a "freight war between Thai and French shipping lines which had the effect of reducing shipping costs to Europe by roughly a third of the normal price" (Phillips, 1974). Attempts to develop cassava export industries in other countries failed. These countries differed from Thailand in that cassava in the former countries is a staple, while in Thailand it is not. The effect of the domestic market seems to be to raise domestic cassava prices above the EEC price of cassava--e.g., when the c.i.f. price of Thai cassava chips was $70/ton the price of farinha de mandioca in Northeast Brazil was approximately $100/ton.

The message for other potential exporters of tropical root crops is that producers, processors, and exporters must be able to respond to the market challenge. A more important lesson to be learned is that cassava and other tropical root crops can be a very important animal feed ingredient in producing countries, particularly as more intensive livestock systems gain popularity in tropical countries. Indigenous crops may be substituted for imported crops; for example, 1 ton of imported maize could be replaced with 850 kg of dried cassava and 150 kg of groundnuts, or 1 ton of soybeans could be replaced with 625 kg fishmeal and 375 kg of cassava.

The starch and animal feed export markets are important because they illustrate the versatility of cassava specifically, and other tropical root crops generally. The existence of export markets for tropical root crops should help to dispel the myth that tropical root crops are subsistence crops or are of importance only for special occasions, and thereby enable producers, entrepreneurs, and policy makers to assess the contributions which tropical root crops may make to the society.

POTENTIAL CASSAVA MARKETS: SINGLE CELL PROTEIN,
AND ETHYL ALCOHOL

The feasibility of using cassava as a substrate
in the production of single cell protein (SCP) has
been demonstrated by recent research carried out at
the Universities of Guelph and Malaya, and the International Center for Tropical Agriculture (CIAT).
The UG-CIAT project[8] has demonstrated that a nearly
balanced pig ration can be produced from fermented
cassava. An advantage of this technique is that it
is relatively small scale, the fermenter has a
capacity of 3000 liters, and it can operate under
nonseptic tropical conditions. Furthermore, if the
SCP production technique is combined with improved
cassava production techniques, it will provide a
means of increasing both the quality and quantity
of pig feed (eventually other types of animal feed)
with the minimum diversion of land from production
of food crops.
Because this research is not crop specific, the
implications are that other tropical root crops
could be used as a substrate. There is, of course,
the need to determine if the enzymes used for
cassava are the best to use with other tropical
root crops. While the SCP can be used directly as
a basis for producing human food products, the
current thinking is that at present the most promising market is the animal feed market.
Another emerging market for cassava is as a
substrate for production of ethyl alcohol (ethanol).
Presently this market is in Brazil and results from
a governmental decision to produce 3,000 million
liters of ethanol by 1980 (Journal, 1978). Brazil
is promoting ethanol production in an attempt to
become nearly self-sufficient in fuel production.
The ethanol is to be added to gasoline (up to 20
percent) to replace imported fuel. As of February
1978, 163 distillery projects were approved by the
Brazilian National Alcohol Committee (Yang, 1978),
expanding annual national capacity by 3,700 million
liters. While many of the distilleries are sugar-
cane based, 8 of the distilleries plan to use
cassava.[9] In fact, the first cassava-based distil-
lery--capacity 60,000 liters/day and erected by
Petrobrás--is now operational.
Again, the use of cassava is indicative of new
markets which could exist for other tropical root
crops. This latter technology is, however, expensive. It is estimated that the investment

requirement for a 49 million liter cassava distillery is $15.8 million.

A QUESTION OF SIZE

There are two components of the size question: the first relates to the size of production units and the second relates to the size of processing units. Regarding the former, the extent of individual or farm plantings are (except for a few plantations) small and probably less than 16 hectares (based on survey of Thai cassava farmers (Phillips, 1977)). Annual processing, on the other hand, can range from family units processing less than one-half ton of roots to produce gari, a human food, to cassava alcohol plants which are scheduled to process 110,000 to 275,000 tons of cassava.

For the human market the author estimates that more than 90 percent of processing of gari and farinha de mandioca is done on a small scale. Of the remaining 10 percent, a proportion of the large scale processing actually is assembling, grading, re-roasting, packaging, and distribution. An example is the operation of some of the large farinha de mandioca plants outside Rio de Janeiro and São Paulo. The other large processors purchase fresh roots and take full responsibility for processing and distribution.

In Thailand the average utilized capacity of starch plants is 96 tons of roots per day operating 150 days per year (Thawee, 1974), while in Malaysia the average utilized capacity appears to be about 30 tons of roots/day (Tan, 1973). The Thais, with their relatively small plant size (and relatively small cassava farms, averaging 3.3 ha), have apparently been able to meet world demand for cassava starch. Thus the implication seems to be that large scale production and processing are not necessary prerequisites for a successful cassava export industry. Regarding the animal feed industry, the first step in processing, chipping, and drying, is small scale with chipping plants having a capacity of approximately 10 tons/day operating 160 days/year (Thawee, 1974). Given this capacity each plant services about 114 hectares. The chips are then either used to produce pellets at the chipping plant or, more commonly, are sold and pelleted at a pelleting plant. It is estimated that 70 percent of the export is produced by imported pelleting equipment, while the remainder is produced by domestically produced equipment (Titapiwatcwakun, 1974). The capacity of the imported equipment ranges from 20,000 to 110,000 tons of pellets per year.[10] Typically, domestically produced equipment has a capacity of only 5,000 tons of pellets per year.

Thus Thailand has developed a dynamic export industry which blends small scale production and initial processing with large scale final processing. It is perhaps worthy of note that both large scale and small scale levels of processing will be preserved, because the large scale producers have indicated that they are not interested in getting involved with chipping activities.

The animal feed market reveals another aspect of size, namely the ability of exporters to provide the volume of product which is required. In the Thai case 85,000 ton ships are constantly being used to ship Thai pellets to Europe. Thus the exporting activities of Thailand are even larger scale than the final processing activities. The implication for other possible exporters of tropical root crops as an animal feed is that, regardless of the scale of the processing plants, there is need to insure that the industry as a whole has the capacity to supply a minimum volume. GATT (1977) has calculated this annual minimum to be 120,000 tons of pellets, while the author estimates that the minimum may be as great as 260,000 tons of pellets depending on shipping costs (Phillips, 1978).

SUMMARY

All tropical root crops have experienced similar increases in production during the past 15 years, but only cassava has had a substantial proportion of production (approximately 12 percent) used in the non-human food markets. The uniqueness of the cassava experience is perhaps fully explained by the opening of the EEC animal feed market to energy-rich feeds, and the Thai production response to this market. The latter was perhaps only possible because the Thais did not consider cassava as a staple crop.

From the cassava experience two inferences for other tropical root crops may be drawn. The first is that markets for other tropical root crops cannot be diversified where the tropical root crops are a staple crop; the second is that tropical root crops have many domestic uses which need to be explored. The export successes of Indonesia and ethyl alcohol plans of Brazil cast some doubt on the first inference, while adding some support to the second inference. The key element is the desire on the part of processors, entrepreneurs and governments to have tropical root crops be more than subsistence crops. The existence of this will be revealed in time.

NOTES

1. Most discussion of the world food situation refers only to availability of food grains, or, in the most mypoc version, to wheat, and completely ignores tropical root crops. Yet tropical root crop production is 40 percent of food grain production (68 percent if rice is excluded).
2. In economics an inferior good is defined as a good for which demand decreases when income increases.
3. It should be noted that the transformation is partially illusory since the term subsistence crop may still be applied to as much as 88 percent of cassava production.
4. It should not be assumed that all tropical root crops will produce equally acceptable flours. The point of the calculation is to indicate the market potential, and hence the need and benefits which could be derived by doing research in this area.
5. T.P. Phillips, "Cassava Utilization and Potential Markets," IDRC-020e, Ottawa, Ontario, 1974.
6. SITC 599.5 excluding vegetable protein products.
7. Sweet potatoes imported from the People's Republic of China and Indonesia have competed with cassava (for animal feed purposes, cassava and sweet potatoes are complete substitutes). However, the importation has never exceeded 150,000 tons/year.
8. Funded by IDRC, Canada.
9. W.N. Milfont, personal communication, 1978.
10. This is equivalent to 50,000 to 250,000 tons of roots per year.

REFERENCES

Bangkok Bank. 1973. Bangkok Bank Monthly Review, Bangkok, Thailand, Vol. 10.

Department of Commerce, U.S. General. Input Bulletin, PT135, Washington, D.C.

Food and Agricultural Organization. 1975. Trade Yearbook 1974, Rome.

_____. 1976. Production Yearbook 1975, Rome.

_____. 1977a. Monthly Bulletin of Agricultural Economics and Statistics, Rome.

_____. 1977b. Trade Yearbook 1976, Rome.

GATT. 1977. Cassava Export Potential and Market Requirements, Geneva, Switzerland.

Halder, S. and Yang, M.C. 1976. Developing country foodgrain projections for 1985. World Bank Staff Working Paper No. 247, Washington, D.C.

Journal do Brasil. 1978. Rio de Janeiro.

Phillips, T.P. 1974. Cassava Utilization and Potential Markets, IDRC-020e, Ottawa, Ontario, Canada.

_____. 1977. A profile of Thai cassava production practices. Proc. Fourth Symp. of Intern. Soc. for Trop. Root Crops, IDRC-080e.

_____. In press. Economic implications of new cassava harvesting and processing techniques. In Cassava Mechanical Harvesting and Post-Production Processing, IDRC.

Tan, K. 1973. The tapioca industry in Malaya. Ph.D. dissertation, University of Malaya.

Thawee, C. 1973. The Production, Marketing and Prices of Tapioca in Thailand in the Year 1971. Agricultural Economics Division, Thai Ministry of Agriculture, Bangkok, Thailand.

Titapiwatatanakun, B. 1974. Cassava industry in Thailand. Master's thesis, Thammasat University, Bangkok Thailand.

Yang, V., Milfont, W.N., Jr., Scigliano, A., Massa, C.O., Sresnewsky, S., and Trindade, S.C. 1978. Cassava Fuel in Brazil. Centro de Technologia Promon: Rio de Janeiro.

Truman P. Phillips: University of Guelph, Guelph, Ontario, Canada

3
Production, Distribution, and Final Uses of Sweet Potato in Taiwan

P. H. Calkins

ABSTRACT

Average annual yields of sweet potatoes have
increased by 184 kg/ha, while the area planted has
decreased drastically. The decrease in area planted
can be attributed to sweet potato's unfavorable comparison with competing products in labor demanded,
economics, and status as a food. Since sweet potato
is a relatively nonperishable, low-cost commodity to
handle, marketing agents like its consistent profit
margin. The most efficient marketing channel was
found to involve only two marketing intermediaries
who preferred to take a smaller profit in the interest of handling larger volumes. Sweet potato uses
are as follows: starch production, of which the
starch is used for food and medicine; the waste
becomes animal feed, citric acid, and other chemical
products; the bulk of the remainder becomes animal
feed, with a small amount going to human consumption.

INTRODUCTION

Taiwan has developed efficient sweet potato
production, distribution, and processing technologies which may well be emulated by other tropical
nations seeking to develop their root crop industries. Over the past twenty years, per hectare
yields have continued to climb in Taiwan, with an

397

annual average increase of 184 kg/ha. A processing industry has been born which efficiently produces starch, feed, and industrial by-products from the crop. Also, the marketing of sweet potatoes has found a major role as a relatively nonperishable commodity with low handling costs. These advantages provide a popular combination in helping to balance the commodity portfolio of vegetable merchants.

Figure 1. shows that the overall area planted to sweet potato in Taiwan has dropped drastically in the last six years. This is mainly because household consumers regard sweet potatoes as a low status food. As their incomes have risen (seven fold since the early 1950s), they have switched to other foods. Other reasons include:

1. Farmers prefer to grow rice since a guaranteed price reduces their risks compared to that of sweet potatoes.
2. Farmers who once gres sweet potatoes as hog feed today prefer to use processed feed which is easier to store and handle. This consists largely of maize, much of it imported from Thailand.
3. Increased irrigation allows cultivation of rice and sugar cane on land once suited only to sweet potato.
4. A growing shortage of farm labor tends to make farmers choose a crop like sugar cane, which can be harvested by factory-owned machines.
5. Processors have found that cassava is cheaper than sweet potato for the production of starch necessary for eel-raising.

Thus, not only have consumer and processing demands declined, the farmers' incentives to produce the crop have lessened as well. Nevertheless, Taiwan's production, marketing, and processing of sweet potatoes constitute a successful supply system of a kind rarely seen in the tropics. But even efficient supply systems are eventually outgrown by society's demands, and this is what is happening in Taiwan. This paper focuses on the aspects of the sweet potato industry in Taiwan which have contributed to its ever-increasing efficiency, despite its loss in size.

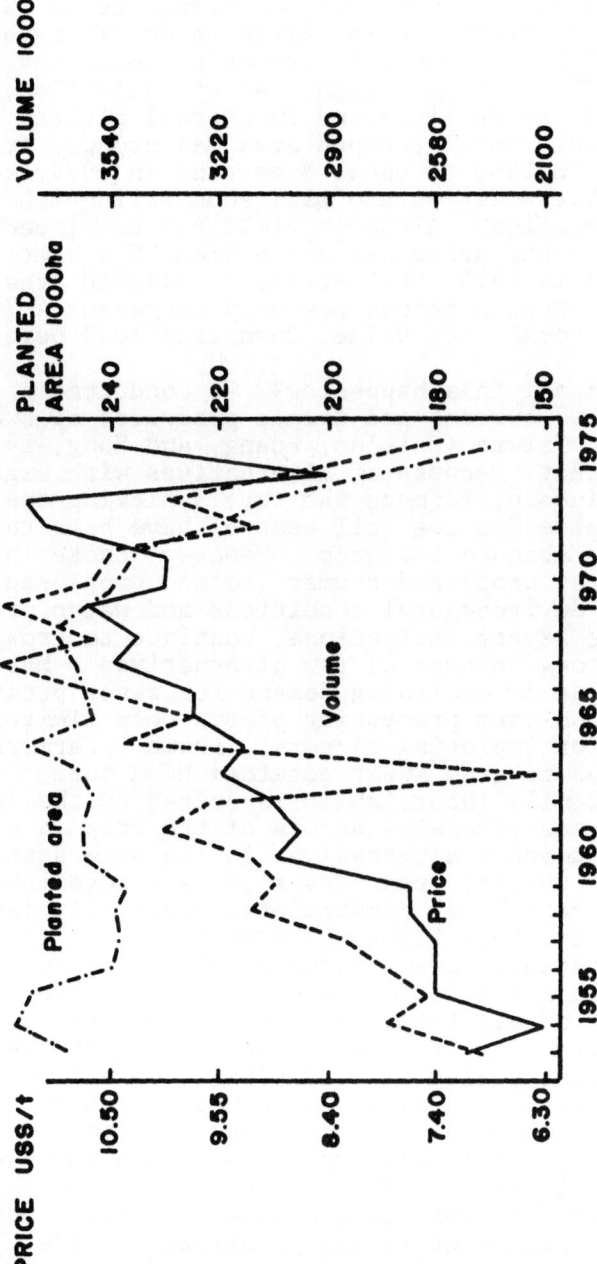

Figure 1. Price, planted area, and volume of sweet potato in Taiwan 1953-1975.

PRODUCTION

According to long-term production trends (Wu,
1977), the planted area in Taiwan is declining by
about 5,177 ha/yr, and the production volume by
about 60,000 mt/yr in Taiwan. At the same time,
sweet potato is still second in overall planted area
only to rice. Total cropped area has dropped from
14 percent in 1966 to under 8 percent in 1976, and
it seems that sweet potato will soon relinquish its
dominant position. Although yield has continued to
increase and the price has risen from US $19/mt in
1966 to $47 in 1976, the decline in planted area has
meant that sweet potatoes now only represent 4.7
percent of total crop value, down from 10.2 percent
in 1966.

Why is all this happening? We conducted a
survey of 317 current and former producers of sweet
potatoes in Taiwan (Calkins, Huang, and Hong, 1977)
and found that, because of alternatives with higher
price and income, farmers who were achieving the
highest yields (in the fall season) have been the
quickest to abandon the crop. However, those in the
spring (first crop) and summer (second crop) seasons, when environmental conditions and water problems impose severe limitations, continue to grow
sweet potatoes because of few alternatives. Moreover, because of declining demand for sweet potatoes
in Taiwan, on-farm processing offers only limited
prospects for improving farmers' income. Farmers
who continue to grow sweet potatoes have more
available family labor, which is suited to the low-
capital, labor-intensive nature of the crop in comparison with other alternatives in the same season
in Taiwan. Current producers also have more land
and face more natural constraints, especially lack
of irrigation, than former producers.

Nevertheless, even producers of spring and
summer crops are able to achieve yields higher than
the average of all tropical countries for which data
are available. After Taiwan, Bangladesh has the
second highest yields with 10 t/ha. Table 1 shows
that relatively high rates of inputs, notably
agricultural chemicals and machinery, contribute to
high levels of yield and profit in Taiwan for sweet
potato as well as for other crops.

It would be wrong to conclude that sweet potato
production technology is inappropriate or undevel-
oped in Taiwan. Far from it; the yields achieved,
particularly during the fall season, are quite

TABLE 1
Production Budgets for Sweet Potato by District 1975-1976

Budget Element	District					
	Changhua (N=20)	Tainan (N=30)	Kaohsiung (N=20)	Miaoli (N=20)	Pingtung (N=17)	Taitung (N=22)
Season	Relay	Fall crop	Fall crop	2nd crop	2nd crop	1st crop
Yield (kg/ha)	13,284	28,163	21,794	14,101	12,262	14,275
Price (US$/100 kg)	2.9	3.7	4.5	3.4	4.7	4.2
Revenue (US$)	385	1,038	975	482	600	601
Expenses (US$)						
A. Cash (US$/ha)						
Human labor	54	91	82	33	55	54
Animal power	14	28	64	14	18	34
Machine power	12	50	67	0	43	15
Chicken manure	0	0	5	1	0	0
N	60	34	42	16	18	32
P	22	12	18	6	5	14
K	21	21	21	8	8	15
Seeding	48	131	83	30	24	32
Pesticide	8	5	2	<1	3	3
Fuel	<1	16	4	0	42	3
Water fee	29	47	0	16	1	11
Land tax	40	42	38	24	8	14
Subtotal	305	480	427	148	226	227
B. Non-cash (US$/ha)						
Family labor	206	141	172	217	115	157
Animal power	70	49	40	128	121	110
Machine power	0	17	22	0	3	3
Compost	166	86	50	24	46	30
Interest on land	139	87	128	71	23	25
Interest on capital	8	14	13	3	5	6
Subtotal	590	394	424	442	313	331
TOTAL	594	874	852	590	539	558
Farm return (US$/ha)	80	558	548	335	374	374
Net profit (US$/ha)	-510	164	123	-108	61	43

Source: Survey results, 1976. US$ = NT$38.

remarkable.. In the economic environment of Taiwan, where processing and human consumption demand have declined so rapidly, this technology has become uncompetitive with alternative crops such as processing tomato, Chinese cabbage, and cauliflower intercropped with lima bean, all horticultural crops with increasing demand patterns.

National and international research institutes in Taiwan could attempt to improve sweet potato production policy in two alternative ways: (1) concentrate on maximizing fall yields and incomes, with possible benefits to urban and industrial consumers but no guarantee of high rates of adoption because of competing crops, and (2) develop better, low-input technologies for low income farmers in the first and second seasons. Present farmers' conditions suggest that the latter technology is more suited to Taiwan, largely in the interests of bettering the rural distribution of income. The production techniques in Table 1 for these seasons could further gain from crop management research to control the production environment and from higher use of stem cuttings, inorganic fertilizer, and farmer-owned animal power to elevate yields and farm income.

When we asked farmers' perception of sweet potato marketing characteristics in comparison with those of some major alternative horticultural commodities, they reported that sweet potatoes have the lowest day-to-day and season-to-season price fluctuations, the lowest season-to-season yield variation except for bamboo shoots, and great ease in transport and storage. Even these advantages, however, are not enough to convince more farmers to produce the crop.

POST-HARVEST HANDLING AND FARMER SALES

In general, sweet potatoes for non-starch purposes are processed on the farm before entering commercial channels. The sweet potatoes are sliced and dried into chips which are used as hog feed. Table 2 shows that most of the cost is for processing, with human labor the main component. The wide range in costs per hectare (US $26 to $112) is primarily due to the variation in amounts processed in the six sample districts.

Table 3 shows how the farmers in each district dispose of their crop. Farmers still plant sweet

TABLE 2
The Structure of On-Farm Post-Harvest Handling Costs of Sweet Potato[a] 1975-1976

Item	District					
	Changhua	Tainan	Kaohsiung	Miaoli	Pingtung	Taitung
			(%)			
Root transportation						
Labor cost	8	0.4	0	7	2	25
Animal cost	28	8.5	3	14	1	35
Machine cost	0	11	22	3	24	3
Subtotal	36	19.9	25	24	27	63
Processing						
Labor cost	64	74	72	75	73	36
Animal cost	0	0	0	0	0	0
Machine cost	0	6	2	0	0	
Subtotal	64	74	74	75	73	36
Chip transportation						
Labor cost	0	0.1	1	0	0	1
Animal cost	0	0	0	0	0	0
Machine cost	0	0	0	1	0	0
Subtotal	0	0.1	1	1	0	1.
Total: %	100	100	100	100	100	100
: US$	26	87	112	67	105	44

[a] Based on one hectare of output. Source: Survey data 1976.

TABLE 3
Sales Patterns of Sweet Potato in Six Districts of Taiwan, 1975–1976

							Sold to					
District	Total Production	Total % Sold	Starch Factory %	Price[a]	Feed Factory %	Price[a]	Local Assemblers/Jobbers %	Price[a]	Hog Producer %	Price[a]	Neighbor and gifts %	Price[a]
Changhua (N=32)	143,683	4	0		0		3	2.2	0		1	2.2
Tainan (N=66)	753,503	53	43	3.5	0		5	4.4[b]	3	3.8[b]	2	3.8[b]
Kaohsiung (N=44)	377,913	24	6	4.4	0		18	4.7	0		0	
Miaoli (N=30)	323,960	36	0		2	2.84	32	3.7	0		2	3.5
Pingtong (N=35)	153,138	3	0		0		1	4.4	1	6.6	1	4.7
Taitung (N=44)	244,106	54	0		0		26	3.8	16	4.5	12	4.4

[a]All prices are in $US/100 kg.

[b]Chip prices: $12.71 local assemblers/jobbers; $13.16 hog producer; and $13.16 neighbor and gifts in Tainan.

Source: Survey data 1976.

potatoes primarily for their value as feed, but the variability in percent retained on the farm is great. Tainan and Taitung sell the most sweet potatoes. Tainan farmers sell primarily to starch factories and Taitung farmers to local shippers, both first stages in the marketing channels to be discussed below.

MARKETING

Sweet potatoes flow mainly in the markets of the central and southern portions of Taiwan, which regulate seasonal demand among themselves (Wu, 1977). For example, fall sweet potatoes are shipped from southern to central Taiwan, and the fall relay crop flows from the central to the south, usually by rail. There is also an insignificant export business to Hong Kong (2.6 mt in 1976).

In a study of perishable commodity marketing in Taiwan in 1977, we found a number of different types of sweet potato marketing agents: local assemblers, jobbers, wholesalers, wholesale-retailers, retailers, and commission agents. A full 50 percent of all marketing agents incurred no handling loss and no marketing agent incurred more than 10 percent. Losses tend to be greatest at the retail level. Table 4 shows that variable marketing costs are extremely low for sweet potato. For most types of agent, transportation is the major expense, followed by labor. Wholesalers have no variable marketing costs, although of course they do have fixed rents and other costs which should be assigned to sweet potato and other crops handled. Therefore, we may conclude that sweet potato is a relatively non-perishable, low-cost commodity to handle. Indeed, when compared with mungbean, bamboo shoots, tomato, cabbage, and Chinese cabbage, sweet potato was the least perishable commodity (next to mungbean) and had the lowest marketing costs. Therefore, marketing agents like to include it in their commodity portfolio.

Given its ease of handling and its relative long-tern storage compared to other crops in question, marketing agents also have more freedom to hold out for higher price markups.[1] Figure 2 shows that there is a broad range of markups, with 50 percent receiving less than 30 percent; 44 percent receiving 30 to 60 percent; and 6 percent receiving

TABLE 4
Variable Marketing Costs by Marketing Agents, Sweet Potato

Unit: $/M.T.

| | | | Variable Marketing Costs | | | |
| | | | | Commission | | |
	Labor[a]	Materials	Transportation	Market	Other Agents	Total
Local Assemblers	1.37	1.71	7.08	–	–	10.16
Jobbers	3.78	2.98	5.68	1.26	0	13.70
Wholesalers	0	0	0	0	0	0
Wholesale-retailers	0	1.28	0.60	1.78	0	3.66
Retailers	0	0.14	2.41	0	0	2.55
Commission Agents	0.43	0	0	0	0	0.43

[a]Imputed value included for own labor.

Source: Survey data, 1977.

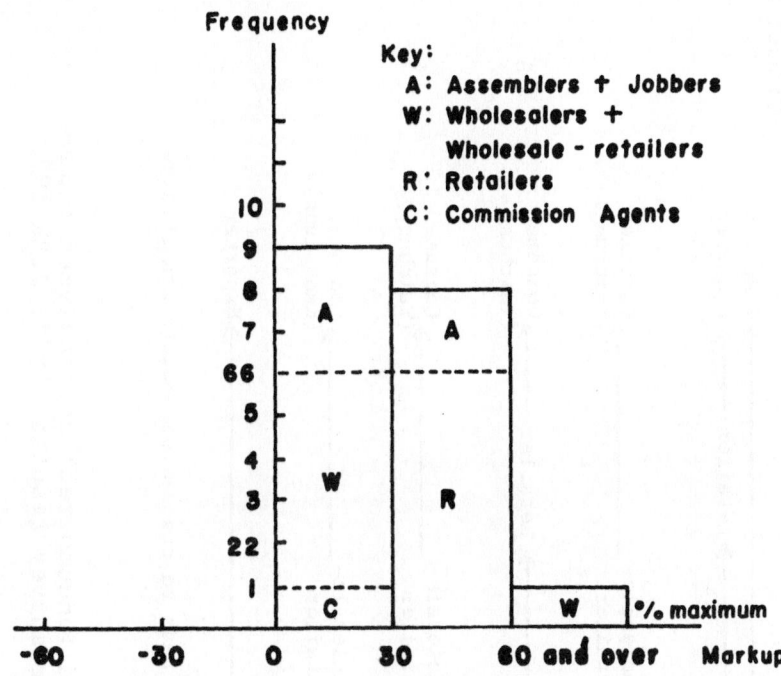

Figure 2. Distribution of markups at maximum prices paid and received by marketing agents for sweet potato (after survey data, 1977).

over 60 percent. There is also a tendency for agents close to the farmer who handle larger volumes to receive less markup.

There was a strong tendency for a given markup at minimum prices to reflect itself in an equal markup at maximum prices. This suggests that agents who deal with sweet potatoes tend to have a very stable percentage profit. The case is quite different with more perishable commodities like Chinese cabbage, where a high markup at minimum prices reflects itself in a sharply reduced markup at maximum prices. Also, the overall range of markups for Chinese cabbage is much lower, for the agent wishes to move the commodity as quickly as possible. Indeed, we found that the more perishable the commodity the lower the markup an agent is willing to accept..

Figure 3 shows the main types of marketing channel for sweet potatoes in Taiwan. Channel 1 takes the product directly from the producer through

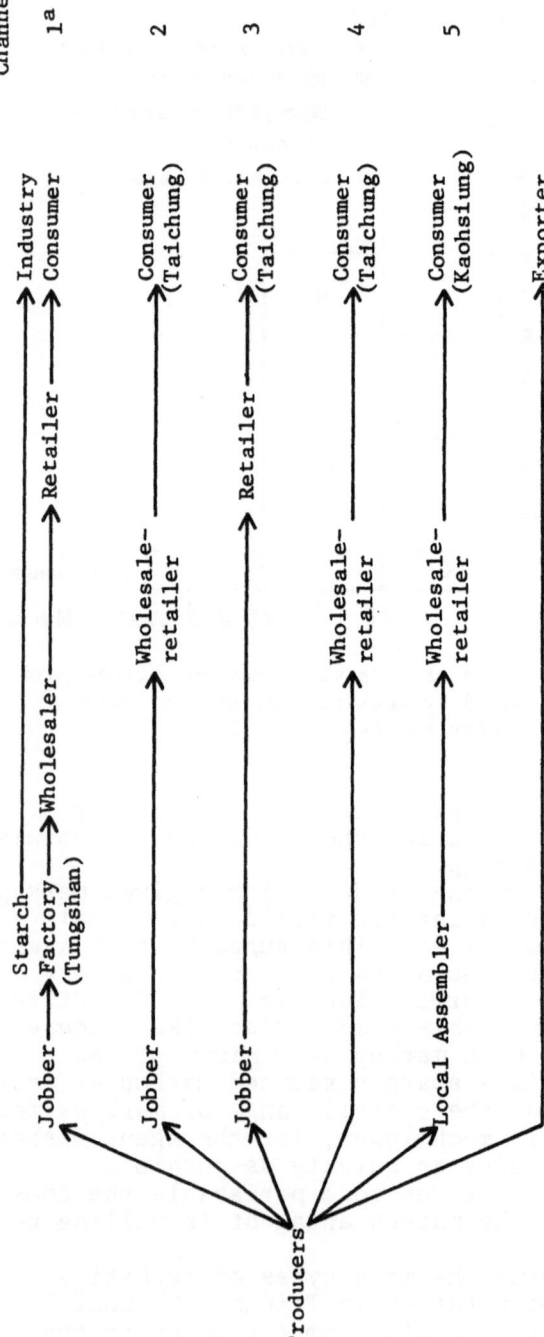

[a]Because of the change in form, no attempt was made to trace sweet potato starch after it left the factory.

Figure 3. Marketing channels identified (unnumbered) and analyzed (numbered) for sweet potato, 1977 (after survey results, 1977; JCRR 1972).

a jobber to the starch factory, while the others involve several combinations of jobbers, wholesale-retailers, and retailers in the route to the consumer. The first four channels are for August data, while the fifth was recorded in December to gauge the effect of different seasons on marketing efficiency. We used cost per metric ton of final product delivered per kilometer to measure technical efficiency, and the sum of marketing agent profits divided by total marketing costs as an index of marketing efficiency. The channel with the lowest values of each index was judged most efficient.

We found that agents in channel 5 were able to achieve both higher technical and economic efficiency than those in other channels. Analysis of their handling procedures showed that the local assemblers in channel 5 had a larger percentage of vegetables in their total revenues (40 percent) than those from other channels (25 to 30 percent). This may give them a slight edge in terms of handling expertise and economies of scale. This is further shown by the fact that channel 5 local assemblers handled about 19 mt/wk and 6 mt/day, higher than in any other channel except channel 1, where distances were so short (3.3 km) that the assembler could handle 342 mt/wk and 21 mt/day. Thus, we may be fairly sure that the channel 5 assembler is taking advantage of economies of scale. Partly as a result, he was able to handle 12.5 kg/min, versus about 3 kg/min in other channels (statistics for channel 1 are not available). He used cheap plastic bags which cost US $0.11 and had a capacity of 60 kg, versus jobbers in other channels who used cloth bags (US $0.34 with a capacity of 72 kg). The plastic bags may be discarded after a single use.

Wholesale-retailers in channel 5 specialized exclusively in sweet potatoes among the target commodities, unlike those in any other channel. Moreover, they had the highest transaction volume per week (4,200 kg) and per day (1,200 kg). They used trucks for transportation, while the channel 2 retailer, for instance, had excessively high transportation costs because he transported his sweet potatoes by motorcycle. Wholesale-retailers in channel 5 used plastic bags, while those in other channels sometimes also used cloth.

A final reason for the greater technical efficiency of channel 5 is that the number of kilograms lost in handling and shipping (52 kg/mt) was the lowest except for that of channel 1, where short distances meant no losses. This is because channel 5 was a winter channel and all varieties of sweet potato were relatively free from rotting. In summer, by contrast, varieites like Early-70 spoil very quickly if exposed to rain. Special varieties, such as Tainan 31 and Tainan 57, have to be used during summer if losses are to be reduced.

The reasons why channel 5 was so economically efficient were straightforward: (1) it involved only two marketing intermediaries, and (2) they seemed to prefer taking a smaller profit in the interest of handling larger volumes.

What quality of sweet potatoes is preferred in Taiwan? Table 5 shows that outer appearance was the most important (this is a more important consideration than price for consumers), with a good red color and crispness next, followed by sweetness, softness, and storability. Crispness, color, softness, and storability were more highly valued in sweet potatoes than in other crops, while outer appearance is less important than in tomato, and sweetness less important than in cabbage, Chinese cabbage, mungbean, and bamboo shoots.

To complete the comparison with other perishable commodities, Table 6 shows additional key physical and economic characteristics. Water content has a strong positive correlation with physical loss and an inverse correlation with both economic efficiency index and farmers' share of consumer price. Sweet potatoes are the exception, with low indices of economic efficiency (showing that marketing agents take unusually low profits) and farmers' share of consumer prices (showing that farmers also contribute to the low price paid by the consumer). Thus, even though sweet potatoes are a low-cost, efficiently handled product, the consumer is still not interested.

PROCESSING

Sweet potatoes have about 75 percent water content (Table 6), in addition to about 1.3 percent protein (Wu, 1977), leaving about 23 percent non-protein solid matter. Most of this may be used for starch production. The problem is that the quality of sweet potato α-starch[2] is not as good as that of white potato and cassava. Thus, although sweet potato was once tried as a source of the sticky base for eel-feed in pond-raising operations in Taiwan,

TABLE 5
Marketing Agents Preferences in Product Quality

Item	Outer Appearance (size, color, shape)	% Responses[a] Inner Content and Taste								
		Sweet-ness	Low Cellu-lose	Keeping Quality	Crisp-ness	Tender-ness	Soft-ness	Color	Thin Skin	Fresh-ness
Chinese Cabbage	50	64	82		5					
Cabbage	8	72	92		20	4				
Tomato	90	10							10	
Bamboo Shoot	17	61	61		6			28		
Sweet Potato	78	33		11	44		22	44		
Mungbean	50	43							21	7

[a]Total percentage may exceed 100 because of multiple answers.
Source: Survey data, 1977.

TABLE 6
The Relationship Among Physical and Economic Characteristics in the Marketing of Selected
Commodities

Commodity	Water Content[a] (%)	Physical Loss (%)	TEI[b]	EEI[c]	Farmers' Share of Consumer Price (%)
Chinese cabbage	95.7	32.0	2.46	0.56	48
Cabbage	94.2	36.3	2.16	1.19	54
Tomato	95.0	12.4	1.38	3.28	56
Bamboo shoot	82.7	8.3	8.59	4.06	58
Sweet potato	75.3	5.0	1.07	1.69	37
Mungbean	11.8	2.7	5.82	31.31	76

[a]Adapted from Food Industry Research and Development Institute, Table of Taiwan Food Composition
(Hsinchu, 1971). All other data are averages of the channels studied in this report.

[b]Technical Efficiency Index.

[c]Economic Efficiency Index.

it was quickly replaced by imports of white potato and cassava starch.[3] Therefore, only about 5 percent of sweet potato production in Taiwan (2.4 million mt/yr) is used in the production of starch, while the bulk going to the production of animal feeds and a small amount for human consumption such as snacks.

Channel 1 in Figure 3 showed the marketing of sweet potatoes to starch factories. There are 95 such factories in Taiwan, most valued at between US $1300 and $4000 and having a ground space of 300 to 450 m^2 (Wu, 1977). Of all the starch produced, only 20 percent is from sweet potatoes (the rest is from cassava and corn) and only 45 factories produce sweet potato starch.

Because of poor market organization and the sporadic nature of sales, it is difficult to construct complete marketing channels for the starch industry (Wu, 1977). Therefore, the Asian Vegetable Research and Development Center (AVRDC) conducted interviews of three representative starch factories in southern Taiwan to determine their production budgets (Table 7), sources of supply, and final outlets. From our survey, we found that 70 percent of the sweet potatoes processed were bought directly from the farmers and 30 percent from wholesalers. These were converted into first grade starch (13.5 percent), second grade starch (1 percent), and refuse (8.8 percent), accounting for almost all the solid matter. First and second grade starch were sold to wholesalers (71 percent), retailers (7 percent), and food processing firms (22 percent). The starch was finally used for food (96.4 percent), and as a base for medicines (3.6 percent).

Of the refuse, 3.5 percent was purchased by farmers, 33.7 percent by the chmical industry, and 62.8 percent by wholesalers. It was finally used for animal feed (45 percent), for the production of citric acid (32 percent), and for making other chemical products (32 percent).

Because nothing goes to waste, current processing technologies seem quite efficient. One possible source of further efficiency would be the introduction of a pressurized extraction process to remove the excessive water produced. In this way, factories could effectively retrieve the components of the liquid waste and reduce the thickness of the effluent (Wu, 1977).

It has been suggested (The China Times, June 9, 1976) that the current low level of 7.5 percent

TABLE 7

Processing Costs and Returns (Totals and Average for Three Starch Factories in Shanhua Area, October 1975 to September 1976)

Item	Factory A US$	A %	B US$	B %	C US$	C %	Average US$	Average %
Value of output:								
S.P. 1st grade starch	163,184	39.6	101,053	26.0	88,863	41.9	117,700	34.8
S.P. 2nd grade starch	10,800	2.6	0	0	5,826	2.8	5,542	1.6
S.P. refuse	31,500	7.6	8,947	2.3	13,263	6.3	17,904	5.3
Cassava 1st grade starch	163,137	39.5	243,420	62.5	87,128	41.1	164,562	48.7
Cassava 2nd grade starch	14,216	3.5	0	0	0	0	4,739	1.4
Cassava refuse	29,804	7.2	36,026	9.2	16,804	7.9	27,545	8.2
Total	412,641	100	389,447	100	211,885	100	337,991	100
Cost of input:								
Variable cost								
Sweet potato	148,737	44.5	84,211	25.9	75,789	44.6	102,912	37.2
Cassava	166,737	49.9	204,474	62.9	76,737	45.2	149,316	54.0
Labor	15,158	4.5	27,221	8.3	12,079	7.1	18,153	6.6
Electricity	947	0.3	3,158	1.0	474	0.3	1,526	0.6
Fuel	0	0	3,526	1.1	0	0	1,175	0.4
Taxes	2,632	0.8	2,632	0.8	3,816	2.2	3,026	1.1
Interest	0	0	0	0	1,071	0.6	357	0.1
Subtotal	334,211	100	325,221	100	169,965	100	276,466	100
Fixed cost								
Maintenance of building	263	1.0	1,053	5.6	263	2.4	526	2.8
Maintenance of machinery	658	2.4	1,579	8.4	316	2.9	851	4.5
Depreciation of building	132	0.5	1,053	5.6	343	3.2	509	2.7
Depreciation of machinery	368	1.4	1,579	8.4	752	7.0	900	4.8
Taxes	789	2.9	571	3.0	448	4.1	603	3.2
Interest on equity	19,368	72.2	10,898	57.8	4,854	45.0	11,707	62.2
Management	5,263	19.6	2,105	11.2	3,816	35.4	3,728	19.8
Subtotal	24,842	100	18,838	100	10,792	100	18,824	100
Total	361,053		344,059		180,757		295,290	
Net return	51,588		45,388		31,128		42,701	

import tax on α-starch for the eel-industry be increased to allow Taiwan's industry to survive and grow. About 67 percent of the current α-starch is imported either as final starch or for further processing in Taiwan, and 33 percent is processed from locally grown cassava (Wu, 1977). If the import duty were increased, there would be no direct benefit to the sweet potato starch industry (and hence to growers) because sweet potato, as mentioned, is no longer used in the production of α-starch. However, there would be an indirect benefit, as suggested in Table 7, because the same factory prefers to produce starch from both sweet potato and cassava. This is because the processing seasons for cassava (October-January) and sweet potato (December-April) are complementary and can extend the annual operation time of the factory.

CONSUMPTION

In contrast to the meteoric rise of per capita availability of other crops such as fruits and vegetables, the availability of sweet potato has declined from 53 kg/caput/annum in 1964 to 13 in 1975 and a projected 10 in 1981 (Chen, 1977). This lack of availability, however, has actually been induced by the falling consumption patterns of consumers. Under the Japanese occupation of World War II, the people of Taiwan had been exhorted to mix sweet potatoes with theri rice. Their real preference, however, was for rice alone. As soon as incomes began to rise, people dropped sweet potato from their diets.

Figure 4 shows that the decline over the past years has been most dramatic in agricultural households, but that they still consume about three times as many sweet potatoes as urban households (Liau, 1977). At the same time, the consumption of rice declined somewhat, while the per capita consumption patterns of wheat flour have risen dramatically in urban households. Thus, mainly sweet potatoes but also rice are being replaced in the diet by wheat.

Unlike many horticultural commodities (Figure 5), sweet potatoes have very little seasonality in consumption and are consumed at very low levels.

To investigate sweet potato consumption in more detail, AVRDC conducted a survey in 1977 of preferences and consumtpion patterns in Kaohsiung City, the second largest city in Taiwan and the closest large city to the major sweet potato producing areas of Kaohsiung and Tainan counties. The responses of the residents of Kaohsiung, therefore, represent the upper limit on sweet potato consumption in the major cities of Taiwan.

We found that only 25 percent of the population consumed sweet potatoes regularly, with 65 percent occasionally and 10 percent never at all. A full 66 percent of the people said that if the price of sweet potatoes fell, they would still not consume any more of teh product, revealing the strong price inelasticity of the crop. Even more dramatic: 76 percent of the people said they would not consume more even with an increase in income. As to whether the current price was acceptable, 19 percent said it was quite inexpensive, 26 percent said just right, 30 percent said they didn't know, and only 25 percent said it was too expensive. The ideal price

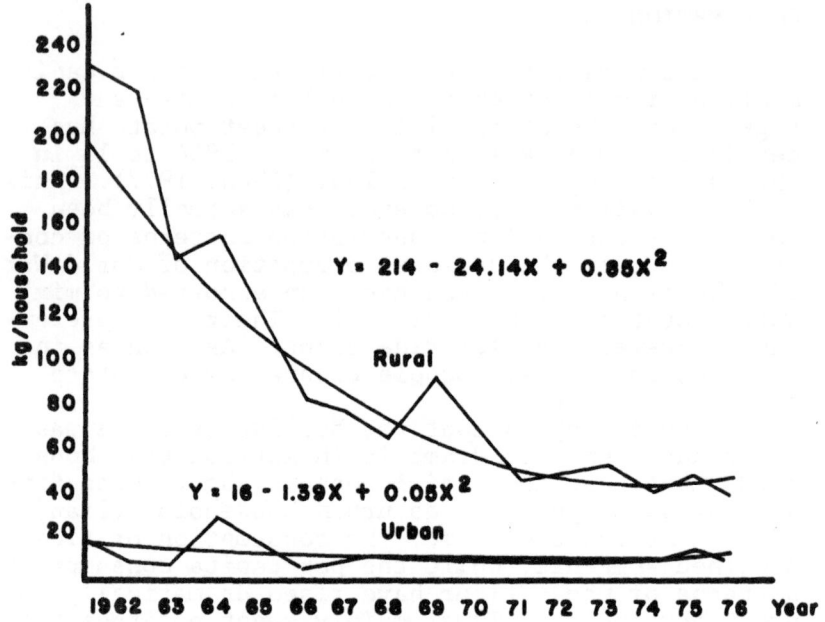

Figure 4. Long-term trends in rural and urban
consumption of sweet potato.
Source: Liau, 1977.

listed was an average of US $0.10/kg, higher than
the actual price of $0.08!

Actual rates of consumption averaged 1.99
kg/household per month. Twenty-four percent of the
households surveyed used it as a staple food, 32
percent as a side dish, and 44 percent as a snack.
In other parts of Taiwan, the predominant use as a
snack is even more pronounced.

What motivates people to eat or not eat sweet
potatoes? The major factors in descending order
were: (1) nutrition, with 87 percent recognizing
the benefits of this bega-carotene-rich food; (2)
taste, with 89 percent enjoying the flavor;
(3) price, with 73 percent reporting acceptable
cost; and (4) visual appeal, with 65 percent claim-
ing to enjoy the appearance of the product.

416

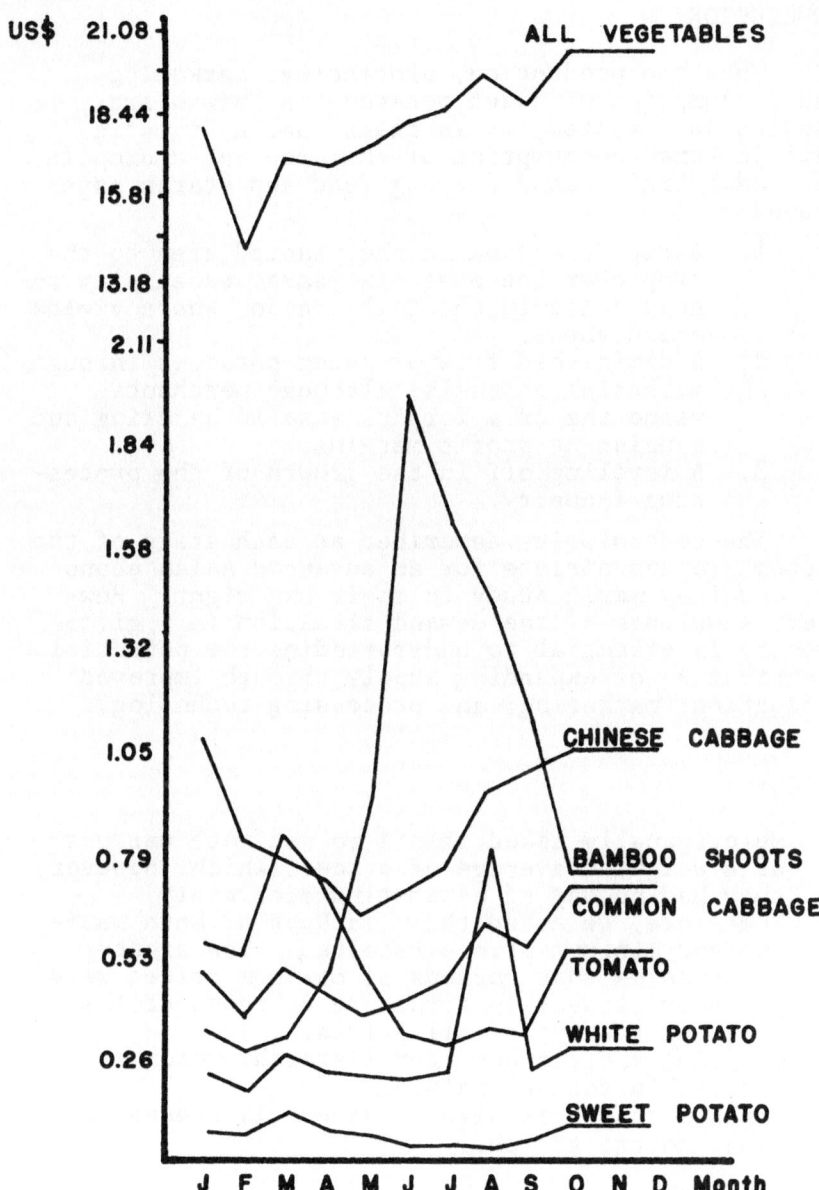

Figure 5. Seasonal patterns in total vegetable purchases compared with those of target vegetable commodities, averages by month, 1973-1976, Taipei City. (Source: Taipei City Household Expenditure Survey Report, 1973-1976. Because of incomplete data, expenditures for Oct.-Dec. are averages.)

417

CONCLUSION

When the production, processing, marketing, and consumption of sweet potatoes in Taiwan are studied as a system, it is clear that a dramatic fall in human consumption of the crop and a drop in the industrial demand for hog feed and starch have caused:

1. A rapid decline in the planted area to the crop over the past six years, especially on good soils in the fall season, where yields are highest.
2. A diminished flow of sweet potatoes through marketing channels, although merchants value the crop for its ease in handling and consistent profit margins.
3. A leveling off in the growth of the processing industry.

The technologies described at each stage of the system are appropriate for an advanced Asian economy, and they merit study in their own right. However, knowledge of the demand situation in a given country is essential to understanding the potential contribution of expanding supply through improved production, marketing, and processing technology.

NOTES

1. We originally asked agents to estimate markups at a weighted average of prices, which, however, they had no way of estimating accurately. Therefore, we asked their markups at both maximum and minimum prices received. The agents informed us that markups at maximum prices were probably closest in structure to those at the weighted average of all prices.
2. Chemically different from β-starch, which is unsuitable for eel-raising.
3. Cassava starch is also considerably cheaper than sweet potato starch.

REFERENCES

Asian Vegetable Research and Development Center. 1976. Sweet Potato Report for 1975. Shanhua, Taiwan, Republic of China.

Calkins, P.H., Huang, S.Y., and Hong, J.F. 1977.
Farmers' Viewpoint of Sweet Potato Production in
Taiwan. AVRDC Technical Bulletin 4, Shanhua,
Taiwan, Republic of China.

Chen, Wu-hsiung. 1977. Application of food balance
sheets in formulation of food policy and agricultural planning--an experience in Taiwan.
J. Agric. Econ. 22.

Liau, Shih-Yih. 1977. Research of Basic Food
Consumption in Taiwan, 1976. Taichung, Taiwan,
Republic of China.

Wu, Chueh-yuan. 1977. Research on the supply and
demand of starch in Taiwan (in Chinese). Taiwan
Bank Periodical 28.

*P. H. Calkins: Asian Vegetable Research and Development
Center, Shanhua, Tainan, Taiwan*

4
Taro Marketing in Hawaii

B. W. Begley
Heinz Spielmann
G. R. Vieth

ABSTRACT

Poi was the major staple of the people of
Hawaii when Captain Cook made his voyages to the
Islands in the 18th century. Since that time, with
the settlement in Hawaii of other ethnic groups from
Asia, Europe, and the mainland United States, other
staples such as rice and Irish potatoes have become
of greater importance than poi. But throughout the
post-contact period of Hawaiian history, poi has at
least in the public mind been associated with
Hawaiians. This is still the case today. But no
attempt has been made to actually measure how much
consumption of poi can be attributed to Hawaiians.
A study was therefore initiated to interview a sufficiently large number of people to derive some
meaningful information. Though deliberate efforts
were made to ensure that Hawaiians would be well
represented, in essence the size of the sample was
such that the central question asked was, who in
Hawaii in 1977 eats poi and how often.

In the first part of the paper, therefore, the
procedures and results of a survey carried out on
the island of Oahu, where Honolulu is located, are
described. In addition to ethnicity, the importance
of variables such as income and education in influencing poi consumption is also shown.

In the second part of this paper, some ideas
are developed on the problems and strategies needed

to market a traditional product in the present era
of supermarkets and fast food restaurant chains.
Given the changes in diet, buyer behavior, and emphasis on convenience foods of today, the need to
consider changes in form, appearance, packaging, and
indeed, the total approaches to marketing of poi,
are discussed briefly.

PART I
CONTEMPORARY CONSUMER PREFERENCES
FOR POI IN HAWAII

INTRODUCTION

To the Hawaiian people, both before and after
Captain James Cook's discovery of the islands, poi
was a staple food (Handy et al., 1972). Unlike
their Polynesian relatives in Tahiti, Tonga, and
Samoa, the Hawaiians steamed and pounded taro until
it became the viscous product we know today as poi.
This was and still is quite a different product from
the baked taro widely used throughout Polynesia or
the South Pacific.
Following Cook's voyages and the coming to
Hawaii of Caucasians, Filipinos, Japanese, Chinese
and other ethnic groups, poi's position as the major
staple in the islands changed radically. Other ethnic groups brought with them their preferred
staples, and today rice or even Irish potatoes are
of far greater importance in general in Hawaii than
poi, which has slipped into the position of being a
favored food for a small, though distinct, number
of people in the State.[1]
While the growth of importance in people's
diets of rice or potatoes was occurring during the
19th century, the displacement of poi from common
usage appears to have speeded up rapidly in recent
times. Forty years ago the quantity of poi being
produced on Oahu was almost double that of today,
and per capita consumption annually was then almost
30 pounds (13.6 kg). Today it is close to 5 pounds
(2.3 kg).[2]
One can speculate about some of the factors
that influenced this change. For example, during
the last forty years modern food merchandizing has
taken mammoth strides with the development of the
supermarket. There has been a thorough integration
of Hawaii into the mainland food distribution and

421

marketing system with its emphasis on packaging and advertising convenience foods. Then, of course, the settling in Hawaii of large numbers of people from the U.S. mainland with palates quite unused to poi, particularly over the last twenty-five years, has meant that the proportion of poi-eating Polynesians in the total population has declined.

Poi has been an important food, yet despite its historical importance, very little has been published on taro or poi, either at the production or consumption stage of its cycle. One has to go back over twenty-five years to the last published report treating poi. This was by Rada (1952) and dealt not with poi in Hawaii but with the potential mainland market for taro products. Some comments were included, however, on local consumption of taro products, one of which noted that "in 1936 Hawaiians ate 70 percent of poi provided." Kubo, writing a paper on the history of taro and taro products in 1970, was clearly conscious of the dearth of scholarly articles on poi and turned to the local newspapers to unearth a considerable volume of material which showed that during the 1930s, 40s, and the Second World War, poi continued to receive a lot of attention in the press. Yet despite its having been in common use, no study was ever carried out which focused on who, in addition to the Hawaiians, ate poi in Hawaii, how much was consumed, and what patterns of consumption, brand preferences, household preferences among family members, existed. In addition, little has been written about foods customarily associated with poi, the manner in which price or nutrition influences consumption, and the degree to which income, occupation or ethnicity influences the amount and frequency of use.

Today, although it is difficult to cite actual figures, there is sufficient circumstantial evidence to suggest that most people in Hawaii do not eat poi, or do so only when they attend special events such as luaus. If such is the case, it becomes rather unproductive to focus on the average person in Honolulu who eats poi perhaps once or twice a year at a luau. More useful results can be gained from focusing on that segment of the population which regularly consumes poi. This was done in this study, which was carried out in 1976-78 on the island of Oahu, where Honolulu is located.

BACKGROUND

Taro and Poi: Production and Consumption in Honolulu

Between 60 and 70 percent of the wetland taro grown on the neighbor islands of Kauai, Maui, and Hawaii is shipped to Oahu. A small quantity of this, perhaps 3 percent[3] is sold as table taro and the balance is processed into poi. In contrast with the situation of 35 years ago, little commercial taro is grown on Oahu today. What were formerly the taro lands of Oahu are today used for hotels, highways, and houselots. But if Oahu is no longer of importance as a production center, such is not the case with consumption. Over 80 percent of the population of the Hawaiian Islands lives on Oahu (the City and County of Honolulu), and the bulk of poi eaten in the State is consumed by people living in Waianae, Kailua, Waimanalo, Kapalama and other suburbs of Honolulu. Clearly, then, this is the logical place to carry out research work on the consumption or marketing of poi.

Where Consumption Fits into the Total Root Crop Delivery System: Some Key Questions

Part of the analysis of the total food delivery system which begins with farmers raising taro in the taro patches of Hanalei on the island of Kauai and concludes with poi being eaten in households in Nanakuli or Kapalama on the island of Oahu, required a number of consumer interviews. Recognizing that even under the heading of consumer utilization there are a large number of topics, this study deals with just a few of them in depth. Taking as our target the poi consumers, a representative sample of people was asked questions designed to determine:

1. Who eats poi in Hawaii;
2. How much, how often and when it is used; and
3. What factors best help explain consumption and purchasing patterns.

THE INTERVIEW FORM AND LOCATIONS FOR INTERVIEWS

In designing the original questionnaire, the time constraints that would be imposed on the inter-

viewer were clearly kept in mind. He or she would
be interviewing the customer close to where poi was
displayed in a supermarket or near the check-out
counter. Brevity and succinctness there became
critical, clearly cutting down on the number of
questions that could be asked.

The questionnaire was roughly divided into
three parts: in part one, information was requested
on frequency of usage, quantities regularly purchased, brand preferences, age of product preferred.
Information was also sought on the time or occasions
when poi was used, who in the household consumed it,
and the various combinations with other foods that
may be customarily prevalent. In part two, information on reactions to such factors as price, color,
nutrition, and brand preferences was sought.
Finally, in part three, data were obtained on the
background of the respondents and in particular on
education, ethnicity, income, occupation, and
household size. A copy of the questionnaire is
enclosed in the appendix to this report.

Conduct of Interviews

Traditionally poi is associated with Hawaiians.
Therefore, after constructing a purposive sample,
interviews were arranged in marketplaces located in
the heart of or close to areas where concentrations
of people with Hawaiian ethnic background were to
be found, including such suburbs of Honolulu as
Waianae and Kalihi. Subsequent results justified
this method, but also pointed to the need of carrying out interviews in areas where the proportion of
Hawaiians was low. This was done and helped gain
an understanding of the importance of poi to such
ethnic groups as the Japanese, Chinese, Caucasian,
and others in Honolulu.

It should be stressed that this cannot be
regarded as an in-depth study of consumption patterns for poi throughout Hawaii, or of the causes of
change in preference over time for rice or potatoes
in comparison to poi. It describes mainly who is
buying and using poi, and not why this is being
done. In essence, this was a short and relatively
inexpensive survey of people well acquainted with
the use of poi, to determine consumption frequencies and some of the socioeconomic characteristics
of poi users on Oahu. It represents a first step in
documenting, rather than a final statement on, the
place of a traditional food in this supermarket age.

424

GENERAL FINDINGS

The first set of questions asked sought information on items such as frequency of use, quantities purchased, occasion of usage, combinations of use with other foods, persons in the household consuming poi, brand preferences, preferences in degree of freshness, and the position of poi with respect to other staples.

General Usage and Frequency of Purchase of Poi

Seven hundred sixteen consumers in four areas of Honolulu were first asked the question: how often do you use poi in the course of a week? The answers are shown in Table 1.

Though those interviewed were people who classified themselves as users of poi, over 60 percent said they used poi only once a week or less. On the other hand, a third of those interviewed indicated that they consumed poi two or more times a week. These are probably individuals who buy several large packets of poi each week. Poi is not a product that is kept in the refrigerator, therefore it could be predicted that there would be an association between frequency of usage and frequency of

TABLE 1
Frequency of Use of Poi

Answer	No. of Buyers	Percentage
Less than once a week	244	34.3
Once a week	202	28.4
Twice or more a week	266	37.3
Total	712[a]	100.0

[a] The total number of people interviewed was 716, but not everybody responded to each question. So each table shows the actual number of people who replied to the question or questions asked and percentages calculated on responses, not on the total number of people interviewed.

425

purchase. This would follow if we accept in principle that consumers do not keep poi in reserve, but
only make a purchase when they plan to use it within
a day. In fact, just less than 60 percent of respondents bought a packet or less a week, which is
close to the figure of slightly over 60 percent who
said they used poi once a week or less.

Actual figures on purchases of poi are shown
in Table 2.

Packets are the sealed plastic bags in which
poi is packaged and sold on Oahu in 14- and 28-ounce
(392 and 784 g) sizes. The figure of slightly less
than 60 percent who bought 1 packet or less, referred to above, is derived by adding the top two
percentages shown on the right of the table.

Frequency of Use by Area

As is shown in Table 3, there is considerable
variation in the frequency of use by area.

The most striking contrast is between the pattern in Waianae and that in the central part of
Honolulu. Eighty-five percent of respondents in
Waianae used poi one or more times a week, compared
with 49 percent in Kalihi and 55 percent in the
eastern section of Honolulu (McCully to Hawaii Kai).

Though the importance of ethnicity in explaining consumption will be taken up in more detail
later, it should be mentioned at this stage that the
proportion of Hawaiians in the total sample of those
interviewed in the central areas was significantly
lower than in Waianae or in the Windward areas. But

TABLE 2
Size of Purchase

Answer	No. of Buyers	Percentage
Less than 1 packet a week (large or small)	196	28.4
1 packet a week (large or small)	211	30.5
2 or more packets a week	284	41.1
Total	691	100.0

TABLE 3
Frequency by Area (Percentages)

Answer	Windward (%)	Central Kalihi (%)	Central East (%)	Waianae (%)
Less than once a week	34.0	51.1	45.6	15.0
Once a week	30.0	27.8	30.8	24.9
2-3 times a week	13.6	14.3	11.8	35.2
Over 3 times a week	22.4	6.8	11.8	24.9
	100.0	100.0	100.0	100.0

in the Windward area where the proportion of Hawaiians in the sample was comparable to that of Waianae,
66 percent of respondents used poi once a week or
more, which is considerably lower than the frequency
of use figure of 85 percent found in the Waianae
sample. This would appear to suggest that the use
of poi among regular users is higher in Waianae than
in the other areas of Honolulu where interviews were
carried out.

When is Poi Used?

    Several of the processors, who are more common-
ly known as millers, expressed interest in the answer to the question: when do you most often use
poi throughout the year? Processors suggested that
there is an increase in demand for poi over the
summer. This is not reflected, however, in the
answers shown in Table 4.
    Clearly the processors know from their sales
figures that more poi is sold in summer than in
winter, but these figures actually suggest that
regular users of poi do not account for the additional sales. Probably special events or activities
commonly occurring in summer which are attended by
tourists account for this phenomenon. On such
occasions poi is served--and hopefully consumed--
by visitors to the Islands.
    From the point of view of the processor, the
increased summer demand for poi is not entirely a
boon. Higher air and water temperatures at that

427

TABLE 4
Time of Year Poi is Used

	Frequency	Percentages
Summer	17	2.5
Winter	4	.6
All year	671	96.9
	692	100.0

time result in disease which often plays havoc with taro yields. Consequently, the quality of taro available is poorer, and the actual quantity available is probably less in summer than in winter.

Consumption by Various Household Members

In over 80 percent of cases, the male head and woman of the household said they ate poi, and over 70 percent of sons and daughters also answered in the affirmative. But when respondents were asked whether relatives or others in the household also ate poi, over 75 percent gave a negative response in each case.

It is difficult to offer an explanation for this phenomenon, except perhaps to suggest that poi is served mainly when immediate members of the family eat together but not when relatives and friends are participating in the meal. This can also lead to the conclusion that--as in the case of milk in Hawaii--households regard poi as food that is not to be served to guests.

INFORMATION ON THE PRODUCT

A series of direct questions was asked concerning preferred "state" or condition of poi, poi brand preferences, combination with other foods, and poi's rank among various staples.

Condition of the Product

Consumers have the option of buying poi fresh, day-old, or two days old, after which it is taken off the shelf. The relative preferences are shown in Table 5.

Brand Preferences

Data collected on brand preferences have to be treated with considerable caution for several reasons. In the first place, a relatively small proportion of all markets carry all three brands of poi sold on Oahu. Even some of the largest markets sell only one specific brand, which appears to be almost always the leading brand. Therefore, though the bulk of the interviews--probably over 90 percent--were carried out in markets displaying at least two brands, it could not be claimed that the figures represented true preferences. The figures therefore can be interpreted to mean that these are the brand preferences, given the limited range of choices available in most cases. The information on brand preferences by area is shown in Table 6.

What emerges from these figures, first of all, is that "Taro" brand, in addition to being the over-all preferred brand leader, leads in all four areas. "Haleiwa" brand, however, is strong in Waianae and Central Kalihi.

Secondly, except in Waianae, there are a large number of consumers who indicate no clear preference. This supports the responses to the question asked later: "When buying poi, is availability of a particular brand very important, unimportant, or

TABLE 5
Age Preferences for Poi

Age	Frequency
Fresh[a]	495
Day-old[a]	240
Two days old	65

[a]Many people marked more than one line, indicating that they were indifferent in their choice between fresh or day-old poi.

429

TABLE 6
Brand Preference by Area (Percentages)

Brand	Windward (%)	Central Kalihi (%)	Central East (%)	Waianae (%)
Taro	58.5	44.3	54.7	45.6
Kalihi	4.7	18.8	2.9	13.5
Haleiwa	9.5	25.6	8.8	37.3
No Preference	27.3	11.3	33.6	3.6
	100.0	100.0	100.0	100.0

of some importance?" Over 50 percent of those res-
ponding indicated that the availability of any
given brand was unimportant.

What Foods are Combined with Poi?

Respondents were asked to indicate whether
they ate poi with certain other specified foods:
shellfish, meat, fish, or other. No clear patterns
emerged, but in general the answers indicated that
poi was most often eaten with fish.

Poi and Other Staples

Both in this and in another survey carried out
on consumer preferences for sweet potatoes, the
overriding importance of rice in the diet of the
people interviewed clearly emerged. In the poi
study 71.4 percent of those interviewed stated that
they usually have rice with their main meal. Only
37.8 percent indicated that they regularly had poi
as part of their main meal. This is a significant
finding when it is kept in mind that these answers
came from people who were consistent poi-users.

Importance of Price and Other Selected Factors to Consumers

In the often crowded conditions of a super-market it is difficult to obtain satisfactory answers requiring reflection and time regarding frequency and quantity of purchase. Yet it was felt important to obtain some information on consumers' attitudes on factors which might be expected to influence their buying behavior. A few factors considered to be of some importance were listed and reactions to them elicited. The responses are shown in Table 7.

How important is price to you? Almost 50 percent of the respondents who were asked this question indicated that price was relatively unimportant. This question can be considered together with the statement, "Poi today costs too much." The first is a general question and looks at the role of price, as against other factors, and set in the context of inflation, recent price increases for food, higher taxes, and the shrinking purchasing value of the dollar. The second is a more specific question determining the consumer's perception of the price of poi today relative to the price of poi in the past. Clearly there are two different sets of considerations involved, but they are related. The question that might be asked in a further study is the extent to which price influences the frequency and quantity of purchases.

Two other interesting responses were obtained to questions on color and nutrition.

1. Almost half of the respondents felt that color was important. In trying to explain this, it is possible to hypothesize that color provides an indication of freshness. But, in fact, it is more likely that color is associated with a particular brand.

2. Many respondents regarded poi as nutritious because of its acclaimed value as a baby food, and particularly because of its importance for children suffering from allergies to cereal foods. That poi may have certain therapeutic uses was discussed by Derstine and Rada (1952).

TABLE 7
Factors Influencing Purchase of Poi

Question: How important to you is PRICE, AVAILABILITY, and COLOR?

	Very Important (%)	Unimportant (%)	Of Some Importance (%)	No Answer (%)
Price	28.5	49.7	21.6	0.1
Availability	32.0	50.7	16.9	0.4
Color	49.3	31.3	19.0	0.4

Question: Do you feel that:

	Agree Strongly (%)	Disagree (%)	Don't Know (%)	No Answer (%)
Poi tastes good	88.0	0.7	1.7	9.6
Poi today costs too much	47.8	27.1	17.3	7.8
Poi is nutritious	85.1	1.3	5.9	7.8
Good poi is readily available	62.3	14.0	15.8	8.0

Finally, on the question of the availability of a particular brand, as mentioned earlier, 50.7 percent stated that it was unimportant, while 32 percent considered it very important, and 16.9 percent affirmed it was of some importance.

SOCIOECONOMIC DATA AND KEY ASSOCIATIONS

Occupation

The first socioeconomic variable on which information was collected, and one which is frequently linked with education and income, is occupation. Data on this variable in the sample are shown in Table 8.

Education

Educational standards attained by men and women interviewed were somewhat similar. Fifty-three and one-tenth percent of men and 54.1 percent of women completed 12 years of school, and 21.9 and 22.4 percent of men and women, respectively, had one or more years of university or post-high school education. Twelve and two-tenths percent of the men, but only 6.4 percent of the women, did not answer this question, suggesting that education was a sensitive issue for some respondents.

TABLE 8
Occupation of Men/Women

Category	Frequency		Frequency	
	Men	Percentage	Women	Percentage
Unemployed	32	5.0	2	.3
Laborer	77	12.3	12	1.7
Blue-collar	204	32.6	92	12.8
White-collar	170	27.1	203	28.4
Retired	55	9.0	11	1.5
Other	88	14.0	46	6.4
Housewife			350	48.9
	626	100.0	716	100.0

Household Income

Some 117 respondents did not answer the question on household income (Table 9). Almost half of those are accounted for by the 50 questionnaires on which the question on income was omitted at the request of the management of one of the participating supermarkets.

Ethnicity

Traditionally poi consumption has been associated with people of Hawaiian or part-Hawaiian ethnic background. Consequently, care was taken to establish detailed identification of respondents with Hawaiian and other ethnic origins. Further, information on the ethnic identification of both male and female members of households was collected. The information on women is summarized in Table 10.

Interviewers in supermarkets quietly approached shoppers and asked them, do you use poi? If the answer was yes, the interviewer continued with the interview. If no, the interview was stopped at this point. This simple question on poi usage led to a rough and ready determination that Hawaiians rather than other ethnic groups were main users of poi. This supported the assertion, made over 40 years ago, that 70 percent of those who ate poi were of Hawaiian extraction (Rada, 1952).[4] People from other ethnic groups in Hawaii also use poi. Our task was to determine how often and how much.

TABLE 9
Household Income

Category (Income)	Frequency	Percentage
Less than 4,999	55	9.2
5,000 - 9,999	87	14.5
10,000 - 14,999	158	26.4
15,000 - 19,999	164	27.4
20,000 or greater	135	22.5
	599	100.0

TABLE 10

Ethnic Composition of Women in Sample (Percentages)

Category	Windward (%)	Central Kalihi (%)	Central East (%)	Waianae (%)	Overall (%)
Full or part Hawaiian	62.4	50.4	40.7	76.8	60.3
Japanese	26.6	30.9	33.3	17.4	26.0
Other	11.0	18.7	26.0	5.8	13.7
	100.0	100.0	100.0	100.0	100.0

Though Table 10 shows the ethnic composition of women interviewed, it parallels very closely the patterns for men included in the survey.

Household Size

It might be assumed that the more people that live in a house--particularly if they are of the immediate family--the more poi will be consumed. Clearly there are exceptions to this rule, but this supposition adds some importance to the household size data collected, which are shown in aggregated form (Table 11).

Over a third of the respondents lived in households of six or more, and the modal value of a household unit was 5 members. The percentage of households with 5 or more members was significantly higher in the sample than is found in the general population of Oahu.

Relationships Between Socioeconomic Variables and Consumption Patterns

The indication that 60 percent of respondents using poi had Hawaiian backgrounds established in a general way that ethnicity was related to consumption patterns. However, the frequency of poi use by Hawaiians needed to be explored. In addition, it was also necessary to establish the importance of other variables such as occupation, education and

435

TABLE 11
Household Size

Number in Household	Absolute Frequency	Percentage
3	204	28.5
4	117	16.3
5	124	17.3
6	115	16.1
7-16	156	21.8

household income and, if possible, analyze their influence on poi consumption. In studying these major socioeconomic variables, it was recognized that these are often linked and operate collectively rather than as a single variable, in influencing consumption behavior.

The first step in establishing the causal relationships was to construct a set of cross-tabulations to compare consumption with occupation, income, education, household size, and ethnicity. In Table 12 the frequency of use of poi by women who are housewives and white collar workers and by men who are blue and white collar workers is shown.

In general, it appeared that housewives and blue collar male workers ate poi more frequently than white collar workers, whether male or female. Also, the former were much more likely to use poi two or more times a week than white collar workers.

Frequency of Use by Income

To facilitate interpretation, the data on income and on frequency of use have been aggregated and are shown in Table 13.

Some association between household income and consumption emerges from the data, but no sharp contrasts are in evidence. At the lower end of the income scale (<$10,000) fewer people buy less than once a week than at the top of the income scale ($20,000 plus). Secondly, those earning $20,000 or more are markedly less likely to use poi two or more

TABLE 12
Frequency of Poi Use by Occupation (Percentages)

Frequency	Women		Men	
	Housewife (%)	White Collar (%)	Blue Collar (%)	White Collar (%)
Less than once a week	27.6	43.3	33.0	47.7
Once a week	24.4	34.8	27.4	29.6
Twice or more a week	48.0	21.9	39.6	22.7
	100.0	100.0	100.0	100.0

TABLE 13
Frequency of Poi Use Associated with Household Income
(Percentages)

Frequency	Household Income		
	Less Than $10,000 (%)	$10,000– $19,999 (%)	$20,000 or More (%)
Less than once a week	28.5	35.6	43.0
Once a week	32.3	29.4	31.8
Two or more times a week	39.2	35.0	25.2
	100.0	100.0	100.0

times a week than those earning less than $10,000,
but are as likely to use poi once a week, as those
on lower incomes.

437

Frequency of Use by Education

Respondents were asked to indicate the number of years they had spent in high school and college. In Table 14, consumption by those who had graduated from high school is contrasted with that of those who had some college. Consumption by females who graduated and those who had some high school education, but who did not graduate, were similar.

Female respondents with college education are more likely to use poi less than once a week than those who are high school graduates. On the other hand, a higher percentage of those with some college education are likely to use poi at least once a week. The sharpest contrast, however, is between those likely to use poi two or more times a week. Over 45 percent of those who are high school graduates--the figures are very similar for those with less than 12 years of education--use poi two or more times a week, as compared with 18.1 percent of those with some college education. These figures, as well as those in Table 12 (frequency and occupation), and to a slightly lesser degree those in Table 13 (household income), are indicative of major contrasts between those respondents who are frequent (two or more times a week) and those who are casual (once or less than once a week) users of poi.

Ethnicity

Poi was the staple food of Hawaiians in the 18th century. Rada (1952) commented that in 1927

TABLE 14
Frequency of Use by Education (Females) in Percentages

Frequency	High School Graduate (%)	Some College (%)
Less than once a week	28.5	45.6
Once a week	25.3	36.3
Two or more times a week	46.2	18.1

70 percent of poi was consumed by Hawaiians. Clearly, the preliminary question asked of shoppers in this survey established that the Hawaiians were still the dominant poi consumers today. But over and beyond this, two important questions yet to be asked were (1) how often is poi used among Hawaiians and others, and (2) how do consumption patterns differ between the ethnic groups using poi? The data on frequency of poi consumption by ethnic groups are shown in Table 15.

The contrasts between consumption among Hawaiians and those in the "other" category (Caucasian, Chinese, Korean, etc.) are very striking. The Japanese appear to occupy a position in the middle. Eighty percent of the Hawaiians in the sample ate poi at least once a week, almost 50 percent used it two or more times a week, whereas 73.6 percent of the "other" group consumed poi less than once a week. Over 27 percent of the Japanese ate poi once, and a similar proportion ate it two or more times a week. The number of those with Japanese ethnic background constituted a much smaller part of the sample than the Hawaiian group, both in percentage and absolute terms. However, clearly among those Japanese who actually eat poi, consumption frequencies are fairly high.

Further Analysis

Though justifying the assertion that ethnicity is an important factor in explaining consumption of

TABLE 15
Frequency of Use by Ethnic Origin (Females) in Percentages

Frequency	Hawaiian and Part-Hawaiian (%)	Japanese (%)	Other (%)
Less than once a week	20.0	46.6	73.6
Once a week	31.2	27.0	17.6
Two or more times a week	48.8	26.4	8.8
	100.0	100.0	100.0

poi, cross-tabulations also indicated that household income, occupation, household size, and education were also statistically important. Some further work using factor analysis[5] was, therefore, carried out in an attempt to differentiate statistically between major and minor variables in explaining consumption behavior.

A single socioeconomic measure was derived from the three variables of occupation, household income and education, using factor analysis. Then analysis of variance was used to determine the major and minor variables explaining consumption. As a result of this analysis it was possible to determine that ethnicity was a major factor in explaining consumption. The combined socioeconomic variable on the other hand was not significant in explaining consumption. Among the three ethnic groupings used, Hawaiians and part-Hawaiians tend to consume more poi than Japanese and others, a conclusion that could be anticipated from data in Table 15. Regression analysis indicated that Hawaiian or part-Hawaiian households are likely to consume poi more often than are non-Hawaiian households.

PART II
ISSUES THAT AFFECT TODAY'S MARKETING
OF TARO-BASED PRODUCTS

INTRODUCTION

In the first part of this paper the case was made that the major consumers of poi are the Hawaiians. As a proportion of the total population in Hawaii, however, the number of Hawaiians has decreased sharply over the last 200 years. At the same time some major changes have taken place in people's diets and in the manner and style in which food is marketed, both on the mainland and in Hawaii. All of these changes have consequences for the marketing of poi.

"If I were starting from scratch and marketing poi I would tackle the job differently from how it is done now. I would find the people in this State and on the mainland who wanted a starchy food with a different taste and heavily promote it to them." Such were the comments of an advertising executive

discussing the marketing of poi recently. Certainly
if the changes that have occurred in the marketing
environment even over the last decade are not taken
into account, then in this era of the McDonald's
hamburger and T.V. dinners a package of poi could
easily become an anachronism if the necessary adap-
tations are not made in the form, packaging and
promotion of poi.

In this section of the paper some of the changes that have contributed to creating the marketing
environment of the later part of the 20th century
are first discussed. Then some possible approaches
and considerations related to the marketing of poi
are outlined.

CHANGES THAT HAVE OCCURRED

Food Habits

Bennett and Pierce (1961) carried out a study
of changes in the American National Diet between
1879 and 1959, and reported that the consumption of
calories had dropped about 25 percent during that
time. In particular, the total food calories
derived from starchy foods had dropped from 52 to
24 percent. The starchy foods studied were wheat
flour, sweet potato, cornmeal and potatoes. In the
words of the authors, "this decline is the most con-
spicuous feature of long-term changes in the composition of the national diet," and they project
further reductions in the use of starchy foods.
Even without the influence of other factors that
were at work, if Hawaii followed national trends, a
sharp decrease could have been anticipated in the
consumption of poi.

During this time--1879 to 1959--the group of
sugary foods and flavored foods increased sharply,
particularly during the early part of the period,
in contrast to the declind in consumption of starchy
foods. But if changes were occurring in food composition and variety of foods available, changes in
the forms, shape and size of food, emphasis on convenience, food marketing methods, and finally, in
the places where food was purchased and consumed,
also became prominent.

In the past 50 years a transformation has taken
place in the handling and presentation of food.
Instead of appearing mainly in their fresh or dried
state, many foods today undergo any one of several

preservation processes designed to extend the life of the product. As a consequence, "basic food" is modified drastically before being packaged into standard sizes.

Convenience and branding are two key concepts that are emphasized in food merchandising today. Some of today's foods reflect the deliberate aim by processors to develop convenient packaging to meet the requirements of households ranging in size from individuals to families of five or more. Hand in hand with the "convenience" concept is that of branding for purposes of advertising and promotion in order to establish definitive association with and differentiation of a given product.

The New Commercial Environment

Food consists of a much more sophisticated range of products today than in the past. Refrigeration, the growth of food processing and preserving industries, and the development of efficient transportation linkages across the globe have ensured that food no longer consists of a few basic staples supplied by local farmers or home grown, with luxuries occasionally available at high prices. Instead, food today has become a range of several thousand lines stocked and colorfully presented in modern, efficient supermarkets and available the year round, except for some locally produced fresh vegetables. Each of the individual items displayed in supermarkets has to compete for shelf space and creates its own demand, based on some goodness, freshness, taste, nutrition or convenience criteria. The cost of promotion and advertising required to create individual or brand awareness is measured in thousands and often millions of dollars. This assumes, of course, that the particular brand can match other brands in price and quality and is packaged attractively enough to command attention. Food today has to be merchandised as effectively as non-food items in the battle for consumers' dollars.

Locational Changes: Where Food is Eaten

Between the 1930s and the 1960s the place in which food was primarily purchased in Middle America was the supermarket, and the place of its consumption was the home. This is now no longer always the case. Fast food restaurants have caused a minor revolution in the last decade in the place where food is eaten and the total quantity consumed away from home. The Progressive Grocer, the magazine of supermarkets, noted in its November 1977 issue that eating away from home has grown by more than 20 percent, and that the fast food portion doubled in the years between 1967 and 1976. Though less than 10 percent of all food purchased is eaten at fast-food restaurants, the number of meals eaten out appears likely to increase in the foreseeable future.

Clearly the national food picture has changed over the last 175 years, and most of these changes have reached Hawaii. Here the building of super-markets began over 30 years ago, and an explosive expansion of fast food franchise operations is taking place at the present time. At this point it appears appropriate to consider certain other facts which contribute directly to the commercial environ-ment, and which touch directly on the marketing of poi in Hawaii.

HAWAII: LOCAL DEMOGRAPHIC TRENDS

In 1853 the Hawaiians comprised 95 percent of the total population in the islands. By 1970 this number had shrunk to 9.3 percent. Meanwhile, Cauca-sians, Japanese and Filipinos--who together account-ed for less than 5 percent a century ago--now constitute 39, 28 and 12 percent, respectively, of Hawaii's population. The influx of Caucasians dur-ing the 30 years between 1940 and 1970, of whom very few appear to have acquired a taste for poi, is particularly striking. Two-thirds, or almost 200,000 of the 300,000 Caucasians now living in Hawaii, have arrived since 1940 (State of Hawaii, 1977).

INTEGRATION OF HAWAII INTO UNITED STATES COMMERCE

Supermarkets, as mentioned before, have been in Hawaii for over 25 years, and fast food stores have been in operation for many years, but the current dramatic expansion in the number of fast food restaurants is a recent phenomenon. What has made these developments entirely logical is the fusion of Hawaii and the U.S. mainland into a single commercial system.

For example, an order for crab placed with a distributor in Los Angeles or Seattle by telephone in the morning can be available to diners the same or the next day in a restaurant in Waikiki. Though bulky items are normally shipped to Hawaii from the mainland, from Australia, or any part of the world, a serious shortage of an important staple can be remedied through air shipments within 48 hours from almost any point around the globe.

If, on the one hand, shipping airlines and the telephone have effectively bound Hawaii to the mainland and enabled the pushbutton delivery of products, these same links have induced Hawaii's population to bring its consumption patterns into line with fashions evolved on the mainland. Further, those who have settled in Hawaii can and do fully indulge the tastes they had acquired prior to their arrival, and may not be induced to explore island foods, except on rare occasions. The Nebraskan settling in Hawaii may eat less potatoes than his father 50 years ago, but he does eat potatoes. On arrival in the islands he may try rice, but he is unlikely to treat poi or <u>aburage</u> (fried soybean curd) as anything other than exotic items to be sampled at a party.

Given this condition, what problems will be faced in the marketing of poi in the 1980s? At this point it should be made clear that no attempt will be made to map out a complete marketing strategy for that product. Rather, some implications of the environment in which a basically traditional, unchanged product is marketed are sketched out with particular attention given to the taste and form of the product, and the need for an altered promotional strategy given today's commercial environment.

POI AS A MARKETABLE ITEM

Taste of Poi

Poi has a taste for which it takes time to acquire a liking. It is distinctive. It is different from anything to which a visitor or a new resident from the mainland or Asia is accustomed. Perhaps its position is somewhat analogous to fresh yogurt, which in its original form has a distinctive flavor. While some consumers may be induced to acquire a taste for it, most have come to accept it in a fruit-flavored form. In this context, earlier remarks on the increase in consumption of products with sugary and fruit flavors seem quite pertinent. It might therefore be inferred that for poi to have

general appeal, fruit, honey, or some such appropriate flavoring would have to be added. There is
a local precedent for this--honey-flavored poi was
sold in the islands during the 1930s.

The example of yogurt would seem to suggest
that if a product has some distinctive characteristic and its flavor is modified to suit today's
preferences, a segment or even the broad mass of
the population can become a suitable market target.
It would seem possible that if, in addition to producing "pure poi" for the "Hawaiians" the bulk of
poi produced could be modified to suit various
sweet palates, its appeal might be broadened considerably.

Poi as a Starchy Food

Because poi is a starchy food it faces problems
that rice and Irish or sweet potatoes have to face.
The potato industry responded to this situation by
selling a brand name product and by developing snack
lines such as chips. The poi industry would probably have to face up to a similar situation and focus
its attention on a segment of the population that
consumes and utilizes large quantities of calories.
This group would include children, manual workers
or body builders. In addition, the industry would
need to exploit the possibilities that development
of taro chips or taro cookies might offer.

The Form of Poi

Clearly it appears that the form in which poi
is to be marketed has to be determined by concentrated research and by carrying out a series of
feasibility studies. Honey or fruit-flavored poi
would appear to be one such form. But are there
other forms? research would be needed to answer
this question and would have to examine the combinations of an array of foods with various taro products. This would involve obtaining reactions of
divergent ethnic groups in Hawaii.

The specific sizes and cartons in which taro
products would need to be packaged should also be
determined. The important element of "convenience"
is linked not only to packaging but also to handling
and preparation. Clearly the whole question of
whether poi is, or can be made into, a convenience
food needs some special research. The final
product the consumer sees must satisfy these basic

criteria, whether it is destined for supermarkets or fast food stores.

Still on the subject of form, _kulolo_ is a sweetened, semihard taro-based product made from coconut, taro and sugar, which is little known even in Hawaii. Kulolo might be representative of a type of product with considerable market potential.

These brief comments are intended to serve as an introduction to the topics of taste and forms. It is perhaps sufficient to say at this juncture that the traditional form of poi does not seem to have universal appeal, and that the actual form or forms should be determined by market and particularly by consumer research.

MARKET RESEARCH

Market research can be viewed as an essential foundation in the development of markets for taro-based products in Hawaii and elsewhere. Modern marketing places great emphasis on the need to assess acceptance of a new product. Given the competitive environment alluded to earlier and the tastes of today, it is therefore essential to predict who, in what quantities and at what price, will eat honey- or strawberry-flavored poi, kulolo, mango-flavored taro chips, or whatever other products are developed. Market research attempts to do this. Assessing what consumer reactions are likely to be products involves much painstaking research and involves considerable expenditure, but is considered essential in order to obtain indications of: (1) product acceptance, and (2) the particular population segment that will buy this new product.

Promotion and Advertising

Promotion and advertising strategies are logically derived from market research findings. Such findings would aid in identifying those taro-based products which appeal to particular groups of people in the islands and on the mainland. Market research would also help to determind the most logical food outlets to be considered and the packaging and size of fruit-flavored poi or macadamia-taro chips that would be most acceptable.

All of this information would be invaluable in

mapping a promotional strategy, at the heart of which would be an educational program aimed at creating an awareness of the "new" taro products.

A particular promotional approach might be to incorporate an ethnic element in the strategy. Using the success achieved by Taco Bell as a model, it might be possible to incorporate the ethnic association linked with poi to the products being promoted. In this case, ethnicity would not imply that the traditional food would be targeted towards "Hawaiians," but rather that a food--or associated foods--suitably modified to fit the palates of the present-day total population of Hawaii, including visitors, would retain the traditional name association.

Advertising and promotion would clearly involve considerable effort and expense to highlight the characteristics and attractive features of some yet-to-be-determined product or lines of products. But in the food marketing environment of today, advertising and promotion are essential. They represent tools which may be used to create a "position" in consumers' minds for poi not currently available.

CONCLUSION

In conclusion, it should be re-emphasized that this is not an exhaustive description of changes in consumer habits, in the operation of the marketplace, and in the potential and problems of marketing poi. Rather, it is an attempt to touch briefly on some of the changes which have caused the food marketing picture in 1978 to differ markedly from that in 1928 and--even more dramatically--from that of 1828 when poi was a staple for perhaps 90 percent of the people living in Hawaii. Given the changes that have occurred in the marketing scene, it is logical to conclude that the form of poi being marketed must also change. The Hawaiians may want to retain poi in its original form. But for poi to have a profitable and lasting future, perahsp the bulk of poi or some other taro-based product must change in order to have wider appeal in Hawaii or on the mainland. Having said that, the success story of yogurt suggests that poi also could have an exciting future--if the investments in research and development are made.

NOTES

1. Such are the preliminary findings which came out
   of some research carried out on sweet potato
   consumption in Hawaii. Interviewing for this
   study took place in 1976-1977.

2. This is an estimate. See: Bryan W. Begley,
   "Taro in Hawaii: Study of a Root Crop Delivery
   System," paper in preparation for the Department
   of Agricultural and Resource Economics University of Hawaii, June 1978.

3. This is the figure suggested by a Department of
   Agriculture Official closely involved in the
   collection of data on production and movement of
   taro between the islands in Hawaii.

4. Overall, 60.3 percent of 716 respondents who
   used poi were Hawaiians or part-Hawaiians.

5. At this point appreciation is expressed for the
   invaluable assistance given in this phase of the
   work by Walter Hudson of the School of Social
   Work, University of Hawaii, Honolulu, Hawaii.

REFERENCES

Bennett, M.K. and Pierce, R.H. 1961. Changes in
   the American national diet, 1958-1959. Food
   Research Institute Studies, No. 2. Stanford
   University.

Derstine, Virginia and Rada, Edward L. 1952. Poi
   in Hawaii. Agricultural Economics Bulletin
   No. 3, University of Hawaii.

Haddock, Daniel and Hernandez, Leslie. 1952. Consumer preferences for taniers (Xanthosoma spp.)
   in Puerto Rico 1949-1950. University of Puerto
   Rico Agricultural Experiment Station, Rio
   Piedras, Puerto Rico.

Handy, E.S. Craighill and Handy, Elizabeth Green,
   with the collaboration of Mary Kawena Pukui.
   1972. Native Planters in Old Hawaii: Their
   Life, Lore and Environment. Bishop Museum
   Bulletin 233, Honolulu, Hawaii.

Kubo, Patricia. 1970. The history of taro and taro
   products in Hawaii. Term paper, Department of
   History, University of Hawaii.

Progressive Grocer. 1977, New York. November:37.

Rada, Edward L. 1952. Mainland Market for Taro Products. Agricultural Economics Report No. 13, University of Hawaii.

State of Hawaii. 1977. Data Book. Department of Planning and Economic Development, Honolulu. November.

Bryan W. Begley, Heinz Spielmann, and Gary R. Vieth:
Department of Agricultural and Resource Economics, College
of Tropical Agriculture, University of Hawaii, Honolulu,
Hawaii

5

The Effects of Transportation and Government Policies on International Trade in Root Crop Products, Especially Cassava

Peter R. Walters

ABSTRACT

Cassava enters international trade in two main forms--as pellets for use in animal feeds and as starch for use in both the food and non-food sectors. Thailand is the major world exporter of both these products. While the export of Thai pellets has increased rapidly due to favorable European Economic Community (EEC) regulations and advances in transportation, starch exports to the two major markets have stagnated due to government policies in Japan and transport difficulties to the United States. The trade in sweet potato slices to EEC is also compared with the Thai cassava pellet trade. These case studies highlight the problems in developing international trade in root crops, the necessity for thorough market research, and the need for adequate transport facilities and clear govern- ment policies in the exporting countries. Because of the difficulties in developing international trade in both the animal feed ingredient and starch markets, producers of tropical root crops should normally concentrate on their domestic markets.

INTRODUCTION

In this paper consideration will first be
given to cassava. This enters international trade
in two forms--as pellets for animal feeds and as
starch for use in both food and non-food sectors.
Thailand is the major world exporter of both these
products, and the effects which transportation and
government policies have had on the varying fortunes of these two products in the export markets
will be examined. As an aside it must be pointed
out that, although this paper concentrates on the
non-food uses of root crops, as stated above,
starch is used in both the food and non-food sectors. However, to avoid unnecessary complication,
for the purposes of this paper no distinction will
be drawn between these two sectors.
Following from this, the use of sweet potatoes
by the European Economic Community (EEC) compound
animal feed industry will be briefly examined.
Finally, the lessons to be drawn and the possibilities for developing international trade in other
tropical root crops will be considered.

EXPORTS OF CASSAVA PELLETS TO THE EEC

The EEC is, as will be described later, the
only market for cassava pellets. These are used
by the compound animal feed industry. Trade
statistics (Table 1) show a rapid increase in the
level of imports, which have nearly doubled
between 1974 and 1977 to reach 3.9 million tons,
and can be expected to go well over 4 million tons
in 1978 (representing more than 10 million tons of
fresh roots). The majority of these supplies comes
from Thailand, with small quantities originating in
Indonesia. The major consuming countries are the
Netherlands, taking about a half of total supplies,
followed by West Germany, Belgium, and France. In
1978 other EEC markets may be opened up for the
first time. Both government policy and transportation have had a major role to play in the development of this trade, and these will now be
considered.

Government Policy

It is important to realize at the outset that
cassava pellets are competitive as an ingredient

451

TABLE 1
Imports of Cassava Pellets to the EEC ('000 tons)

Destination	1974	1975	1976	1977
The Netherlands	1088	1233	1514	2027
West Germany	431	484	666	961
Belgium	394	449	680	733
France	164	146	174	201
TOTAL	2077	2312	3034	3922
of which from:				
Thailand	1713	1874	2768	3647
Indonesia	259	313	172	109

Source: National trade statistics.

in EEC compound animal feedstuffs only because of the EEC's Common Agricultural Policy (CAP) price structure and the regulations governing cassava imports. An understanding of the operation of the CAP is essential when studying the details of the EEC cassava market. However, for the purposes of this paper, I intend to keep discussion of this to a minimum and merely highlight certain relevant factors. (A full discussion of EEC agricultural policy can be found in the ITC monograph--see Footnote 1).

Cereal prices in the EEC are supported by establishing target, intervention, and threshold prices. In order to safeguard farm incomes, threshold prices (which are minimum import prices) have been set at levels that have usually resulted in EEC cereal prices being above world market prices. A variable levy, equal to the difference between the c.i.f. Rotterdam price and the threshold price, is charged on all cereal imports. The maintenance of high cereal prices through the CAP has encouraged animal feed compounders to search for cheaper sources of protein and energy to substitute for cereals in their rations.

One such cereal substitute is cassava, imports of which are governed by a different regulation. Under this regulation no threshold price has been

452

fixed, and instead a levy amounting to only 18 percent of the barley levy is charged on imports of cassava pellets. Duties on cassava pellets are also bound under GATT rules whereby they may not exceed 6 percent of the value of the goods.

Thus the maintenance of high cereal prices and a maximum import duty of only 6 percent on cassava imports, has made cassava economically attractive to compounders in the EEC. While the price structure of the CAP has allowed the market for cassava to develop, it must also be realized that the market is highly vulnerable to any changes in these regulations. In this context it is pertinent to note that France, as the largest cereal surplus producing country in the EEC, periodically presses the community to take action to reduce the level of imports of cereal substitutes. The importance of EEC agricultural policy in allowing the market for cassava pellets to be developed is further illustrated by the fact that this is the only significant market to have developed. It is probably true to say that, had the EEC anticipated the level that cereal substitute imports (and subsequently cassava) would attain, the original regulations may well have been more restrictive in their nature.

Another important EEC ruling affecting cassava imports provides that, to be eligible for the maximum 6 percent import duty, pellets may not contain more than 3 percent by weight of binding agents. If the level is higher, the pellet is considered as a compounded product and is subject to a higher levy. Used at the 3 percent level no really satisfactory binding agent has been found; therefore, pellets often break up during transportation, causing one of the major drawbacks to cassava usage--dust and handling problems. This regulation therefore adversely affects consumption levels. In the Netherlands, for example, factories situated in or near towns are often unable to use cassava in large quantities. In France, handling problems have led some compounders to discontinue its use. Trial shipments to Italy, the United Kingdom, and Denmark, where port facilities are not adapted to handling cassava, have all caused considerable unloading problems. Because of these problems, compounders only include cassava in rations when it has a clear price advantage over other raw materials.

At the national level, a policy which constrained cassava utilization in Germany until

recently was the requirement that compounders
declare the "open formula" of their feed mixes, i.e.
that they list the percentages of raw materials used
in the compound. Farmers showed considerable resistance to buying compounds with a high cassava content. More recently, however, the "open formula"
labelling requirement has been replaced by the
"closed formula" system under which only the nutrient contents of the final compound have to be
declared by the manufacturer, thus lessening the
farmers' resistance. So far the levels of bacteria
and fungi in cassava are not controlled by official
EEC standards, but attention is increasingly being
focused in this direction. The warm, moist cassava,
if not properly dried, forms an ideal medium for
bacterial and fungal growth that may cause subsequent health problems to livestock. If standards
are introduced these too can be expected to have an
influence on trade.

The above discussion has focused on the
critical role EEC policy has had in developing the
cassava market. It must also be appreciated, however, that government policy in the exporting
country can have an equally important role to play
in trade development. The Thai government, for
example, assisted in the initial development of the
cassava pellet industry by allowing duty-free
importation of machinery and granting tax holidays
in the early years of operation, with no export duty
charged on pellet exports. At the same time, on the
transport side, the government has encouraged the
building of roads and port facilities.

Transportation

While EEC policies allowed the market for
cassava to develop, advances in transportation, more
than anything else, enabled such large volumes to be
shipped to Europe.

Originally cassava was shipped in the form of
dried chips and meal. The first development was the
advent of pelletization in the mid-1960s which
enabled significant reductions to be made in the
shipping volume required. Trade in chips and meal
subsequently ceased. At the same time, a good network of well-surfaced roads were constructed
throughout Thailand. This considerably facilitated
the movement of cassava from farms to chippers and
pelletizers, and subsequently to the exporters.

It is, however, developments in shipping size and loading methods which have caused the major reductions in freight costs to be achieved. Formerly cassava was shipped in cargoes of up to about 15,000 tons. However, in the mid-1970s consignments of 50,000 to 80,000 tons became the norm, with some cargoes exceedings 100,000 tons. This considerable increase in the size of cargoes resulted from the excess capacity which developed in the bulk carrier fleet around this time. Shippers were able to charter these carriers and use them to transport cassava at a saving of some 20 percent on freight costs. With the larger vessels developments were also needed in loading operations. Cargoes loaded in Bangkok are limited to about 15,000 tons owing to harbor depth. The larger vessels are now loaded in the Gulf of Thailand by one of two methods. Firstly, the bags of pellets may be taken by lighter from the exporter's wharves, lifted on board by mobile cranes--which are themselves lifted on deck by a floating crane--and bled into the holds (it is necessary to use mobile cranes since the bulk carriers have no lifting equipment of their own). Secondly, and more recently, a 3-km jetty has been built in the Gulf of Thailand, down which the pellets are loaded by conveyer belt. Together, these various shipping developments have enabled a large volume of cassava to be transported cheaply, thus enhancing the economic advantage of cereal substitution in the EEC.

The development of the Thai cassava pellet industry has been a remarkable success story, and there can be no questioning the vital role which government policy and transportation had to play in this.

EXPORTS OF CASSAVA STARCH FROM THAILAND

While cassava pellet exports have boomed in Thailand, starch exports have faced a generally static market. The principal importing countries are Japan and the United States, and imports of cassava starch to these markets are shown in Table 2. Thailand is the only significant supplier to the Japanese market and the major supplier to the U.S. market, followed by Brazil. Apart from 1974, the Japanese market has only increased slowly over recent years, while the U.S. cassava starch market

has gradually declined. The situation for exporters
to these two markets will be considered separately.

Japan

Annual consumption of starch in Japan amounts
to about 1.2 million tons. Of this only about
100,000 tons are normally imported, the remainder
being manufactured domestically from locally grown
sweet and white potatoes, and corn imported from
the United States. Japanese agricultural support
policies have encouraged the use of potatoes, and
substantial investments have been made in the cornstarch industry. On imports of starch, Japan
operates a variable import quota system and will
increase the level of imports only if demand exceeds
domestic supply. Thai cassava starch is competitive
with other starches in Japan in terms of both price
and quality. However, because of Japan's policy of
sheltering the domestic starch industry, the prospects for increasing the market for cassava starch
are not favorable. Any permanent increase will
depend on political pressure to increase the import
quota.

In 1974, when a shortage of starch occurred in
Japan due to a drop in United States corn supplies,
imports from Thailand doubled. In 1975, however,
imports returned to previous levels. This temporary
export boom led to considerable investment being
undertaken by the Thai starch industry in the hope
of continuing a higher Japanese import quota. Since
this was not the case there is now considerable

TABLE 2
Imports of Cassava Starch to Japan and the United States
('000 tons)

Destination	1970	1971	1972	1973	1974	1975	1976	1977
Japan	50	47	51	72	140	71	82	94
United States	94	68	64	49	74	39	43	38
of which from:								
Thailand	73	68	57	43	58	30	38	36

Source: National trade statistics.

excess starch extraction capacity in Thailand,
indicating how vulnerable the industry is to changes
in government import policies.

United States

Most of the starch used in the United States
is produced domestically from corn, and only small
quantities are imported. Cassava starch is the
dominant import, the volume of trade having gradually declined over recent years.

The reasons for the fall in imports are mainly
the increase in shipping costs and the longer
delivery time needed for imports compared with
domestic sources of supply. Shipping costs from
Thailand to the United States have trebled since the
early 1970s, mainly owing to high unloading costs
in the United States. Despite the fact that cassava
starch is imported duty free, this increase has
eroded its competitive price advantage over cornstarch. On the supply side, manufacturers are
reluctant to hold stocks, and whereas cornstarch can
be bought locally, cassava starch must be ordered
some 6 months in advance.

In summary, Thai starch exports to the two
main markets have stagnated largely due to government import policies in Japan and transport difficulties to the United States.

It is also interesting to note that in 1976
Thailand was able to diversify her export markets
and send sizeable quantities of starch to Indonesia.
However, this market was only short-lived before
prohibitive import duties were applied by the
Indonesians.

Before drawing conclusions from these case
studies for the prospects for international trade
in other root crops, the trade in sweet potato
slices to the EEC merits brief attention.

EEC IMPORTS OF SWEET POTATO SLICES

Sweet potato slices have similar nutrient composition to cassava. Of all the cereal substitutes,
sweet potato provides the most direct competition
for cassava in the compound animal feed ingredient
market. Table 3 illustrates the extent of this
trade, almost all supplies coming from China. The
marked drop in imports has been due to unavailability of supplies. The main drawbacks with sweet

TABLE 3
Imports of Sweet Potatoes to the EEC ('000 tons)

Importing Country	1974	1975	1976	1977
Belgium	125.6	91.5	32.7	2.2
Germany, Fed. Rep.	46.2	22.2	21.1	2.2
France	1.2	0.7	1.2	1.5
The Netherlands	0.5	0.1	0.1	0.2
Total	173.5	114.5	55.1	6.1

Source: National trade statistics.

potato slices are that they are more difficult to handle than cassava pellets, and supplies are irregular. Nevertheless their quality has been good, and it is this factor that has enabled them to find a sizeable market in Belgium in the past. Belgian compounders are very quality conscious and conservative, and, despite the greater handling problems involved, prefer the better quality product.

PROSPECTS FOR ROOT CROPS OTHER THAN CASSAVA ENTERING INTERNATIONAL TRADE

The example of cassava has illustrated the critical role played by both government policy and transportation in the fortunes of the cassava pellet and cassava starch trade. If other root crops are to be traded internationally for use in the non-food sector, i.e., as an animal feed ingredient or as starch, these two factors can again be expected to have an important role to play. This highlights the necessity to carry out market research in the import markets with particular regard to government policies and regulations; the need for regular and competitive transport services; and a commitment to export-oriented policies by the governments of the exporting countries before embarking on international trade. Furthermore, it is often the case that government incentives favor cereal production, which tends to work against root crop production.

Having said this, however, it is appropriate to question just how good the prospects are for tropical root crops to enter international trade and compete in the animal feed ingredient and starch markets.

The Animal Feed Ingredient Market

The only export market where tropical root crops are likely to find a worthwhile demand as an animal feed ingredient is the EEC where policies encourage cereal substitution. In the case of sweet potato this would mean competing with cassava. In the ITC monograph on cassava a list of the requirements for the establishment of a cassava pellet export industry have been drawn up. Any root crop competing with cassava will also have to meet these requirements, which are as follows:

1. The availability of regular supplies from one year to the next. Since most tropical root crops are consumed by the local population, only surplus amounts are released for export. Unless a regular surplus can be ensured, markets will be difficult to develop.
2. European compounders like to receive regular supplies from any one origin to ensure consistent quality. Cargoes should therefore be sent every 4 to 6 weeks, and each should not be less than 10,000 tons, making an annual requirement in the region of 100,000 tons. This requirement could be reduced if other animal feed ingredients of interest to the shipper were also available.
3. Minimum quality requirements obviously have to be met. Equally important, consistency of quality is high on the list of compounders' priorities. The European compound animal feed industry is highly sophisticated, and in selecting ingredients to use, nutrient content must be known in advance. Too great a variation will require changes in feed formulation.
4. Producing countries must have an adequate infrastructure, in particular a large enough harbor and satisfactory roads to producing areas. Owing to the bulky nature of the crop, distance from producing areas to the port is also a factor to bear in mind.

459

5. Production and freight costs would need to be competitive. For cassava the ability to sun-dry the crop and the availability of cheap labor in the producing countries are particularly important.
6. Government incentives and assistance may be needed to encourage the development of the export industry and to provide the necessary infrastructural demands.

As far as cassava is concerned, apart from Thailand, only Indonesia has developed a pellet export industry. The prospects for other cassava-producing countries being able to meet enough of the above requirements and enter the market do not look promising at present. No other country is likely to build up as large an industry as that in Thailand, and certainly in the early stages it would not be able to achieve the same economy in ocean freight costs. The example of Chinese sweet potatoes, however, does show that smaller markets do exist, particularly for high quality products, if their price is right. Overall, however, the market is likely to become increasingly competitive and difficult to break into.

The Starch Market

Starches can be substituted for each other in most end uses. No one starch can therefore be considered in isolation from the overall starch market. The large excess capacity in the Thai starch industry, and the limited success the industry has had so far in finding new and satisfactory long-term export markets, highlights the problems that other countries considering starting a starch export industry are likely to face. Again, this is not to say that small markets may not develop, but overall the prospects for international trade are again not good. Trade between the developing producer countries themselves might be a possibility in both these markets, but import substitution policies often hinder such trade developments, as was illustrated by the Thai-Indonesia starch trade.

DEVELOPMENT OF DOMESTIC MARKETS

The above discussion has gone wider than the government policy and transport aspects of developing international trade. However, in light of all the aspects covered and the difficulties in developing international trade in both the animal feed ingredient and starch markets, producers of tropical root crops should consider the prospects for developing their domestic markets. The fact that the development of a viable export market is usually based on a flourishing domestic market strengthens this conclusion. Increasingly, developing countries are beginning to invest in their own animal feed plants which often rely on expensive raw materials. The opportunities for substituting these with locally available raw materials, e.g., root crops, should constantly be borne in mind.

Similarly, with industrialization there is likely to be a rising demand for starch and its derivatives in developing countries. If starch extraction industries are launched, they will probably be best oriented toward meeting these domestic requirements.

Peter R. Walters: Tropical Products Institute, 56-62 Gray's Inn Road, London, WC1X 8LU, England